T0255357

The Victorian Palace of Science

The Palace of Westminster, home to Britain's Houses of Parliament, is one of the most studied buildings in the world. What is less well known is that while Parliament was primarily a political building, when built between 1834 and 1860, it was also a place of scientific activity. The construction of Britain's legislature presents an extraordinary story in which politicians and officials laboured to make their new Parliament the most radical, modern building of its time by using the very latest scientific knowledge. Experimentalists employed the House of Commons as a chemistry laboratory, geologists argued over the Palace's stone, natural philosophers hung meat around the building to measure air purity, and mathematicians schemed to make Parliament the first public space where every room would have electrically controlled time. Through such dramatic projects, Edward J. Gillin redefines our understanding of the Palace of Westminster and explores the politically troublesome character of Victorian science.

Edward J. Gillin completed a D.Phil. at the University of Oxford in 2015 and is now a Research Fellow at the University of Cambridge. He specialises in British science, technology, architecture and politics in the nineteenth century, with his current work focussing on the role of sound in the production of Victorian scientific knowledge. Previous works cover topics such as the Cunard Steamship Company, early twentieth-century political protest, and Isambard Kingdom Brunel's *Great Eastern* steamship. He received the 2015 Society of Architectural Historians of Great Britain's Hawksmoor Medal, and in 2016 was awarded the Usher Prize from the Society for the History of Technology.

SCIENCE IN HISTORY

Series Editors

Science in History is a major series of ambitious books on the history of the sciences from the mid-eighteenth century through the mid-twentieth century, highlighting work that interprets the sciences from perspectives drawn from across the discipline of history. The focus on the major epoch of global economic, industrial, and social transformations is intended to encourage the use of sophisticated historical models to make sense of the ways in which the sciences have developed and changed. The series encourages the exploration of a wide range of scientific traditions and the interrelations between them. It particularly welcomes work that takes seriously the material practices of the sciences and is broad in geographical scope.

The Victorian Palace of Science

Scientific Knowledge and the Building of the Houses of Parliament

Edward J. Gillin
University of Cambridge

CAMBRIDGE
UNIVERSITY PRESS

University Printing House, Cambridge CB2 8BS, United Kingdom

One Liberty Plaza, 20th Floor, New York, NY 10006, USA

477 Williamstown Road, Port Melbourne, VIC 3207, Australia

314-321, 3rd Floor, Plot 3, Splendor Forum, Jasola District Centre, New Delhi - 110025, India

79 Anson Road, #06-04/06, Singapore 079906

Cambridge University Press is part of the University of Cambridge.

It furthers the University's mission by disseminating knowledge in the pursuit of education, learning and research at the highest international levels of excellence.

www.cambridge.org
Information on this title: www.cambridge.org/9781108411615
DOI: 10.1017/9781108303873

First published 2017
First paperback edition 2019

A catalogue record for this publication is available from the British Library

Library of Congress Cataloging in Publication data
Names: Gillin, Edward John, 1990– author.
Title: The Victorian palace of science : scientific knowledge and the building of the Houses of Parliament / Edward J. Gillin.
Description: New York : Cambridge University Press, 2017. | Series: Science in history | Includes bibliographical references and index.
Identifiers: LCCN 2017040737 | ISBN 9781108419666 (hardback)
Subjects: LCSH: Westminster Palace (London, England) | Barry, Charles, 1795–1860. | Architecture and science – England – London. | Science – Great Britain – History – 19th century. | Westminster (London, England) – Buildings, structures, etc. | London (England) – Buildings, structures, etc.
Classification: LCC DA687.W6 G55 2017 | DDC 942.1/32–dc23
LC record available at https://lccn.loc.gov/2017040737

ISBN 978-1-108-41966-6 Hardback
ISBN 978-1-108-41161-5 Paperback

Contents

Figures

Preface

The Palace of Westminster is probably the most recognizable nineteenth-century building in the world. Home to Britain's Houses of Parliament, it is the seat of the nation's democratically elected political representatives. As far as architecture goes, the Palace is the ultimate symbol of political power. Though to look at the Houses of Parliament building is not to see a bastion of democracy, but a fantastic shrine to the medieval powers of monarchy, church, and aristocracy. Adorned with glorious sculptures of English kings and queens (obviously Oliver Cromwell has to wait outside) and decorated with Queen Victoria's monogram and frescoes depicting idealized episodes in British history, this is every bit a royal palace. Even the style exudes aristocratic privilege. Although a Classical silhouette with obvious Italian influences, it is unmistakably Gothic, and while the Classical holds notions of republicanism (both from classical Rome and revolutionary France), the Gothic is feudal and ecclesiastical. Step inside and all the fantasy of Augustus Pugin's Roman Catholic imagination is overwhelming. One cannot help think that this romanticized mock-medieval illusion is a rather inappropriate venue from which to govern a modern nation. If this is Britain's representative legislature, then it is unclear which particular bit of Britain it represents.

It is the premise of this book that to interpret the Palace of Westminster in this way is to misconstrue it. Indeed, it is to fall into an architectural trap set long ago; the building is meant to seem traditional, medieval, and emphasizing of a British constitution in which the Crown is central. After all, this was a Parliament building to allay fears of radical political change in an age of reform. Yet beyond the reassuring symbols of ancient authority lies a far more modern, radical, and perhaps dangerous power: science. Parliament's architecture might appear medieval, but it was designed as a startlingly modern legislature, embracing the latest scientific learning. The Palace is a vast network of early Victorian science and was, in the mid-nineteenth century, a site of knowledge production.

Historical comparisons are always risky, but it is hard not to feel that now is an appropriate time to revisit the 1830s with a focus on the new

Parliament building. Built in an age of great social and political uncertainty, where existing institutions struggled to survive in the face of radical and popular pressures, this was a time when both the physical form of Parliament and the British state were recast. And at the centre of all this turmoil was the problem of knowledge. With religious teachings and aristocratic powers challenged, new bodies of scientific knowledge, such as geology and chemistry, provided alternate cultural authorities. If 2016 was the year in which substantial elements of society 'had enough of experts' and the Oxford Dictionaries opted for 'post-truth' as their Word of the Year, then the 1830s was a time when purveyors of science worked hard to fashion their knowledge as 'truth' and build cultural authority for themselves.

The political upheavals seen recently will engage historians for years to come as they devise explanations and historical comparisons. Yet in the background, Britain's parliament building has attracted headlines. In the context of immense political uncertainty, the future of the Palace has been a quiet, yet important, cause for discussion. With its physical state dilapidated, its appearance and practicality in question, and the costs of renovating and maintaining the property divisive, the Palace of Westminster has itself become controversial again. A recent committee of MPs and Lords have warned that without a restoration costing at least £4billion, the future of the building is in peril.

Politicians have often commented on Parliament's physical setting, either as a cause for celebration, ridicule, or disgust. Perhaps most recently, in her book *Honourable Friends?* (2015), the MP Caroline Lucas condemned the physical setting of Parliament and described it as a place of privilege and indolence. She proposed a dramatic solution. Electronic voting instead of division lobbies, iPads replacing paper, meetings beyond Westminster, office space, and an architectural redevelopment to include glass walls through which the public might see the machinations of their government, as can be done in Berlin and The Hague. What is so interesting about such proposals is that they echo some of the more radical calls for a new Parliament made in the 1830s. Attention to efficiency, architecture, division of space, public access, and even the location of the building were all central to the debates surrounding the new Palace. The problems faced in the 1830s are still resonant today.

Popular and architectural historians have contributed to this growing interest in the Palace, providing new interpretations of the building. While Henrik Schoenefeldt has drawn attention to the elaborate ventilation schemes and their significance within architectural and design history, Caroline Shenton's *Mr Barry's War* (2016) has provided an entertaining account of the Palace's construction from architect Charles Barry's perspective. His enigmatic partner and co-designer of the Palace, Augustus Pugin, was the subject of Rosemary Hill's hugely successful biography, *God's Architect* (2007). At the University of York, John Cooper has been leading a considerable research project on the architecture of St Stephen's Chapel (the original home of the Commons) between 1292 and 1941. In London, at the Institute of Historical Research, Rebekah Moore has investigated the temporary accommodation used for Parliament during the 1830s and 1840s. All this activity has reminded us that the architecture of Parliament presents us with important questions over the history of our political system. The form a nation's legislature takes is not straightforward, but reflects the ideas and values a society holds dear. What past studies have failed to show is the extent to which the Houses of Parliament embody science and the degree to which Victorian society wanted their legislature to be a bastion of scientific knowledge.

I first came across this relationship between science and the new Parliament building by accident in early 2012. While undertaking research on the cultural significance of the Houses of Parliament's architecture during the nineteenth century, I had to read through a host of parliamentary debates, minutes from select committee meetings, and commission reports. Throughout these, the names of individuals well known within the history of science, but perhaps obscure in architectural circles, kept appearing. The Astronomer Royal, George Biddell Airy, was a frequent contributor to royal commissions, while experimentalists Michael Faraday, Goldsworthy Gurney, David Brewster, and David Boswell Reid regularly provided evidence to committees. Charles Wheatstone, John Frederic Daniell, and John Tyndall were all officially employed to perform chemical investigations into various building materials. William Whewell and William Buckland were unofficially consulted for advice, and geologists Henry De la Beche and William Smith worked on a royal commission to select stone for the building. The naval architect John Scott Russell, railway engineer George Stephenson, and several of Isambard Kingdom Brunel's relatives sat on committees to provide technical knowledge. In short, it was clear that the celebrities of early-Victorian science were drawn together to help build the nation a new Parliament. On looking through their personal papers and newspaper articles, it became apparent that scientific knowledge played a central part in the building's construction. As I delved deeper it was also obvious that this role was far from simple. The use of such new and untried knowledge was full of epistemological and political problems.

There is no volume which closely examines all of this scientific activity and places it in its political and social contexts. My study unites histories of science, architecture, and politics into a cultural history of the building. This book argues that knowledge and architecture were interlinked through the act of governing in mid-nineteenth-century Britain. Unpacking exactly how this happened is the aim of this book.

Acknowledgements

Thanks go first and foremost to William Whyte, without whom this project would simply not have been possible. He has provided kindness and support beyond all expectations, and has been the most fantastic mentor throughout the last five years. I must express my thanks to Crosbie Smith who has been a constant inspiration and trustworthy guide in all I do. Without him, I would not have had the confidence and knowledge to produce this work. I am grateful and honoured to have had critical advice from Simon Schaffer and Geoffrey Tyack. Simon's guidance earlier in the project was invaluable, and later on he was instrumental in transforming what was a D.Phil. thesis into a book. Geoffrey's enthusiasm has been a great source of energy throughout; especially his recommendation to focus on the architecture itself when this volume was at an embryonic stage. I have been privileged to have had conversations and feedback from Christina de Bellaigue and Jane Garnett. This project began with a chat at Girton College with Ben Griffin, and I have been grateful to him ever since. In recent months I have benefitted from excellent advice from Lucy Rhymer at Cambridge University Press, as well as Jim Secord and two immensely constructive reviewers; together they have made the publishing process exceedingly smooth. Indeed I am grateful to all the staff at Cambridge University Press who have made the experience most enjoyable, especially Julie Hrischeva and Allan Alphonse. A very special thanks goes to Silke Muylaert for all her insights and support. Not only is she one of the most intelligent critics I've been able to discuss ideas with, but also she has been constantly at my side throughout this work.

For their extensive and generous support, I should like to thank Robert White, Laura Treloar, Alexander Teague, Peter Exell, Steve Teague, Montgomery Spencer, and the late Estelle White. Thanks go also to Tim Marshall, Mark Curthoys, David Boswell, Harry Rodgers, Peter Mandler, Graham Lott, Lyndal Roper, Oz Jungic, Oliver Zimmer, Horatio Joyce, Mark Collins, Craig Muldrew, Grayson Ditchfield, Diego Rubio, Perry Gauci, Graham Harding, Penny Tyack, Richard

Parfitt, Susan Brigden, Alana Harris, David Ormrod, Sam Burgess, Ben Marsden, Mark Collins, Philip Boobbyer, Robert Parker, Yakup Bektas, Jon Agar, Don Leggett, Sir Mark Jones, Judith Pfeiffer, Frank James, Stephen Courtney, Oliver Carpenter, Rachel North, Jenny Bulstrode, Keith Shephard, Ben King, Robert Hall, Harry Mace, and the one and only Chris Yabsley. I am immensely grateful to Henrik Schoenefeldt, Geoffrey Cantor, Graeme Gooday, and Peter Roberts. It has also been a privilege to discuss this project with Robert Thorne and Robert Bowles, who have an incredible knowledge of the buildings. The Modern British History and the Architectural History seminars in Oxford have provided an incredibly constructive atmosphere in which to work. Late direction was provided by David Trippett, for which I am very grateful. Thanks go also to my students, especially Ting Ma, Ayush Mehta, Sarah Nagle, Katharine Ogburn, and Yi Xie. The help and assistance of the inspirational David Lewis has been fantastic and will not be forgotten.

I owe immense debts to Andrew Morrison and Gill Nixon at the British Geological Survey, Annie Pinder and all the staff at the Parliamentary Archives, and Kathleen Santry at the archives of Oxford University Museum of Natural History. Thanks also go to the staff at Tate Britain, the British Library, King's College London Archives, Kent University Templeman Library, the National Archives at Kew, the Royal Observatory at Greenwich, the Sackler Library, Edinburgh University Centre for Research Collections, University of Edinburgh New College Library, Cambridge University Library, the Radcliffe Science Library, the Duke Humphrey Room, and the Weston Library. The staff at the Bude Castle Heritage Museum were lovely and enthusiastic in helping me to research Goldsworthy Gurney's career. For funding I thank the Arts Humanities Research Council, St John's College Oxford, St Cross College, the Oxford Wellcome Unit for the History of Medicine, and the University of Oxford History Faculty. The map work is entirely due to Michael Athanson at the Bodleian Library. I should like to thank the Bodleian for offering me reduced cost image permissions, while Cambridge University Library kindly waived these altogether. At the 2015 'Making Constitutions, Building Parliaments' conference, Joseph Coohill, Rebekah Moore, Caroline Shenton, Martin Spychal, Rosemary Hill, and Philip Salman were all fantastic and provided some essential late inspiration. A final word of thanks is due to Paul Roberson for personally introducing me to the Westminster Clock. Behind this project stand two unseen but incredible figures, without whom it would not have been possible, namely my parents Steve and Louise Spencer, and it is to them that I dedicate this work.

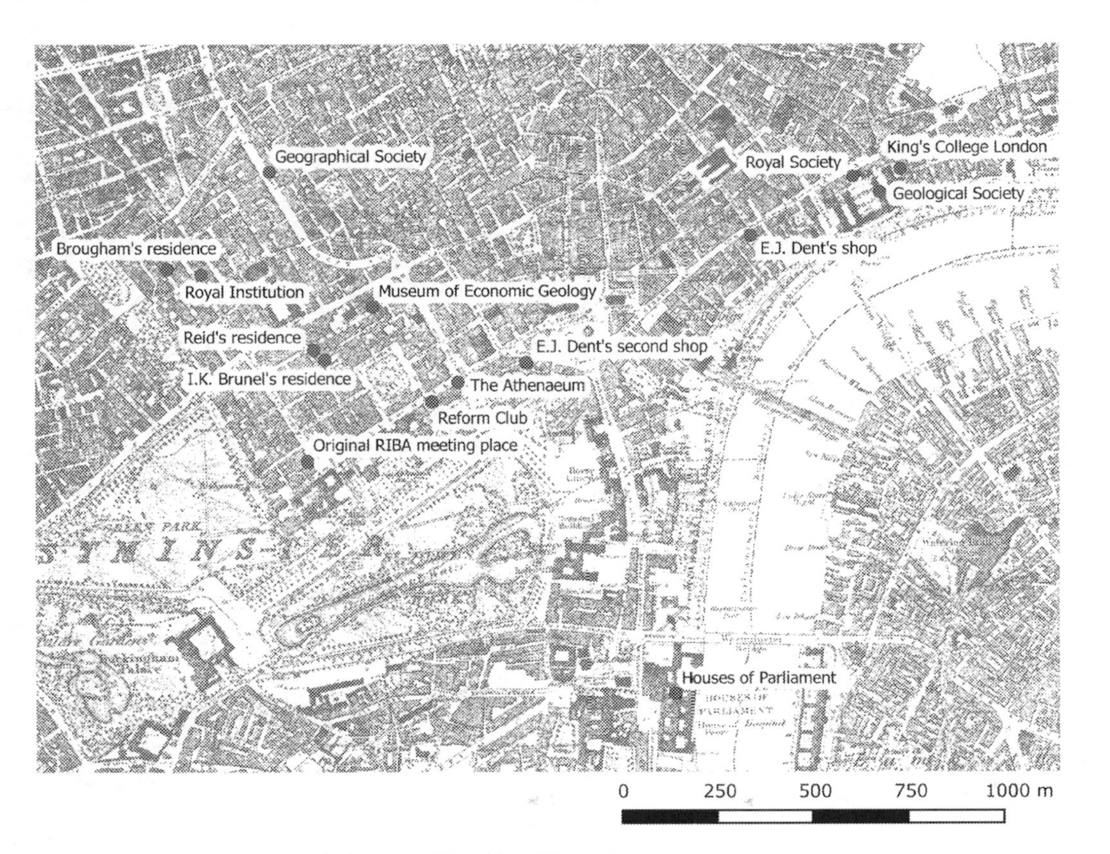

Map A The venues of science in early-Victorian Westminster

Introduction

If you wanted to find out where science happened in early Victorian Britain, you could have done worse than head to the bustling metropolis of Westminster. There you would discover a hive of scientific investigation amidst a densely packed selection of institutions, laboratories, shops, and museums. Along the Strand at Somerset House was the meeting place of the Royal Society, dating its origins back to 1660, and the eminent Geological Society, next door to the newly opened laboratory of King's College London. Just a few minutes away, scientific instrument makers and clock builders offered the latest mechanical contrivances in shops around Charing Cross and Trafalgar Square. About the gentlemen's clubs of St James's and Pall Mall you could bump into learned men of science, bustling to-and-fro between work and dinner. Along Duke Street lived celebrated engineers, such as Isambard Kingdom Brunel, and on Jermyn Street the Museum of Economic Geology displayed the latest findings of the capital's most fashionable science. Walk a little further and you might catch Michael Faraday performing a demonstration of his latest experimental findings at the Royal Institution.[1] Perhaps less obvious though, there was another place of intense scientific activity, and this was the Palace of Westminster, home to Britain's Houses of Parliament (see Map A).

The Palace of Westminster might seem an unusual focus for a history of science, but it is the purpose of this book to revise fundamentally the way we see the new Parliament building, and to show that scientific knowledge was at the centre of its construction. Furthermore, this use of scientific knowledge at Westminster, which included geology, chemistry, and mathematics, matters to our understanding of the relationship between politics and science in Victorian Britain. My account is not only a history

[1] Bernard Lightman, 'Refashioning the spaces of London science: elite epistemes in the nineteenth century', in David N. Livingstone and Charles W. J. Withers (eds.), *Geographies of Nineteenth-Century Science*, (Chicago, 2011), pp. 25–50; Iwan Morus, Simon Schaffer, and Jim Secord, 'Scientific London', in Celina Fox (ed.), *London – World City, 1800–1840*, (New Haven, 1992), pp. 129–42.

of architecture, then, but a study of the interaction between government and specialist knowledge. To contemporary audiences, science appeared as a powerful resource which, if carefully employed, might build credibility for the nation's political elites. Yet it was also a time in which science was not a self-evident, rarefied product; it lacked any clear definition or inherent authority. Varying ideas over how to use it, how to make it, and who could be relied on to undertake it, meant that trustworthy science was a hard thing to find. What constituted scientific knowledge was not just a question of what was 'true' and who was 'right', but rather, what was credible and who could be trusted. At the same time, while science might contribute to political credibility, it could also constitute a dangerous new knowledge, which challenged existing forms of social authority.

No better example of science's troublesome character appeared than during the rebuilding of the Houses of Parliament. On the night of 16 October 1834, a terrifying fire destroyed the old, largely medieval, Palace of Westminster. This apocalyptic event initiated almost three decades of work to rebuild the nation's legislative assembly. While the ruins of the old Palace were converted into temporary debating chambers for the Lords and Commons in early-1835, the Whig government selected architect Charles Barry's designs for a new Gothic legislature in 1836 following a controversial architectural competition. Until his death in 1860, Barry oversaw the new Palace's construction, completing the House of Lords in 1847 and the House of Commons in 1852. This undertaking was a massive architectural challenge, but building the permanent new Parliament also involved the production and employment of scientific knowledge on an immense scale. The invoking of knowledge asserted to be scientific for the new Parliament building was hugely ambitious. It was not only about resolving technical problems, but part of a broader moment where politics looked to science for authority.

The Politics of Science in Early Victorian Britain

The 1830s and 1840s were uncertain times for British society, with the omnipresent threat of revolution never far from the minds of the nation's political elites. Beyond the corridors of Whitehall and Westminster, rapid population growth and industrialization threatened the existing political hierarchy.[2] This was most dramatically realized through the series of

[2] Utilitarian philosopher John Stuart Mill (1806–1873) believed there was a crisis of political authority, see Richard Yeo, 'Science and intellectual authority in mid-nineteenth-century Britain: Robert Chambers and vestiges of the natural history of creation', *Victorian Studies*, Vol. 28, No. 1 (Autumn, 1984), pp. 5–31, 7.

Chartist gatherings in the northern industrial cities, including meetings in Glasgow of 150,000 and Manchester of 250,000 attendants in 1838.[3] This was an age in which old institutions were not safe. The Crown and Church were obvious targets, but reformers also directed their efforts against Parliament, local government, the legal system, slavery, religious discrimination, and all the trappings of a society riddled with aristocratic patronage. The year 1829 witnessed Catholic emancipation in the form of the Roman Catholic Relief Act, followed by the abolition of slavery in 1833, the replacement of self-appointed boroughs with local councils in 1835, and a continual reduction in the Church of England's privileges. These were proceeded by more than ten years of trade tariff reductions which followed the ever-popular demand for 'Free Trade'. The greatest scalp of all, the reform of Parliament, came in 1832 with the Great Reform Act which extended the franchise to just under a fifth of the adult-male population of England and Wales, gave greater political representation to Britain's booming industrial towns, and largely remedied the corruption of rotten boroughs. Arguably all of these upheavals were attempts to build trust in the state and secure its legitimacy, but they were upheavals all the same.[4] From 1828 until the abolition of the Corn Laws in 1846, the form of the British state changed dramatically as it faced immense pressure from reformers and revolutionaries. It is perhaps telling that around seventy-two radical MPs sat in Parliament between 1832 and 1835. It was amidst all this drama that the construction of the new Houses of Parliament took place.

Britain in the 1830s was a place of reform, and though the extent of this reform has been debated, what is unmistakable is that science was inseparable from this morass of social and political change. Science, or at least varying programmes of science, were central to British reform culture. Most obviously, the traditional institutions of science made tempting targets for restructuring. Perhaps more significantly though, scientific knowledge provided an intellectual basis for those intent on reordering the nation's existing social and economic orders. Science not only appeared simultaneously as an illuminator of divine providence, a form of moral improvement, an inducer of social mobility, and a legitimizer of

[3] Jack Morrell and Arnold Thackray, *Gentlemen of Science: Early Years of the British Association for the Advancement of Science*, (Oxford, 1981), p. 10; among other demands, the Chartists called for universal male suffrage, secret ballots, and paid MPs, see Martin Daunton, *Progress and Poverty: An Economic and Social History of Britain, 1700–1850*, (Oxford, 1995), p. 499.

[4] Martin Daunton, *Trusting Leviathan: The Politics of Taxation in Britain, 1799–1914*, (Cambridge, 2001), p. 61.

political hierarchies, but also a radical body of knowledge.[5] Historians of science have frequently shown how disordered knowledge can be linked to social disruption. As soon as a significant portion of society loses belief in existing forms of knowledge, politics can quickly become chaotic and unpredictable, with populism a troublesome thing to control or reason with. Historians have also maintained that solutions to problems of knowledge are often related to restoring social order. Controlling and regulating new bodies of knowledge can be ways of recovering political order.[6]

At no time has this relationship between society and knowledge been more obvious in Britain than during the 1830s. With the promotion of reading and a surge of cheap publications, broad society had never had such access to scientific knowledge. George Combe's *The Constitution of Man* (1828) and Charles Lyell's *Principles of Geology* (1830–1833) established controversial new subjects, and works such as John Herschel's *A Preliminary Discourse on the Study of Natural Philosophy* (1831) linked the cultivation of science to good character and respectable social behaviour.[7] For the first time, science was disseminated beyond privileged elites to middle class and, occasionally, working class audiences. This expansion in knowledge fermented utopian hopes of a new age of improvement. But it also threatened existing knowledge, namely the traditional Newtonian model of the universe which fit so neatly with Anglican theology, and which seemingly functioned as a bulwark of social and political stability. New chemical and mathematical theories fuelled terrifying interpretations in which all in the universe, including the mental and spiritual, might be the result of matter in motion.[8] It was widely feared that to abandon old accepted truths about nature would see a rise in the belief of a mathematical universe, free from divine interference; science could in this form support materialism and atheism. This knowledge was

[5] On the use of science in programmes for social improvement, see Lawrence Goldman, *Science, Reform, and Politics in Victorian Britain: The Social Science Association, 1857–1886*, (Cambridge University Press: Cambridge, 2002); on politics and medicine in the 1840s, see Anne Hardy, 'Lyon Playfair and the idea of progress: science and medicine in Victorian parliamentary politics', in Dorothy Porter and Roy Porter (eds.), *Doctors, Politics and Society: Historical Essays*, (Rodopi: Amsterdam, 1993), pp. 81–106.

[6] As shown with respect to the Royal Society during the 1660s, in Steven Shapin and Simon Schaffer, *Leviathan and the Air-Pump: Hobbes, Boyle, and the Experimental Life*, (Princeton, 1985), pp. 15, 283, and 344.

[7] James A. Secord, *Visions of Science: Books and Readers at the Dawn of the Victorian Age*, (Oxford, 2014), p. 3; for examples of the radical impact of reading science, see James A. Secord, *Victorian Sensation: The Extraordinary Publication, Reception, and Secret Authorship of Vestiges of the Natural History of Creation*, (Chicago, 2000), pp. 11–13.

[8] Secord, *Visions of Science*, pp. 7–8.

inherently dangerous and destabilizing, but it found a wide range of receptive audiences.

Evolutionary theories over the development of living organisms presented just such a dangerous body of knowledge in the 1830s. While traditional histories have concentrated on Anglican Oxbridge science and the later acceptance of Charles Darwin's evolutionary theory following his 1859 *On the Origin of Species*, Adrian Desmond has argued that beyond these elite circles, earlier evolutionary theories were not disbelieved. Evolutionary ideas offered new models of social organization which, though rejected in polite scientific communities, were embraced in radical circles. There were myriad audiences eager for new knowledge with which to challenge existing social and economic orders; groups which were keen to proclaim science as a basis for reform. These groups not only looked to new forms of science, but sought to reorganize its institutions. With London displacing Edinburgh as the empire's capital of medicine, young students targeted the Oxbridge dominated Royal Colleges of Physicians and Surgeons. At the forefront of these moves to seize control over anatomical and medical knowledge were students and radical MPs, such as Thomas Wakely and Henry Warburton.[9] Along with medical reformers, London was home to radical artisans, reforming Whigs, and materialist atheists. These groups conscripted new forms of science in their attempts to reform legal, medical, political, and scientific establishments.[10] Equally radical and distinguished were the Utilitarians who promoted a more professional approach to social organization. Mostly of middle class stock, the Utilitarians staked their own claims for reforming both society and science. Rather than polite learning for a hereditary elite, Utilitarian science was a tool for manufacturing knowledge of value to reformers within government and the professions.[11] From 1826 this Utilitarian science found a home at the newly established University of London, as well as on the Royal Institution's governing body.

Not all programmes for reforming science came from radicals. While science provided intellectual foundations for middle-class calls for meritocracy and working-class desires for revolution, it could also be used to build social stability. Amid all the social disaffection of the 1830s, appeals to nature were actively employed to maintain political order. For political and religious authorities, science was rhetorically valuable in explaining the natural place of man, while for manufacturers such knowledge appeared

[9] Adrian Desmond, *The Politics of Evolution: Morphology, Medicine, and Reform in Radical London*, (University of Chicago Press: Chicago, 1989), pp. 12, 101–02.

[10] Ibid., pp. 5, 101. [11] Lightman, 'Refashioning the spaces of London science', p. 29.

to justify their work economically.[12] The formation of the British Association for the Advancement of Science (BAAS) in 1831 was just such an endeavour, declaring its principal objective of giving 'a stronger impulse and more systematic direction to scientific enquiry . . . and to turn the national attention to the objects of science and obtain a removal of any disadvantages of a public kind which impede its progress'.[13] Importantly, the BAAS's first meeting in York was held amidst the national controversy of the two failed reform bills to extend the franchise, while its decision to hold annual meetings in different cities throughout the country was an effort to bring science to regions of social unrest.[14] The majority of the BAAS's leadership favoured moderate centrist political reform and were usually Whigs, Liberals, or Peelite Conservatives, while in religion they were often liberal Broad Church Anglicans. These collective values ensured that the BAAS was an instrument of public order and social cohesion.[15] Efforts like those of the BAAS emphasize how the study of nature could be employed to legitimize existing political and economic order. Yet the BAAS was not merely an attempt to order knowledge amidst political turmoil. It was also a response to traditional aristocratic-dominated science, which the Royal Society embodied. After the death of Joseph Bank (1743–1820) who had served as the dominating President of the Royal Society for some forty-two years, London science gradually shed aspects of its aristocratic nature. Indeed, the BAAS was at the centre of this move away from the patronage of the landed Anglican gentry and the Crown.

The building of the new Houses of Parliament unfolded in the context of this intimate relationship between science and society. While it is well-known that Victorian science was deeply political, the intense scientific activity surrounding the project shows the extent to which this relationship shaped the character of Britain's government. Science could appear a valuable commodity for politicians, carrying implications of modernity and enlightened governance. In an increasingly urbanized and industrialized nation, it was important for Parliament to appear as the legitimate organ of political power.[16] Political reforms and institutional restructurings were central to this

[12] Morrell and Thackray, *Gentlemen of Science*, p. 33; on the economy as natural order, see Daunton, *Progress and Poverty*, p. 495.

[13] Bodleian Library, Oxford (BOD) Ms Dep. Papers of the British Association for the Advancement of Science (BAAS) 5, Miscellaneous Papers, 1831–1869, Folio 39, 'First Resolution of the York Scientific Meeting', (1831).

[14] Morrell and Thackray, *Gentlemen of Science*, p. 10, 98–99; on this mobile character, see Charles W. J. Withers, *Geography and Science in Britain, 1831–1939: A Study of the British Association for the Advancement of Science*, (Manchester, 2010), pp. 24–65.

[15] Morrell and Thackray, *Gentlemen of Science*, pp. 20–22.

[16] Vernon's answer to this is that from 1832 to 1867 the English state increasingly disciplined radical political forms, replacing them with a progressively less democratic political system, see James Vernon, *Politics and the People: A Study in English Political*

programme, but the rebuilding of the Palace of Westminster was also part of these efforts to secure political credibility. When it came to building a new legislature, this entailed the selection of specific scientific instruments and practices, and individuals possessing technical know-how. In short, the fashioning of Parliament as a physical network of scientific knowledge provides insights into nineteenth-century understandings of how to be modern and progressive in government.

The prominence of scientific knowledge in building the new legislature involved broader questions of governance. The challenge of how to exercise power and gain the respect of subjects is one which governments constantly face, and this is a central concern for my work.[17] Recent histories have asserted that the defining characteristic of the nineteenth-century British state was that it governed subtly through liberalism, and that this governance often took material forms. Patrick Joyce argues that the state, consisting of a collection of institutions and practices, exerted power by securing society's apparent liberties.[18] Liberalism engendered a commitment to maximizing individual freedoms, but was accompanied by 'political technology' for control, including surveillance and increased specialist expertise: in other words, forms of knowledge which were also techniques for ruling.[19] For Joyce, this exercising of state power was manifest through the material world, such as infrastructure networks. Physical objects, often guided by extensive bodies of knowledge, were engineered as forceful, but subtle ways of governing.[20] As Joyce puts it, 'it quickly becomes apparent that the seemingly neutral world of

Culture, c.1815–1867, (Cambridge, 1993), p. 9; for the political workings of Parliament, see T. A. Jenkins, *Parliament, Party and Politics in Victorian Britain*, (Manchester, 1996); on the post-1832 rise of centralized power to manage an expansion of liberty, see David Vincent, 'Government and the management of information, 1844–2009', in Simon Gunn and James Vernon (eds.), *The Peculiarities of Liberal Modernity in Imperial Britain*, (Berkeley, 2011), pp. 165–81, 181.

[17] Michel Foucault, 'Governmentality', in Graham Burchell, Colin Gordon, and Peter Miller (eds.), *The Foucault Effect: Studies in Governmentality with Two Lectures by and an Interview with Michel Foucault*, (Chicago, 1991), pp. 87–104, 87–88; for example, eighteenth-century mechanical automata inspired metaphors for social order and were also directly connected to Enlightenment understandings of government, industry, management, and labour, see Simon Schaffer, 'Enlightened automata', in William Clark, Jan Golinski, and Simon Schaffer (eds.), *The Sciences in Enlightened Europe*, (Chicago, 1999), pp. 126–65, 131, and 164.

[18] Patrick Joyce, *The State of Freedom: A Social History of the British State Since 1800*, (Cambridge, 2013), pp. 10, 28; this builds on Foucault's work on liberal governance and the exertion of control through spatial apparatus, see Michel Foucault, *Discipline and Punish: The Birth of the Prison*, (Trans.) Alan Sheridan, (Harmondsworth, 1977), pp. 200–06.

[19] Patrick Joyce, *The Rule of Freedom: Liberalism and the Modern City*, (London, 2003), p. 1; Simon Gunn and James Vernon, 'Introduction: what was liberal modernity and why was it peculiar in imperial Britain', in Gunn and Vernon (eds.), *The Peculiarities of Liberal Modernity in Imperial Britain*, pp. 1–18, 9.

[20] Joyce, *The State of Freedom*, p. 30.

science and technology is eminently political, just as the political world partakes of science and technology'.[21] Government then, does not just take place through law and legislation, but through material cultures. Material networks are forms of governmentality, or rather they are rationales and technologies involved in governing human actions.[22] It is not that objects have a deterministic power of their own, but that human agency can be embedded in material systems, and that this can exert a governance of its own.[23] If we think about the implications of this understanding of the Victorian state for a study on the architecture of the Palace of Westminster, then it seems clear that the building was itself intended to fulfil a purpose in the business of governance. In this interpretation of the state, there is an apparent distinction between the government, in a more traditional sense of centralized bureaucratic power consisting of MPs in Parliament and administrators in Whitehall, and governance through material networks.[24]

By looking at the ways science featured in the building of the new Parliament building, my study unites these two notions of governance. On the one hand the Palace was itself a material network of technologies, including lighting and clocks, which could shape the actions of politicians. However, this was also a way for these politicians to teach lessons to society. By being seen to build and work in a building embodying scientific knowledge they could appear as enlightened statesmen. The question then, is how did science contribute to Parliament's credentials as a legitimate venue of political power? Part of the answer to this lies in

[21] Ibid., p. 10; for examples, see pp. 53–99; Patrick Joyce, 'Filing the Raj: political technologies of the imperial British state', in Tony Bennett and Patrick Joyce (eds.), *Material Powers: Cultural Studies, History and the Material Turn*, (London, 2010), pp. 102–23; Jo Guldi, *Roads to Power: Britain Invents the Infrastructure State*, (Cambridge, Massachusetts, 2012), pp. 5, 21; also see, Stuart Oliver, 'The Thames embankment and the discipline of nature in modernity', *The Geographical Journal*, Vol. 166, No. 3 (September, 2000), pp. 227–38; on the government's use of scientific knowledge more generally, see R. A. Buchanan, 'Engineers and government in nineteenth-century Britain', in Roy MacLeod (ed.), *Government and Expertise: Specialists, Administrators and Professionals, 1860–1919*, (Cambridge, 1988), pp. 41–58; on the government and lighthouses, see Roy M. MacLeod, 'Science and government in Victorian England: lighthouse illumination and the Board of Trade, 1866–1886', *Isis*, Vol. 60, No. 1 (Spring, 1969), pp. 4–38; Timothy Mitchell, *Rule of Experts: Egypt, Techno-Politics, Modernity*, (Berkeley, 2002), pp. 21, 34–35.

[22] Patrick Joyce and Tony Bennett, 'Material powers: introduction', in Bennett and Joyce (eds.), *Material Powers*, pp. 1–21, 3.

[23] Ibid., pp. 3–7; Otter recognizes that while technological networks do not determine human actions, they do condition behaviour, see Chris Otter, *The Victorian Eye: A Political History of Light and Vision in Britain, 1800–1910*, (Chicago, 2008), pp. 258–63.

[24] Joyce and Bennett argue that the state is not an entity, but a site of transient powers, see Joyce and Bennett, 'Material powers: introduction', pp. 2–3.

science's apparently unbiased, empirical nature; it appears free of politics.[25] To mobilize knowledge of the natural sciences, seemingly apolitical, to maintain social order was a persuasive technique for governance.[26] That this was done for the rebuilding of the Palace of Westminster presents a very material example of how scientific knowledge contributed to the character of the nineteenth-century British state.[27]

Spectacular Credibility

Through wonderful new technologies, promises of ceaseless progress, and utopian hopes for the future, Victorian science captivated a broad range of society. Recent studies have shown science to have been an increasingly popular subject, open to non-elite groups which included the working classes and women, and which increasingly diverse and sophisticated performers popularized.[28] The politics of knowledge stretched much further than ideas over social reform. It was inseparable from the authority and content of science itself. Different understandings over how science should be performed and demonstrated embodied contrasting political ideologies. It was the performance of scientific practitioners and their relation with audiences which shaped their authority.[29] These alternate models of what was proper and appropriate scientific behaviour became embedded through different venues of scientific display. Artefacts and experiments were performed and understood differently in the Royal Institution and the Royal Society than they were in public galleries. Performances of science in these differing locations were tailored to specific audiences and carried distinctive epistemological and cultural messages.[30] As Iwan Morus observed, to really understand Victorian science

[25] Latour explains that 'modernity' consists of identifying nature and culture as two distinct beings, and then designating two ontological zones, in Bruno Latour, *We Have Never Been Modern*, (Trans.) Catherine Porter, (Cambridge, Massachusetts, 1993), pp. 10–11.

[26] Argued in, ibid., p. 37.

[27] Dan Hicks, 'The material-cultural turn: event and effect', in Dan Hicks and Mary C. Beaudry (eds.), *The Oxford Handbook of Material Culture Studies*, (Oxford, 2010), pp. 25–98, 28–9.

[28] Bernard Lightman, *Victorian Popularizers of Science: Designing Nature for New Audiences*, (Chicago, 2007), pp. 8, 18; also see Bernard Lightman, 'The visual theology of Victorian popularisers of science: from reverent eye to chemical retina', *Isis*, Vol. 91, No. 4 (December, 2000), pp. 651–80.

[29] Consider, for example, Humphry Davy's chemical displays, in Jan Golinski, *Science as Public Culture: Chemistry and Enlightenment in Britain, 1760–1820*, (Cambridge, 1992), 188–235; on the development of spaces of chemical knowledge, see Robert Bud and Gerrylynn K. Roberts, *Science Versus Practice: Chemistry in Victorian Britain*, (Manchester, 1984).

[30] Iwan Rhys Morus, 'Worlds of wonder: sensation and the Victorian scientific performance', *Isis*, Vol. 101, No. 4 (December, 2010), pp. 806–816, 810–11.

we have to look at this relationship between audience and performer. If scientific performances embodied particular approaches to knowledge, then examining which audiences gathered where, what they witnessed, and how performers tailored their display to each audience are all important for revealing the dynamics of Victorian science.[31]

To understand how science was used at Parliament, we have to look at these relationships between producers of scientific knowledge and their audiences. It is not enough to examine the content of the knowledge conveyed; we must instead examine how such knowledge was construed to be creditworthy. This is because in early-Victorian Britain, it was unclear exactly what constituted science.[32] It was a challenge for protagonists to construct their knowledge as different to alternative, non-scientific bodies of knowledge.[33] At different times during Parliament's construction, governing ministries, committees, royal commissions, bureaucrats, the Office of Woods and Forests, independent MPs, and the architect Barry all made efforts to select scientific knowledge for the building. These were the audiences, then, to whom scientific practitioners sought to promote their work. But they held differing notions of what constituted science. Accounting for the knowledge chosen for the Palace can best be achieved by analysing how science was believed to be credible within these audiences. How knowledge was produced was important to how it might be designated scientific and, ultimately, creditworthy.[34] The trouble at Westminster in the 1830s and 1840s was that there was no unanimously agreed framework for producing credible science and this was a source of constant controversy which dogged the building of Parliament. In particular, the notion of 'experiment' proved extremely difficult to categorize. Experiment could make knowledge scientific, sustaining innovation and legitimizing failure, but it could also be risky if it threatened to undermine the physical structure of

[31] Ibid., p. 815.

[32] On demarcating what was science, see Thomas F. Gieryn, 'Boundary-work and the demarcation of science from non-science: strains and interests in professional ideologies of scientists', *American Sociological Review*, Vol. 48, No. 6 (December, 1983), pp. 781–95, 782; scientific authority did not correspond neatly with social hierarchies or particular training, see Graeme J. N. Gooday, 'Liars, experts and authorities', *History of Science*, Vol. 46, No. 4 (2008), pp. 431–56; the categories authority and expertise are easily, and often wrongly used interchangeably, see Don Leggett, 'Naval architecture, expertise and navigating authority in the British Admiralty, c.1885–1906', *Journal for Maritime Research*, Vol. 16, No. 1 (May, 2014), pp. 73–88.

[33] Yeo, 'Science and intellectual authority', p. 8; a point well made in Alison *Winter, Mesmerized: Powers of Mind in Victorian Britain*, (Chicago, 1998), pp. 6–8.

[34] Argued in Bruno Latour and Steve Woolgar, *Laboratory Life: The Construction of Scientific Facts*, (Princeton, 1986), pp. 19–21, 24.

Parliament or the health of MPs. Even when not dangerous, at the time of Parliament's construction the correct way of conducting experiment was a controversial subject. Disagreements over where hypothesis should feature in experiments and how to interpret results made experimental practices particularly troublesome.[35] In the mid-nineteenth century it was not clear if experiment should involve making an initial hypothesis and then investigating if it was true, or if knowledge should consist purely of observations made during experimental investigations. The credibility of experiment could be fragile if not well managed.[36] If experiments could be shown to be replicable in many places and observable by trustworthy witnesses, then this could enhance their validity, but achieving such replicability was a great challenge.[37] Indeed the location of experiments also contributed to their troubling nature. If it was uneasy to secure a consensus over how to perform an experiment, then getting an agreement on where to do it was difficult. As much as the ambiguities surrounding science and experiment, what constituted a laboratory was not straightforward, with boundaries over what was an appropriate place for scientific research often vague.[38] Even the House of Commons could serve as a 'laboratory' if placed in the hands of certain experimentalists.

While methods and locations were important to what science was considered appropriate, credibility involved a lot more than experimentation. Scientific credibility was not purely dependent on a body of knowledge's validity, but was built through persuasion and rhetoric. Being credible meant being believable; to be an accepted conveyer of reliable knowledge.[39] Building on this, Ben Marsden and Crosbie Smith propose that we must not classify technologies, or rather practical applications of knowledge, simply as 'failures' or 'successes'.[40] Rather than assume the

[35] For example, see G. N. Cantor, 'Henry Brougham and the Scottish methodological tradition', *Studies in History and Philosophy of Science*, 2, no. 1 (1971), pp. 69–89.

[36] Ben Marsden and Crosbie Smith, *Engineering Empires: A Cultural History of Technology in Nineteenth-Century Britain*, (Basingstoke, 2005), p. 10; see Steven Shapin, 'The invisible technician', *American Scientist*, Vol. 77, No. 6 (November–December, 1989), pp. 554–63.

[37] Simon Schaffer, 'Glass works: Newton's prisms and the uses of experiment', in David Gooding, Trevor Pinch, and Simon Schaffer (eds.), *The Uses of experiment: studies in the natural sciences*, (Cambridge, 1989), pp. 67–104, 70.

[38] Argued in, Graeme Gooday, 'Placing or replacing the laboratory in the history of science?', *Isis*, Vol. 99, No. 4 (December, 2008), pp. 783–95.

[39] Steven Shapin, *Never Pure: Historical Studies of Science as if It was Produced by People with Bodies, Situated in Time, Space, Culture, and Society, and Struggling for Credibility and Authority*, (Baltimore, 2010), p. 18.

[40] For discussions over the terms 'technology' and 'practical science', see Marsden and Smith, *Engineering Empires*, p. 3; Graeme Gooday, '"Vague and artificial": the historically illusive distinction between pure and applied science', *Isis*, Vol. 103, No. 3 (September, 2012), pp. 546–54.

adoption of a particular scientific idea or technology was determined by its inherent value, we should analyse how its selection was contingent on more complex cultural interpretations. The success of a scientific scheme involved its social, as well as material, worth. How knowledge and apparatus were represented was crucial to their selection and this involved much more than deterministic notions of success or failure.[41]

When competently orchestrated, public displays of new science could be a powerful way to build credibility, but such spectacles could prove embarrassing if the performance was underwhelming. At Parliament, this was all too evident. This was the very heart of Britain's political system, and the performance of science at this location was both dramatic and public. The management of spectacle was crucial work for practitioners looking to fashion themselves as scientific authorities, and no spectacle was grander in mid-nineteenth-century London than the new Palace of Westminster. Previous studies have explored spectacle as a powerful tool in the manufacture of new knowledge. Electricity, for example, provided a valuable cultural resource in post-1820 London. Different protagonists could manipulate the spectacle that electricity provided for their audiences. While at the Royal Institution Michael Faraday could present his latest findings to genteel patrons, to popular audiences at the National Gallery of Practical Science experimentalists provided electrical wonders.[42] Faraday offered precisely rehearsed lectures and took care over who to invite to present papers, but beyond such elite spaces, electricity was displayed through popular amusements and practical applications. Spectacles were presented in different ways in order to secure authority with specific audiences. Displays in themselves were not enough to secure credibility, but required much work to ensure a productive interpretation and portrayal through eyewitness accounts and specialist discussions.[43]

Spectacle could be unpredictable, however. Displaying was crucial to building credibility, but there was the potential for public sights to go

[41] Marsden and Smith, Engineering Empires, p. 7; also see, Ben Marsden, 'Blowing hot and cold: reports and retorts on the status of the air-engine as success or failure, 1830–1855', *History of Science*, Vol. 36, No. 4 (1998), pp. 373–420.

[42] Iwan Rhys Morus, *Frankenstein's Children: Electricity, Exhibition, and Experiment in Early-Nineteenth-Century London*, (Princeton, 1998), p. 5; see pp. 13–42 on Faraday; compare to popular presentations, pp. 70–98; on audiences and science in popular culture, see Roger Cooter and Stephen Pumfrey, 'Separate spheres and public places: reflections on the history of science popularization and science in popular culture', *History of Science*, Vol. 32, No. 3 (1994), pp. 237–67, 243–44.

[43] Don Leggett, 'Spectacle and witnessing: constructing readings of Charles Parsons's marine turbine', *Technology and Culture*, Vol. 52, No. 2 (April, 2011), pp. 287–309, 291; Simon Schaffer, 'Natural philosophy and public spectacle in the eighteenth century', *History of Science*, Vol. 21, No. 1 (1983), pp. 1–43.

wrong, while to be seen to manage a spectacle could also damage an individual's credentials. Isambard Kingdom Brunel's *Great Eastern* steamship, for example, was certainly a spectacle of science and engineering, but as the monstrous vessel trundled through delays, scandals, failed launches, and catastrophic disasters, its spectacular nature did not enhance the standing of those who built her. It instead diminished their reputations. Spectacles attracted a wide range of commentators and the right to define scientific spectacles was not limited to self-styled 'men of science'. They made apt subjects for politicians, journalists, religious ministers, and a whole variety of cultural authorities.[44] Perhaps no single project attracted quite the variety of scientific interest as the new Houses of Parliament. Here was not just a spectacle confined to a laboratory or shipyard, but one apparently representative of the nation. While electricity could be simultaneously manipulated in different laboratories, and different practices could be adopted for alternate engineering projects, there was only one national Parliament. That this was the single architectural embodiment of the nation's political system meant that it was a unique spectacle. One that stirred the most intense of scientific controversies and tried claims of credibility to the absolute limit.

The Architecture of the State

This book is about science and architecture in Victorian Britain, but it is also about the state and governmentality. It is about the relationship between knowledge and the business of governing. At stake was the problem of what constituted an appropriate legislature for a modern industrial society. As much as style, sculptures, and aesthetics, the use of scientific knowledge was a deeply political choice. Previous histories of Parliament's architecture have been almost obsessive in their efforts to determine the meaning of the Gothic and interpret its appropriateness for a nineteenth-century legislature.[45] Was the style representative of an English political constitution, inherently associated with a narrative of English liberty, or was it indicative of a nation unable to detach itself from its aristocratic and ecclesiastic traditions? For art historians, the question

[44] Edward Gillin, 'Prophets of progress: authority in the scientific projections and religious realizations of the *Great Eastern* steamship', *Technology and Culture*, Vol. 56, No. 4 (October, 2015), pp. 928–56, 929.

[45] For example, see Roland Quinault, 'Westminster and the Victorian constitution', *Transactions of the Royal Historical Society*, Vol. 2 (1992), pp. 79–104; Andrea Fredericksen, 'Parliament's genius loci: the politics of place after the 1834 fire', in Christine Riding and Jacqueline Riding (eds.), *The Houses of Parliament: History, Art, Architecture*, (London, 2000), pp. 99–111; on style see, J. Mordaunt Crook, *The Dilemma of Style: Architectural Ideas from the Picturesque to the Post-Modern*, (London, 1987).

of style seems to have been the fundamental problem of building the new Palace and this has justified the extensive volume of historical investigation into Barry's collaboration with the Gothic artist, Augustus Pugin.[46] Others have placed the choice of the Gothic in a political context, with George Weitzman focusing on the affront the style caused to radicals and W. J. Rorabaugh declaring the Gothic to be consistent with Whig and Tory notions of English history.[47]

The question of style is assumed, quite rightly, to be a political subject. However, the building's technical aspects are usually treated as mundane self-evident necessities. Practical questions do not warrant attention because they seem to require no explanation other than that they were required efficiencies. Even when historians have considered the practicalities of building Parliament, they have done little to link this work with wider political and scientific contexts. M. H. Port's investigation of the engineering aspects of the architecture is insightful, but fails to integrate this with contemporary understandings of science and politics.[48] Henrik Schoenefeldt's highly detailed analysis of Parliament's ventilation has shown how much scientific knowledge was used for the building's design, but remains primarily concerned with Parliament's place in a broader development of atmospheric control in architecture.[49] Most recently, Caroline Shenton has explored the technical challenges of building the Palace from the perspective of Charles Barry. While such a study has been long overdue, problems of technical knowledge are treated not as political and epistemological controversies, but as personal disputes, usually with Barry as the aggrieved party.[50] What we have then, is a tacit agreement that while questions of style are inherently political, technical problems do not involve politics, or at least, that they do not need to be explained in a political context.

This distinction is underlined in Eric Hobsbawm's concept of an 'invented tradition', which he defined as practices, usually ritualistic or

[46] For example, see Phoebe B. Stanton, 'Barry and Pugin: a collaboration', in M. H. Port (ed.), *The Houses of Parliament*, (New Haven, 1976), pp. 53–72; Rosemary Hill, *God's Architect: Pugin and the Building of Romantic Britain*, (London, 2008).

[47] George H. Weitzman, 'The Utilitarians and the Houses of Parliament', *Journal of the Society of Architectural Historians*, Vol. 20, No. 3 (October, 1961), pp. 99–107; W. J. Rorabaugh, 'Politics and the architectural competition for the Houses of Parliament, 1834–1837', *Victorian Studies*, Vol. 17, No. 2 (1973), pp. 155–75.

[48] M. H. Port, 'Problems of building in the 1840s', in Port (ed.), *The Houses of Parliament*, pp. 97–121; Denis Smith, 'The building services', in Port (ed.), *The Houses of Parliament*, pp. 218–31.

[49] Henrik Schoenefeldt, 'The temporary Houses of Parliament and David Boswell Reid's architecture of experimentation', *Architectural History*, Vol. 57 (2014), pp. 173–213.

[50] For example, see Caroline Shenton, *Mr Barry's War: Rebuilding the Houses of Parliament after the Great Fire of 1834*, (Oxford, 2016), pp. 96, 145, and 166.

symbolic, 'which seek to inculcate certain values and norms of behaviour by repetition, which automatically implies continuity with the past'.[51] In fact, Hobsbawm considered the rebuilding of Parliament in a Gothic style to be the most obvious example of this, conjuring up romanticized notions of English history and a fictive constitution.[52] Parliament's architecture was, in this understanding, a way of constructing an imagined past to ensure the building appeared a place of political continuity and legitimate governance. But this leaves us with a problem. Hobsbawm distinguished between unnecessary traditions which assert social order, and 'pragmatically based norms', essential for convenience and efficiency. His description of horse riding illustrates this. While wearing hunting pink is a tradition, wearing a hard hat can be explained as 'practical sense'. Such measures for efficiency and safety are not invented traditions, 'since their functions, and therefore their justifications, are technical rather than ideological'.[53] This is a division which establishes traditions, such as matters of style, as inherently political, in contrast to applications of practical knowledge, which seem fundamentally rational. In this dichotomy, the Gothic appears ideological while attention to lighting, building, and ventilation seems only technical. Yet, as we shall see, science's value in the building of Britain's new Parliament was also ideological.

To understand how politics and science were physically united in the new Parliament, this book considers its architecture very differently to previous histories of the building. It treats the building of architecture as an act of governance in its own right. Once we see the extent to which science was employed in this architectural project, it becomes clear that an architectural history in which we read the Palace to decipher its meaning and to explain the structure's symbolism will not suffice. Architectural works might have symbolic meaning, but they are also built to house, protect, sustain, and sometimes to teach.[54] Buildings perform functions. The Parliament building was constructed to house politicians and hold debates, but it was also designed as a seat of authority and to appear a credible site of governance. The building of the structure involved a great deal more than attention to Gothic detail to build continuity with the medieval Palace of Westminster. In short, it entailed the adoption of the latest scientific knowledge and apparatus because these were deemed appropriate for a modern legislature.

[51] Eric Hobsbawm, 'Introduction: inventing traditions', in Eric Hobsbawm and Terence Ranger (eds.), *The Invention of Tradition*, (Cambridge, 1983), pp. 1–14, 1.

[52] Ibid., p. 1. [53] Ibid., p. 3.

[54] William Whyte, 'How do buildings mean? Some issues of interpretation in the history of architecture', *History and Theory*, Vol. 45, No. 2 (May, 2006), pp.153–77, 158–89, and 170.

Among all the functions of buildings to be examined in recent years, the role of architecture in the history of science has been a particularly fruitful subject. Scholars have explored the ways nineteenth-century scientific buildings were fashioned to promote specific understandings of science. The architecture of scientific institutions projected varying ideologies, as well as the aspirations and intentions of their promoters.[55] The internal structure of museums, lecture-theatres, laboratories, and meeting rooms represents an architectural dimension to the organization and presentation of scientific knowledge.[56] Buildings such as London's Natural History Museum and the Oxford University Museum of Natural History were themselves tools for presenting knowledge of nature to public audiences. Architecture could be instructive in showing the correct way of producing science. Buildings played a central role in the making and legitimizing of knowledge.[57]

Arguably though, no building in Victorian Britain mattered more for science than the Palace of Westminster. Parliament was a place that contemporaries were eager to shape as a site of science.[58] Promoters of different science worked hard to have their work employed for the Palace's construction and have it embodied permanently in the building's architecture.[59] There was a consensus among these individuals that

[55] Sophie Forgan, 'Context, image and function: a preliminary enquiry into the architecture of scientific societies', *British Journal for the History of Science*, Vol. 19, No. 1, (March, 1986), pp. 89–113, 91; on museums see Sophie Forgan, 'The architecture of display: museums, universities and objects in nineteenth-century Britain', *History of Science*, Vol. 32, No. 2 (1994), pp. 139–62; on architecture and science, see Peter Galison and Emily Thompson (eds.), *The Architecture of Science*, (Cambridge, Massachusetts, 1999); the Cavendish Laboratory's architecture is considered in, Simon Schaffer, 'Physics laboratories and the Victorian country house', in Crosbie Smith and Jon Agar (eds.), *Making Space for Science: Territorial Themes in the Shaping of Knowledge*, (Basingstoke, 1998), pp. 149–80, 160, and 179.

[56] Forgan, 'Context, Image and Function', pp. 106–07; contemporaries made political analogies out of their scientific institutions, for example, the Royal Society appeared similar to the House of Lords, and the Geological Society, which held its meetings in a cross-benched debating chamber, was comparable to the House of Commons, see Bernard H. Becker, *Scientific London*, (London, 1874), p. 23; Martin J. S. Rudwick, *The Great Devonian Controversy: The Shaping of Scientific Knowledge among Gentlemanly Specialists*, (Chicago, 1985), p. 19.

[57] Carla Yanni, *Nature's Museums: Victorian Science and the Architecture of Display*, (Baltimore, 1999), pp. 6–7, 10; also see Sophie Forgan, 'Building the museum: knowledge, conflict and the power of place', *Isis*, Vol. 96, No. 4 (2005), pp. 572–85; Samuel J. M. M. Alberti, 'The Status of museums: authority, identity, and material culture', in Livingstone and Withers (eds.), *Geographies of Nineteenth-Century Science*, pp. 51–72.

[58] For the analytical importance of 'place', see Edward S. Casey, 'How to get from space to place in a fairly short stretch of time: phenomenological prolegomena', in Steven Feld and Keith H. Basso (eds.), *Senses of Place*, (Santa Fe, 1996), pp. 13–52, 14, 19.

[59] Lefebvre argues that all ideas and values are trialled as they travel over space, see Henri Lefebvre, *The Production of Space*, (Trans.) Donald Nicholson-Smith, (Oxford, 1991), pp. 416–17; built on in Crosbie Smith, '"The 'crinoline' of our steam engineers":

having one's work used in this way was important. The inclusion of an instrument or a body of knowledge in the physical home of Britain's Parliament was perceived to be a huge endorsement. If a system of lighting or ventilation was good enough for Parliament, then where was it not good enough for? So, while this book considers how scientific knowledge featured in the building of Parliament, it also reveals the ways in which the construction of the building contributed to the manufacturing of new science. As will be shown, be it the use of geology, horology, chemistry, or optics, the application of science to the Parliament building, virtually without exception, provoked profound disagreement, and stoked fiery passions. It would be no exaggeration to say that the Palace was the architectural opportunity of the century. But it was also a unique chance for scientific practitioners. It was a project every bit as dramatic as other London spectacles in this period, including the Thames Tunnel in 1843, the Great Exhibition in 1851, and the *Great Eastern* steamship in 1858. Perhaps predictably, as the nation grew more accustomed to the Palace as the home of its political system, the attention these works of science attracted diminished. Both the architecture and the science quickly became thought of as old-fashioned. The challenge for this book is to recapture the original meaning of the building. When we do this, it quickly becomes apparent that questions of style represent just one part of a much bigger project to build a Parliament fitting for an increasingly industrial society.

Following the 1832 Reform Act, what Parliament meant was open to diverse interpretations. For some, the building ought to operate like a machine, objectively manufacturing legislation; for others, the Palace embodied the British constitution.[60] If we think about the new Palace's architecture in terms of its role in securing Parliament an identity as enlightened or progressive, then this has implications for our understanding of science and politics in mid-nineteenth-century Britain. Throughout this book, we see how the credibility of science and its spectacular character were directly related to politics at the place of Parliament. Although politicians, architects, conveyors of specialist knowledge, and commentators all shared in a consensus that the new building should be 'scientific', what exactly this entailed was not easy to answer. Disagreements over what constituted good science plagued the work at Parliament. What is interesting about the building of Parliament is that the process drew directly on

reinventing the marine compound engine, 1850–1885', in Livingstone and Withers (eds.), *Geographies of Nineteenth-Century Science*, pp. 229–54.

[60] This resonates with sociologist Bourdieu's conception that symbolic capital can translate into political power, see Pierre Bourdieu, *Language and Symbolic Power*, (trans.) Gino Raymond and Matthew Adamson, (Cambridge, 1991), pp. 166–70, 192.

knowledge from the most prominent scientific disciplines of the 1830s. While the BAAS promoted science broadly, it also organized science into a hierarchy of disciplines. Mathematics, chemistry, and geology were prioritized in the BAAS's division of the sciences. Section A of the BAAS covered mathematics and physics, while Section B was devoted to chemistry and mineralogy, and Section C was responsible for geology. These three sections represented the premier knowledges of their day. That they were also the sciences most relied on during Parliament's construction is significant and merits attention. To focus on how they guided Parliament's construction shows that Parliament did not just conscript knowledge generally defined as scientific. It enrolled the most prized of all sciences. The question then, is why did contemporaries believe that such knowledge could secure Parliament a reputation as enlightened? What was it about the practices of geology which made them appropriate to guide the selection of Parliament's stone? What was it about mathematically devised galvanic instruments which made them appear reliable for time-keeping in the Commons? Why were experimentalists and chemists sought to purify the Palace's air? And what was it about natural philosophers that made them suitable to take responsibility for questions over optics and illumination? It is one thing to say that science might contribute to the identity of Parliament, but it is quite another to determine what makes a science politically reputable. At Parliament, how knowledge and apparatus fared was contingent on its credibility, both scientific and political.

1 A Radical Building: The Science of Politics and the New Palace of Westminster

Those who have kingdoms to govern have understandings to cultivate.
Samuel Johnson, *The History of Rasselas*

Both the old and the new Houses of Parliament at Westminster embodied profound concerns over science and politics in the 1830s. After the 1832 Reform Act, which extended the British franchise from about 500,000 to around 813,000 voters, there followed a period of uncertainty over what direction politics was heading in and the roles of scientific thinking and objectivity in political philosophy. Before the ancient Palace of Westminster burned down in 1834 such questions, in relation to the architecture of Parliament, were somewhat limited to politically radical Utilitarian circles. Yet with the problem of rebuilding a new home for Britain's legislature, architecture became a subject of radical, Whig, and even Tory rhetoric. How Parliament was to be rebuilt involved architectural questions of style and engineering, yet contrasting approaches to government informed differing understandings over what was appropriate for the new Palace. This chapter examines the discourse which surrounded the new Palace of Westminster, and places the rebuilding of Parliament in the context of post-1832 politics.

The 1832 Reform Act was the political event of the decade. In this atmosphere of reform the physical Houses of Parliament had been subject to much attention, even before the fire. Varying suggestions for what form Parliament should take and what style or practical concerns should be prioritized were built on contrasting conceptions of what Parliament actually was. This was not self-evident. The Utilitarian *Westminster Review* portrayed Parliament as a machine for the manufacture of legislation. The radical Utilitarian MP Joseph Hume (1777–1858) felt Parliament to be a site of government, in which MPs were to act almost like automata in making objective decisions. Others considered Parliament not as a machine, but as an exemplar of the British constitution, embodying the government's right to rule. Robert Peel, for instance, asserted that this character was inseparably bound from the history which surrounded the location of Westminster as the seat of government.[1] These varying notions of what Parliament was continued throughout the nineteenth century. Conservatives, such as the three-time Prime Minister the Earl of Derby (1799–1869), viewed Parliament as 'a sovereign deliberative assembly, where parliamentary parties defined an authoritative national interest, distinct from ... the clamour of the populace'.[2] For Derby, Parliament's autonomy was essential to the British political system. By comparison journalist Walter Bagehot (1826–1877), writing in 1867, was cautious over the power contained within the walls of the Palace of Westminster, considering the Cabinet to be the true site of governmental authority. Bagehot believed that the true purpose of the House of Commons was to appoint a Cabinet to rule. As he put it, the 'cabinet, in a word, is a board of control chosen by the legislature, out of persons whom it trusts and knows, to rule the nation'. In this analysis, the House of Lords acted as 'a *reservoir* of cabinet ministers' to be chosen by the Commons.[3] Parliament was thus both a chooser of, and a reservoir for, executive governments. At the same time, Scottish philosopher Thomas Carlyle (1795–1881) informed his readers that Parliament was redundant due to the increasing prominence of debate in the pages of *The Times*.[4] With such a varied interpretation of Parliament's meaning, it is hardly surprising that the question of what form the building should take descended into controversy.

While the meaning of Parliament varied, there was a widespread conviction that the new building should embody enlightened government. In the 1830s politics and science were closely interconnected and this relationship was central to much thought on the new Palace. If politics could be progressive by being scientific, then displaying this through architecture was conceived as a powerful form of constructing political credibility for Parliament. The style adopted, materials of construction, attention to lighting and ventilating, and allocation of space to parliamentary business were all central concerns. They were ways of showing how the government was rational and committed to ruling effectively, but science itself was problematic. While reform characterized the 1830s, the threat of revolution was an ever-present spectre. Within this, science had an important role. Moderate Whigs and liberal Tories shared in the utopian hope that the spread of knowledge would bring about social progress. Science revealed divine providence, underpinned the material and

[1] House of Commons debate, 9 February 1836, *Hansard*, 3rd Series, Vol. 31, p. 243.
[2] Angus Hawkins, *The Forgotten Prime Minister: The 14th Earl of Derby. Volume II: Achievement, 1851–1869*, (Oxford, 2008), p. 422.
[3] Walter Bagehot, *The English Constitution*, (ed.) Miles Taylor, (Oxford, 2001), pp. 12–13.
[4] Thomas Carlyle, 'Parliaments', in Thomas Carlyle (ed.), *Latter-Day Pamphlets*, (New York, 1850), p. 10.

spiritual advance of civilization, and justified established political hierarchies. However, existing social orders faced challenges from science's new findings. Geology's teachings over deep time challenged the literal truth of the Bible, while chemistry was feared as a harbinger of materialism, and new astronomical theories proposing that suns evolved from gaseous nebulae shook traditional beliefs over the origins of the solar system. The idea that all in the universe, mental and spiritual, resulted from matter in motion, was a dangerous one socially.[5] Although science carried with it idealistic hopes of a society organized around nature, there were fears that such potent forms of knowledge could be misused. So although science assumed authority in the mid-nineteenth century, it was controversial in the 1830s. Not only were its own credentials unclear, but also it threatened existing authorities.[6]

Past histories have tended to overlook this context for Parliament's architecture. George Weitzman argued that the use of science and attention to utility in Parliament's construction could secure an image of enlightened governance.[7] To make the building practically efficient for conducting public business was to enhance government's credibility. Weitzman explained how Joseph Hume led radical Utilitarian calls for a new Parliament, built in reference to recent natural philosophy focused on optics, blood circulation, and respiration.[8] Hume's Utilitarian emphasis of function was accompanied by a commitment to the neoclassical style as reflective of scientific enlightenment and republicanism. Rorabaugh developed this by explaining how Whigs and Tories both favoured Gothic architecture as a means of implying government continuity in a post-reformed political landscape.[9] In this account, post-1834 Whig and Tory commitments to ecclesiastical Gothic architecture stood in the way of radical demands for a scientific, republican, and neoclassical Parliament.[10] Quinault's work on Parliament and the Victorian constitution extended

[5] James A. Secord, *Visions of Science: Books and Readers at the Dawn of the Victorian Age*, (Oxford, 2014), pp. 243, 7.

[6] Ibid., p. 242; on astronomy and the controversy of the nebular hypothesis, see Iwan Rhys Morus, *When Physics Became King*, (Chicago, 2005), p. 202.

[7] George H. Weitzman, 'The Utilitarians and the Houses of Parliament', *Journal of the Society of Architectural Historians*, Vol. 20, No. 3 (October, 1961), pp. 99–107; also see M. H. Port, 'The old Houses of Parliament', in M. H. Port (ed.), *The Houses of Parliament*, (New Haven, 1976), pp. 5–19, 9–13.

[8] Weitzman, pp. 103–04; on organic conceptions of city architecture, see Graeme Davison, 'The city as a natural system: theories of urban society in early nineteenth-century Britain', in Derek Fraser and Anthony Sutcliffe (eds.), *The Pursuit of Urban History*, (London, 1983), pp. 349–70.

[9] W. J. Rorabaugh, 'Politics and the architectural competition for the Houses of Parliament, 1834–1837', *Victorian Studies*, Vol. 17, No. 2 (1973), pp. 155–75, 156.

[10] Ibid., pp. 160–61.

this interpretation. He demonstrated that the pre-1834 Parliament was a Royal Palace, and that Barry's replacement structure continued this 'regal flavour'.[11] Parliament's longest room was the Victoria Gallery, its tallest structure was the Victoria Tower, and its decoration consisted of royal crests and sculptures. As Quinault put it, Parliament was evidently not 'a temple to Whiggism and Parliamentary sovereignty'.[12]

Philip Aspin has shown that a crude assessment of style as subject to party lines is unconvincing.[13] Attempts to label styles such as Gothic and neoclassical either Whig or Tory do not work. This was also true of making the Parliament building scientific. The value of science was not limited to any specific group, but was broadly shared. There was a sustained conviction that politics could be treated in a way systematic and objective, employing irrefutable knowledge.[14] This approach was often characteristic of Whig political philosophy, but shared similar intellectual foundations with Utilitarianism. The post-Newtonian concept that a good statesman should be adept at retaining and applying knowledge to social problems was, in the 1830s, a common conviction.[15] Joe Bord has demonstrated that this idea of politics as a science was central to Whig political culture.[16] Rather than limit science to Whig political philosophy, Bord cogently argues that science mattered to how a politician acted and displayed himself as a Whig. Illustrating natural philosophy to be at the heart of this 'Whig World', Bord defined four Whig values (liberality, statesmanship, cultivation, and rational sociability) through which scientific understanding could be articulated in the daily life of Whig politics. Building on Bord's conception of Whig manners, or rather the 'ways of being a Whig statesman', this chapter shows how in the 1830s there was an architectural dimension to Whig politics and science, expressed through the literature surrounding the new Parliament.

To do this, I first examine the architectural competition for the new Houses of Parliament and the selection of Charles Barry. I then go on to show how this decision was made at a time of political uncertainty. In 1834, the Whig government was far from stable following the Great Reform Act. Along with calling for further political reforms, radicals targeted the Palace as in need of rebuilding along functional lines. To

[11] Roland Quinault, 'Westminster and the Victorian constitution', *Transactions of the Royal Historical Society*, Vol. 2 (1992), pp. 79–104, 103.

[12] Ibid., pp. 82–86.

[13] Philip Aspin, 'Architecture and identity in the English Gothic Revival, 1800–1850', (Oxford University DPhil, 2013), p. 127.

[14] Stefan Collini, Donald Winch, and John Burrow, *That Noble Science of Politics: A Study in Nineteenth-Century Intellectual History*, (Cambridge, 1983), p. 3.

[15] Ibid., p. 13.

[16] Joe Bord, *Science and Whig Manners: Science and Political Style in Britain, c.1790–1850*, (Basingstoke, 2009), p. 2.

accompany the new political system, they demanded an efficient, more machine-like organ of governance. Such arguments were informed by a commitment to making politics scientific. However, I conclude by showing that while science could guide radical demands for a new Parliament, it could also shape broader interpretations of the new Palace. Amid the debates surrounding the new building, there was a rhetorical consensus that the Palace should reflect enlightened government, and this was widely believed to entail attention to science. This chapter then, shows how rebuilding Parliament was both a political, and an engineering challenge.

Catastrophe and Competition

On 16 October 1834, the Palace of Westminster was destroyed in a fire (Figure 1.1). While some observers feared the apocalyptic disaster to be a divine judgement for the overthrow of God's natural order in 1832, *The Times* chose to capture the drama and 'terrific splendour' of the spectacle.[17] After the flames had finished their destructive work, the newspaper recorded the moment as a political epoch. The question of whether a new political system warranted a new building had long plagued Parliament but, as *The Times* reported, the 'motion for a new house' was now 'carried without a division'.[18] The morning of 17 October marked the dawn of a controversy over how and who should build the new home of Britain's legislature. The Whig Prime Minister Lord Melbourne (1779–1848) instructed Robert Smirke (1780–1867), architect of the British Museum and influential leader of Greek revival architecture, to construct temporary accommodation for the Commons and Lords. Smirke reserved the remains of the old Painted Chamber for the Lords, and the walls of the old House of Lords for the new temporary House of Commons. After the addition of a roof and much refurbishment, the fire-damaged Upper Chamber provided the Commons with a larger space than the original Lower Chamber in St Stephen's Chapel. This conversion from fire-damaged ruins into a temporary Parliament was completed on 17 February 1835, by the adoption of prefabricated iron girders and timbering (Figure 1.2).[19]

[17] Caroline Shenton, *The Day Parliament Burned Down*, (Oxford, 2012), pp. 3, 33; (Anon.), 'Destruction of both Houses of Parliament by fire', *The Times*, (London, England), 17 October, 1834; p. 3; Issue 15611.

[18] (Anon.), 'Destruction of both Houses of Parliament by fire', *The Times*, (London, England), 18 October, 1834; p. 5; Issue 15612.

[19] M. H. Port, 'The new Houses of Parliament', in J. Mordaunt Crook and M. H. Port (eds.), *The History of the King's Works*, Vol. VI: 1782–1851, (London, 1973), pp. 573–626, 574;

Figure 1.1 Joseph Mallord William Turner's 'The Burning of the Houses of Parliament' (1834). This is by permission of the Tate Britain, 2017

Figure 1.2 Committee rooms forming part of the temporary accommodation for the Houses of Parliament

Parliamentary Archives (PA) ARC/PRO/WORK11/26/6, 'Report on St Stephen's Chapel, 1835–37' (16 July, 1835), p. 15; Rebekah Moore is at present researching the first detailed study of the temporary accommodation in her PhD at the University of London, entitled 'Rehousing Parliament: temporary Houses of Parliament and the New Palace of Westminster, c.1830–c.1860'.

The question of the permanent new Parliament's architecture was one which reached the very pinnacle of government. Melbourne wrote to King William IV of his fears should 'debate and diversity of opinion' engulf the question of the new Palace. Melbourne was strongly committed to preserving the historical associations of Westminster's 'ancient and established place of assembly'.[20] Moving sites, he was sure, would encourage a larger accommodation for spectators which had in the past had 'fatal effects', such as during the 1789 French Revolution, when 'large galleries filled with the multitude' had exerted pressure on the 'deliberation of public assemblies, and consequently upon the laws and institutions of nations'. The dangers of large debating chambers were to be avoided. Melbourne also favoured preserving the site in order to continue the 'character, form, and extent' of the old Parliament. He hoped that this continuity would secure political stability. Melbourne requested Smirke rebuild Parliament quickly but his temporary fall from government in November 1834 jeopardized this appointment. Robert Peel's (1788–1850) brief Conservative administration was immediately confronted with the same problem of rebuilding Parliament. Like Melbourne, Peel defended the historical associations of the old site and favoured Melbourne's choice of architect.[21] Smirke had worked on Peel's country seat, Drayton Manor, and his neoclassical townhouse near Parliament. In 1835, Smirke's Conservative credentials were underlined by his membership of the Carlton Club.[22]

It was the ex-Tory MP and master of King Leopold of the Belgians' household, Edward Cust (1794–1878), who campaigned for a 'market' competition, judged by a King's Commission of men interested in architecture, to determine the design of the new Parliament.[23] He wanted individuals, who were not professional architects, to specify the requirements of the competition and evaluate the entries. This 'experiment' of a competition would, Cust reckoned, secure an appropriate building and a

[20] 'Letter from Lord Melbourne to the King, 1 Nov., 1834', in Lloyd C. Sanders (ed.), *Lord Melbourne's Papers*, (London, 1889), pp. 213–14.

[21] Rorabaugh, pp. 161–62; House of Commons debate, 9 Feb., 1836, *Hansard*, 3rd Series, Vol. 31, p. 243; Richard Riddell, 'Smirke, Sir Robert (1780–1867)', *Oxford Dictionary of National Biography*, Oxford University Press, 2004; online edn, May 2010 [www.oxforddnb.com/view/article/25763, accessed 12 September 2014].

[22] Rorabaugh, p. 169.

[23] (Anon.), 'Cust, Sir Edward, baronet (1794–1878)', rev. James Lunt, *Oxford Dictionary of National Biography*, Oxford University Press, 2004 [www.oxforddnb.com/view/article/6973, accessed 7 March 2014]; Edward Cust, *A letter to the Right Honourable Sir Robert Peel, Bart. M. P. on the expedience of a better system of control over buildings erected at the public expense; and on the subject of rebuilding the Houses of Parliament*, (London, 1835), pp. 12, 15.

competent architect.[24] Over the preceding decade Cust had been gravely concerned by the state of public architecture. The 1820s had been a tumultuous time for government works. Under George IV's government, expenditure spiralled out of control, with architects appointed without competition and financial estimates rarely proving accurate. For example, the architect John Nash (1752–1835) initially estimated his designs for Regent's Street would cost £385,000 to build, but by 1826 the expense exceeded £1,533,000.[25] The most embarrassing of these public scandals was Nash's work at Buckingham Palace. Aside from his design's lack of aesthetic unity due to the connection of the main block to terminal pavilions with low wings, the architect's spending was so lavish that it became the subject of a select committee in 1828. It was later revealed that Nash did not believe architects should be held responsible for exceeding the costs of their estimates.[26] Calculating the cost of tradesmen and building materials did not, Nash contended, fall within an architect's remit. Furthermore, architects who were dependent on state patronage, such as Nash and Smirke, were hostile to open competitions, regarding them as a waste of labour.[27] While these government architects had the favour of George IV they could avoid competition, but with his death in 1830 and mounting public anger, there was a growing consensus that a modern professional architect had to be competent at calculating estimates and know the costs of materials, understand the skills and labour required in construction, and produce accurate drawings, good enough to win competitions.[28]

Until losing his seat for Lostwithiel when it was disenfranchised in 1832, Cust had argued in Parliament that the government needed to exert greater authority over architectural projects.[29] He was sure that a competition to find a professional architect, who could deliver skilled work without extravagant expenditure, would best secure the nation's

[24] Cust, *A letter to the Right Honourable Sir Robert Peel*, pp. 16–17; compare with criticism of patronage, in T. Juvara, *Strictures on architectural monstrosities, and suggestions for an improvement in the direction of public works*, (London, 1835).

[25] Geoffrey Tyack, 'Nash, John (1752–1835)', *Oxford Dictionary of National Biography*, Oxford University Press, 2004; online edn, May 2009 [www.oxforddnb.com/view/article/19786, accessed 22 February 2015]; M. H. Port, 'Parliamentary scrutiny and treasury stringency', in Crook and Port (eds.), *The History of the King's Works*, pp. 157–78, 157.

[26] Port, 'Parliamentary scrutiny and treasury stringency', p. 161; on Buckingham Palace, see M. H. Port, 'Buckingham Palace', in Crook and Port (eds.), *The History of the King's Works*, pp. 263–93.

[27] Port, 'Parliamentary scrutiny and treasury stringency', p. 162.

[28] Ibid., pp. 173–74; an architect's status was imprecise in the 1830s, positioned somewhere between a professional and a tradesman, see Christopher Webster, *R. D. Chantrell (1793–1872) and the Architecture of a Lost Generation*, (Reading, 2010), p. 181.

[29] Port, 'Parliamentary scrutiny and treasury stringency', p. 176.

interests. However, he was also eager to ensure Parliament was aesthetically pleasing. Cust did not trust the generation of neoclassical architects, which included Smirke, to deliver acceptable designs. While Buckingham Palace had been generally derided, Cust had personally worked alongside Barry to improve William Wilkins' (1778–1839) designs for the National Gallery which were criticized publicly for lacking aesthetic quality.[30] Cust's calls for a competition in 1835 were part of a widespread public campaign against the old practice of monopolizing architectural projects.[31]

When Peel briefly became Prime Minister after the election of late-1834, he maintained that the choice of new building would remain with those who would pay for the new Palace, the Commons.[32] However, due to the increasing public hostility towards the government's architectural patronage, Peel embraced Cust's calls for an open competition. In March 1835, the Conservative government established a select committee to consider the form of the new building.[33] This committee paid particular attention to practical questions, including the physical arrangements of the Commons and means for preserving MPs' health. Along with matters of practicality, it limited the architectural style to Gothic or Elizabethan, believing that these would be consistent with Parliament's medieval heritage and surrounding architecture, particularly the Henry VII Chapel of Westminster Abbey.[34]

On Melbourne's return to power in April 1835, this open approach was continued. The Whig First Commissioner of Woods and Forests from 1835 to 1841 and brother-in-law of Melbourne, Lord Duncannon (1781–1847), appointed five judges to a royal commission adjudicating on an open competition for the new building. This scheme was part of a wider project to reform the practices of the Office of Woods and Forests, which administered the Office of Works from 1832 and was responsible for public architecture. Duncannon's role of responsibility for public and

[30] M. H. Port, 'Barry, Sir Charles (1795–1860)', *Oxford Dictionary of National Biography*, Oxford University Press, 2004; online edn, October 2008 [www.oxforddnb.com/view/article/1550, accessed 4 December 2013].

[31] M. H. Port, 'The Houses of Parliament competition', in Port (ed.), *The Houses of Parliament*, pp. 20–52, 21–23.

[32] British Library (BL) British Museum, Add. Ms. 40,413, Peel Papers, Vol. CCXXXIII, folio 134, 'Letter from Sir Robert Peel to Edward Cust', (8 February 1835).

[33] M. H. Port, 'The new Houses of Parliament', in Crook and Port (eds.), *The History of the King's Works*, pp. 573–626, 575; also see House of Commons debate, 2 March 1835, *Hansard*, 3rd Series, Vol. 26, pp. 469–71.

[34] *Report from the Select Committee on Rebuilding Houses of Parliament; with the minutes of evidence, and an appendix*, PP. 1835 (262), pp. 3–4; Kenneth Clark, *The Gothic Revival: An Essay in the History of Taste*, (Harmondsworth, 1962), p. 99.

government buildings was both administrative and also highly political.[35] Duncannon and Melbourne delegated the choice of Parliament's style and form to the royal commission, made up of Cust, Thomas Liddell, George Vivian, and as chairman, the Whig MP for Tewkesbury and amateur architect Charles Hanbury-Tracy (1778–1858). With the closing date set for 1 December 1835, ninety-seven proposed designs were submitted from which the commissioners shortlisted four.[36] From these they declared entry sixty-four to be outstanding and announced it as the winner. This entry was that of Charles Barry.

Barry's entry conformed to the competition specification of a Gothic structure. He proposed a Palace rich in medieval Gothic detail, typical of the fourteenth, fifteenth, and sixteenth centuries. At the House of Lords' end of the building, a colossal square tower, known as the Victoria Tower, was proposed over the Royal Entrance. His Gothic plans were to be built around surviving medieval structures, such as Westminster Hall and the Undercroft Chapel, beneath St Stephen's Chapel. In this way, the past was incorporated with the new. Barry's designs, although lavishly covered in elaborate Gothic carvings, were actually rather classical in style, based on the principle of symmetry. Much of the Gothic detail, as will be seen in Chapter 2, was the contribution of Augustus Pugin (1812–1852). For many Tory and Whig commentators, these Gothic plans appeared to reaffirm traditional and institutional power, implying continuity with the past in a narrative of gradual progress.[37] However, Barry's entry was hugely controversial in post-1832 British politics.

Radical Proposals: The Politics of Barry's Appointment

The choice of Barry's designs came at a crucial moment in British politics. Even before the destruction of the old Houses of Parliament the question of what form the legislature should take was controversial, with radical Utilitarians demanding a new building more befitting of a modern, enlightened, political system. These calls for a physical Parliament to accompany the 1832 Reform Act were grounded in a conviction that politics should be made more scientific. An enlightened government required an appropriate building, and radicals were sure that this was not to be found in a Gothic shrine to traditional notions of a Constitution,

[35] Dorothy Howell-Thomas, *Duncannon: Reformer and Reconciler, 1781–1847*, (Norwich, 1992), pp. 198–248, and on Parliament see pp. 225–34; K. Theodore Hoppen, 'Ponsonby, John William, fourth earl of Bessborough (1781–1847)', *Oxford Dictionary of National Biography*, Oxford University Press, 2004; online edn, January 2008 [www.oxforddnb.com/view/article/22500, accessed 4 December 2013].

[36] Port, 'The Houses of Parliament competition', p. 41. [37] Rorabaugh, p. 156.

but in a structure which prioritized function and efficiency. While a Classical style would conjure up images of republican democracy, attention to lighting, ventilation, and suitable space for business would ensure Parliament became a more mechanistic, rational organ of governance.

Before 1834 Joseph Hume led this campaign for a building of efficiency and utility. Born in Scotland, the son of a shipmaster, Hume's education included natural science. Apprenticed to a surgeon in 1790, he then studied medicine at the Universities of Aberdeen and Edinburgh. After gaining experience as a naval surgeon's mate and seeing service with the East India Company from 1799, he took his first seat in Parliament in 1812. A confirmed radical, the MP for Kilkenny had been a fervent advocate of the Reform Bills of 1831 and 1832.[38] He campaigned for the secret ballot, triennial parliaments, further franchise extension, and religious toleration. He also worked alongside Edwin Chadwick to draft public health legislation, employing his experience as a physician. In all his endeavours Hume was guided by his Utilitarian philosophy. During his time at the Montrose Academy, Hume's closest friend had been James Mill (1773–1836), and in later life he shared the philosophic radicalism of Jeremy Bentham (1748–1832) and John Stuart Mill (1806–1873). This radicalism held that all actions should be judged morally right or wrong based on their creation or diminution of pleasure, and that this held implications for legal and social institutions.[39] Utilitarian philosophical radicalism argued that the maximizing of happiness was the moral standard by which all human choices should be measured.

Even before the fire, Hume raised the question of the overcrowded reformed Parliament in the Commons. Debates were conducted, sometimes continually for over seventeen hours, in an atmosphere which Hume deemed unhealthy. In 1833, he declared the House unfit for more than 400 MPs, but due to the 1832 Reform Act, this number was now at over 600.[40] Hume's concerns were grounded in his readings of medical texts, such as physician William Harvey's (1578–1657) teachings, which focused on blood flow and respiration.[41] He found support

[38] V. E. Chancellor, 'Hume, Joseph (1777–1855)', *Oxford Dictionary of National Biography*, Oxford University Press, 2004; online edn, January 2008 [www.oxforddnb.com/view/article/14148, accessed 7 March 2014]; also see Valerie Chancellor, *The Political Life of Joseph Hume, 1777–1855*, (London, 1986).

[39] Ronald K. Huch and Paul R. Ziegler, *Joseph Hume: The People's M.P.*, (Philadelphia, 1985), p. 3; on Utilitarian philosophy and politics, see James E. Crimmins, *Utilitarian Philosophy and Politics: Bentham's Later Years*, (London, 2011).

[40] House of Commons debate, 2 July 1833, *Hansard*, 3rd Series, Vol. 19, p. 61.

[41] Andrea Fredericksen, 'Parliament's genius loci: the politics of place after the 1834 fire', in Christine Riding and Jacqueline Riding (eds.), *The Houses of Parliament: History, Art, Architecture*, (London, 2000), pp. 99–111, 108; on Harvey's teachings and city planning,

from the MP for Bridport, Henry Warburton (1784–1858), a fellow doctor who believed sitting in the Commons to be a 'state of bodily torture'.[42] Hume cited evidence collected during a recent select committee which he had chaired. The problem of Parliament's physical structure had been assessed before the Reform Act, in 1831. During this select committee the architect Benjamin Wyatt (1775–1852) provided technical information on sound and ventilation in the Commons. Wyatt considered possible alterations to the chamber's roof to assist the acoustics. He asserted that raising the Commons' roof would improve hearing, having weighed up the potential for sound vibrations to irregularly collide with 'the particles of the fluid which constitutes the vehicle of sound'.[43] Despite such optimism, the committee judged the House to be both inadequate for enacting efficient public business, and also incapable of undergoing any improvement.

Hume's 1833 committee placed increased emphasis on practical matters and judged that a new Parliament was essential for a reformed legislature. It reviewed the form a new Parliament should take. This committee, which included Peel, Warburton, and the MP for Monmouth and civil engineer, Benjamin Hall (1802–1867), declared ventilation to be very imperfect, and the threats to the health of MPs to be extensive.[44] Hume's questioning focused on the transmission of sound, debating chamber shape, and means of ventilation. The architect John Soane (1753–1837) felt Parliament's existing site secured good air circulation, but recommended a debating chamber based on the Olympic Theatre at Vicenza.[45] This would mean a rectangular room with a semi-circular apse for the transmission of sound. Smirke agreed with Soane's straight parallel sides, but favoured a semi-hexagonal apse.[46] Decimus Burton (1800–1881), protégé of Nash and architect of the Athenaeum, advised Hume that a new Parliament should

see Richard Sennett, *Flesh and Stone: The Body and the City in Western Civilization*, (London, 1994), pp. 255–70.

[42] House of Commons debate, 2 July 1833, *Hansard*, 3rd Series, Vol. 19, pp. 63–64; H. C. G. Matthew, 'Warburton, Henry (1784–1858)', *Oxford Dictionary of National Biography*, Oxford University Press, 2004; online edn, May 2009 [http://ezproxy.ouls.ox.ac.uk:2117/view/article/28672, accessed 8 April 2013].

[43] *Report from the Select Committee on House of Commons Buildings; together with the minutes of evidence taken before them*, PP. 1833 (17), p. 12.

[44] G. F. R. Barker, 'Hall, Benjamin, Baron Llanover (1802–1867)', rev. H. C. G. Matthew, *Oxford Dictionary of National Biography*, Oxford University Press, 2004; online edn, January 2012 [www.oxforddnb.com/view/article/11945, accessed 22 February 2015].

[45] *Report from the Select Committee on the House of Commons' Buildings; with the minutes of evidence taken before them*, PP. 1833 (269), pp. 11–12; David Watkin, 'Soane, Sir John (1753–1837)', *Oxford Dictionary of National Biography*, Oxford University Press, 2004; online edn, January 2008 [www.oxforddnb.com/view/article/25983, accessed 22 February 2015].

[46] PP. 1833 (269), p. 13.

be 'advantageously placed for light and ventilation' and have a rectangular debating chamber with semi-circular ends.[47] This was based on experiments he had made on sound in theatres and churches. Nevertheless, he felt parallel benches were vital to securing the intense cross-bench debate characteristic of English politics.[48] Hume also interrogated Burton over his newly designed system of lighting, conceived so as to avoid stress on the eyes of members.[49] To subsequent witnesses, including architects James Savage (1779–1852), Thomas Hopper, and Edward Blore (1787–1879), Hume pursued questions over the relation between sound and the shape of walls and height of roofs, considering the advantages and drawbacks of a circular debating chamber like that of the Chamber of Deputies in Paris. His commitment to utility shaped his questioning. However, the witnesses' evidence presented little accord over what was a practical form and shape for a new chamber.

Hume's efforts won praise in the *Westminster Review*. Sir Henry Cole (1808–1882), the inventor of the commercial Christmas card and an establishing member of Imperial College London, heralded Hume's campaign for a suitable building for Britain's democratic representatives.[50] As a friend of John Stuart Mill, Cole was close to the Philosophical Radicals and along with promoting improved urban drainage and water supply, would go on to play an instrumental role in the 1851 Great Exhibition before becoming the first director of the Victoria and Albert Museum.[51] He believed the physical structure, or 'machinery', of Parliament to be an important matter. To pursue reformed politics in the 'barbarous' House of Commons was impractical. The building was too small for 658 MPs and presented an unhealthy atmosphere where members were subjected to noxious vapours and typhus fever.[52] Cole felt the Whig and Tory commitment to the existing Commons was a scheme to undermine the impact of the Reform Act. It was 'a regular system for driving out the people's agents by making the house too hot to hold them'.[53]

Hume's commitment to a new Parliament of utility was typical of a broader Utilitarian agenda. Both James Mill and Joseph Hume had Scottish roots, as did their political philosophy. The Utilitarian philosophical

[47] Ibid., p. 28; Dana Arnold, 'Burton, Decimus (1800–1881)', *Oxford Dictionary of National Biography*, Oxford University Press, 2004; online edn, May 2012 [www.oxforddnb.com/view/article/4125, accessed 22 February 2015].
[48] PP. 1833 (269), pp. 28–29. [49] Ibid., p. 30.
[50] Ann Cooper, 'Cole, Sir Henry (1808–1882)', *Oxford Dictionary of National Biography*, Oxford University Press, 2004; online edn, January 2008 [www.oxforddnb.com/view/article/5852, accessed 7 March 2014]; Henry Cole, 'Parliaments of our ancestors', *Westminster Review*, Vol. XXI (October, 1834), pp. 319–34, 319.
[51] Elizabeth Bonython, *King Cole: A Picture Portrait of Sir Henry Cole, KCB 1808–1882*, (London, 1982), pp. 3–6.
[52] Cole, 'Parliaments of our ancestors', p. 320. [53] Ibid., p. 334.

radicalism of Bentham and Mill placed emphasis on the science of legislation; Bentham's political science was a set of utility-based principles guiding the choices of the legislator.[54] Mill and his son, John Stuart, both believed that for the science of politics to be truly enlightened, increasing the franchise was required, and argued that government had to focus on principles rather than party.[55] John Stuart Mill argued that the principle of utility in government was essential because it was the purest measurement of society's progress to high civilization.[56] Mill's emphasis on intellectualism in governance built on the work of the University of Edinburgh's professor of moral philosophy from 1785 to 1810, Dugald Stewart (1753–1828), who placed value on science in politics as a means to progress. Like Stewart, Mill believed that the nature of reasoning in science carried an authority which was transferable to politics. Mill explained that induction, by which he meant the operation of discovering general propositions and rules from observable evidence, was vital in both science and politics. He was confident that induction was at the heart of logic in science, and that 'a complete logic of the sciences would be also a complete logic of practical business and common life'.[57] In politics, as in science, it was important to establish general truths through induction, and then make choices based on these known principles. Mill was thus concerned with establishing a system of epistemological methodology applicable to the physical world, which could be transferred to questions of morality and science.[58] This concept was at the heart of the Utilitarian philosophical radicalism which Hume shared. So not only would a Parliament of utility be practically good for an efficient government, but for the Parliament to itself espouse utility had broader implications. It would demonstrate to observers that Parliament was an organ of objective, scientific government and that here was an authority both capable and credible of ruling the nation.

These radical overtones made architecture and the prioritizing of practical concerns like lighting and ventilation contentious at Westminster. Peel summarized Whig-Tory concerns regarding the Utilitarian drive for a new Parliament in his response to Hume's 1833 select committee report. Peel believed that of all reports he had ever read, it was 'the most imperfect' and

[54] Collini, Winch, and Burrow, p. 95. [55] Ibid., pp. 102, 109. [56] Ibid., p. 14.

[57] John Stuart Mill, *System of Logic Ratiocinative and Inductive: Being a Connected View of the Principles of Evidence and the Methods of Scientific Investigation*, (London, 1886), pp. 185–86.

[58] Collini, Winch, and Burrow, p. 130; also see William Thomas, *The Philosophic Radicals: Nine Studies in Theory and Practice, 1817–1841*, (Oxford, 1979); on politics and science, see Harvey W. Becher, 'Radicals, Whigs and Conservatives: the middle and lower classes in the analytical revolution at Cambridge in the age of aristocracy', *British Journal for the History of Science*, Vol. 28, No. 4 (December, 1995), pp. 405–26.

the 'most discreditable'.[59] He damned the report's inability to produce a decisive opinion on any question and failure to select any of the twenty-two proposed plans for a new House of Commons. He warned that a larger debating chamber would be ineffectual for improving the discharge of public business, and that it was foolish to blame the building for all the faults of health and hearing among MPs.[60] Hume's motion to build a new Parliament was subsequently defeated in a division by 154 votes to seventy. Peel's reservations over a new Utilitarian Parliament were part of wider political uncertainties within Parliament immediately after the 1832 Reform Act. As leader of the Tory party Peel had, after the Duke of Wellington's loss of the leadership following his refusal to consider political reform, opposed Prime Minister Earl Grey's (1764–1845) reform programme between 1830 and 1832.[61] Both in opposition, and during his brief ministry between 1834 and 1835, Peel was cautious over making concessions to radicals. Many Whigs shared these fears. Following Grey's resignation in July 1834, Melbourne was confronted with navigating his ministry through a newly reformed political world.[62] Like Peel, Melbourne was apathetic towards the Utilitarian calls for a new Parliament.

This context of the post-reformed political system is important for understanding the Whig-Tory ambiguity regarding Hume's calls for a Parliament built to embody values central to Utilitarianism. Although it expanded the electorate to men who occupied premises worth at least £10 per annum in towns and £50 per annum in counties, abolished rotten boroughs, and created new representation for large towns like Birmingham and Manchester, the 1832 Reform Act changed very little of the existing political system at Westminster.[63] From 1832 until 1852, only Melbourne's cabinet between 1834 and 1835 contained more commoners than peers.[64] Despite this, Peter Mandler has shown that while the Reform Act had little effect on Parliament, it had a great impact on the

[59] House of Commons debate, 2 July, 1833, *Hansard*, 3rd Series, Vol. 19, pp. 64–65.

[60] Ibid., p. 65.

[61] John Prest, 'Peel, Sir Robert, second baronet (1788–1850)', *Oxford Dictionary of National Biography*, Oxford University Press, 2004; online edn, May 2009 [www.oxforddnb.com/view/article/21764, accessed 7 March 2014].

[62] Peter Mandler, 'Lamb, William, second Viscount Melbourne (1779–1848)', *Oxford Dictionary of National Biography*, Oxford University Press, 2004; online edn, Jan 2008 [www.oxforddnb.com/view/article/15920, accessed 7 March 2014].

[63] Robert Stewart, *Party and Politics, 1830–1852*, (Basingstoke, 1989), p. 32; John A. Phillips and Charles Wetherell, 'The Great Reform Act of 1832 and the political modernization of England', *American Historical Review*, Vol. 100, No. 2 (April, 1995), pp. 411–36, 414.

[64] Stewart, *Party and Politics*, pp. 33, 35; for post-1832 context, see Ian Newbould, *Whiggery and Reform, 1830–41: The Politics of Government*, (Basingstoke, 1990), pp. 81–101.

unenfranchised, provoking popular expectations of reforms to come.[65] Mandler demonstrated that this agitation concerned the post-reformed ministries of Grey and Melbourne. The big question for Grey and Melbourne between 1832 and 1834 was how could governments assert ministerial authority over popularly elected seats and provide direction to legislation? There were two managerial problems for post-1832 governments. Primarily, Parliament witnessed an increased activity by MPs, threatening the government's control over Parliament's legislative timetable, and secondly, an increase in petitioning.[66] The Whig ministry's response in 1833 was to reserve Mondays and Fridays exclusively for government business. This was, as the Speaker of the Commons noted, a reaction to 'popular feeling'. Grey and Melbourne feared that any election would return a government unable to govern; the Reform Act appeared to have created a 'rudderless' administration.[67] By the summer of 1834, Melbourne had endured two exhausting sessions of reformed Parliament. The following decade was a period of consolidating and reaffirming stability in a reformed system. As Mandler put it, 'Whigs and Tories are seen to be rallying together against the threat from below and groping towards the "Victorian Compromise" of moderate Liberalism'.[68] So although Parliament did not appear radically different in 1834 from its 1831 state, the Reform Act provided a context of uncertainty which was important in debates over Parliament's architecture both before and after the fire of 1834. At the heart of these debates was a concern not only with the form of a new building, but what Parliament should be, and what its business was. Challenges from radicals such as Hume were not new after 1833, but drew increasing Whig and Tory caution.[69] As Jonathan Parry concluded, the Whig government, which was above all a 'Reforming' government, never seemed to secure stability.[70]

In this context of uncertainty, the Utilitarian agenda for rebuilding Parliament aroused concern. Hume was proposing a radically new building, conceptualized as a mechanistic organ of government at the very moment the Whig administration sought to secure solidarity. Following Parliament's destruction, the Utilitarian imperative for a building of science and practical utility intensified. Hume assisted Arthur Symonds in publishing a Utilitarian

[65] Peter Mandler, *Aristocratic Government in the Age of Reform: Whigs and Liberals, 1830–1852*, (Oxford, 1990), p. 151; on Whig values, see Abraham D. Kriegel, 'Liberty and Whiggery in early nineteenth-century England', *Journal of Modern History*, Vol. 52, No. 2 (June, 1980), pp. 253–78; J. W. Burrow, *Whigs and Liberals: Continuity and Change in English Political Thought*, (Oxford, 1988).

[66] Jonathan Parry, *The Rise and Fall of Liberal Government in Victorian Britain*, (New Haven, 1993), pp. 95, 102.

[67] Ibid., p. 106. [68] Mandler, *Aristocratic Government*, pp. 155–57. [69] Parry, p. 99.

[70] Ibid., p. 128.

perspective of the opportunity the fire presented in the *Westminster Review*. He believed that the fire had removed the 'aching tooth' of government, eradicating all sentimental attachments to the old building. As he crudely put it, referring to the negligence of the old Parliament's caretaker, all Whig and Tory arguments had been 'dissolved by the carelessness of some slut that forgot to sweep her hearth-stone'. Symonds hoped the new Parliament would be functional, with space for committee rooms, accommodation, and all the appropriate 'machinery' for individual MPs to work to their full potential.[71]

This analogy of an ideal Parliament as a machine ran through Symonds's article. A government ruling by political science and clear principles demanded a mechanistic building of Utilitarian function. He believed the new structure should be 'A powerful machine, of nicest force . . . of wondrous power . . . adjusted to a thousand special functions, yet combining for the production of one grand general effect'.[72] This description shows what was the ideal government building in Utilitarian philosophy, but also illustrates how subjective the meaning of Parliament was. For the Utilitarian journal, the question of a new House of Commons was not one of four walls, but 'by what machinery shall the legislative functions be best performed?' To work mechanistically, Symonds described how the new Parliament must have good access, sound, warmth, ventilation, and efficient means of division, as well as space for reporters, refreshment rooms, records, while also being fireproof. Interestingly he explained that a decrease in representative members would allow for a smaller debating chamber and make questions of heating and ventilating easier to address. The new Parliament could, he suggested, entail considerable political reform. Reducing the number of MPs would increase the legislature's efficiency. Furthermore, he declared that the new Parliament could only be truly enlightened if women were permitted inside the Commons. He believed a female audience within the Commons itself would reduce impolite debates and encourage MPs to be diligent on the benches.[73] A later committee on the admission of ladies to the Commons, which included Hume, appeased the *Westminster Review*'s calls for female spectators; as of July 1835, up to twenty-four spaces in the Strangers Gallery would be reserved for women.[74]

It was not just questions of utility that aroused radical passions. The use of Gothic was immensely alienating. For the likes of Hume, the Gothic did not

[71] Arthur Symonds, 'New House of Commons', *Westminster Review*, Vol. XXII (January, 1835), pp. 163–72, 164–65.
[72] Ibid., p. 165. [73] Ibid., pp. 169, 171.
[74] *Report from the Select Committee on the Admission of Ladies to the Strangers' Gallery; with the minutes of evidence*, PP. 1835 (437), p. 3; on this 'privileged space', see Kathryn Gleadle, *Borderline Citizens: Women, Gender, and Political Culture in Britain, 1815–1867*, (Oxford, 2009), pp. 57–58.

appear an appropriate style for an enlightened legislature. This question of style had actually been a divisive topic for over forty years before the fire. Attempts to reconceive the old Parliament as a neoclassical senate house, including those of architects Soane and Wyatt, had come to nothing following the increased association of the style with republicanism after the 1789 French Revolution. In 1799 Wyatt embraced the Gothic in his plans to rework the Lords' chamber and in subsequent years focused on reconstruction work which showed off the medieval splendour of the palace.[75] The nineteenth century witnessed a commitment to maintaining Parliament's Gothic character as part of an anti-revolutionary affirmation of English politics.[76] Radicals claimed the neoclassical as reminiscent of ancient democracies and emulative of continental republics.[77] Hume also believed that while the Gothic was especially prone to the ill-effects of weathering, neoclassical structures would defy decay better due to their lack of ornament.[78] The *Architectural Magazine* shared in these concerns with style and practicality, asserting 'the importance of science and engineering' in the new Parliament building.[79] The journal's founding editor, John Claudius Loudon (1783–1843), had met Bentham in 1803 after graduating from the University of Edinburgh in 1802. Through his *Architectural Magazine*, Loudon oversaw John Ruskin's first publication. Despite this, the journal only ran from 1834 to 1839.[80] Loudon was concerned with utility in architecture and promoted what he was sure was a rational approach to building which focused on the materials and practices of construction. His journal provided a commentary on the 'pragmatic functionalism' of the 'benefits of modern technology', including heating and ventilation.[81]

Loudon agreed that Hume was correct to be guided by 'fundamental principles of utility', avoiding damp and bad air, and focusing on facilitating efficient public business.[82] The new Parliament should be fireproof, lit by gas lamps, heated by steam boilers, and well ventilated. He felt its construction should build on 'human knowledge', which was

[75] Sean Sawyer, 'Delusions of national grandeur: reflections on the intersection of architecture and history at the Palace of Westminster, 1789–1834', *Transactions of the Royal Historical Society*, Vol. 13 (2003), pp. 237–50, 242.

[76] Ibid., pp. 242–45. [77] Rorabaugh, p. 157. [78] Weitzman, p. 105.

[79] Howard Leathlean, 'Loudon's architectural magazine and the Houses of Parliament competition', *Victorian Periodicals Review*, Vol. 26, No. 3 (Fall, 1993), pp. 145–53, 151.

[80] Brent Elliott, 'Loudon, John Claudius (1783–1843)', *Oxford Dictionary of National Biography*, Oxford University Press, 2004; online edn, May 2010 [www.oxforddnb.com/view/article/17031, accessed 7 March 2014].

[81] Leathlean, pp. 146–47; for an early call for improved ventilation, see Alfred Ainger, *On Ventilation, in reference to the Houses of Parliament*, (London, 1835).

[82] 'The Conductor', 'A new site for the Houses of Parliament suggested, and the fundamental Principles on which they ought to be designed pointed out', *Architectural Magazine*, Vol. III, No. 25 (March, 1836), pp. 100–03, 100–01.

'always progressive'. Loudon favoured open competition and a change of site. Although Loudon wanted a new competition for a new site without style restrictions, he felt Cust's existing competition, far from being unscientific had, in revealing the talent of Britain's architects, already contributed much 'to the progress of architecture as a science'.[83] The attention surrounding the new building was thus seen to itself be compiling and advancing architectural knowledge. The journal especially feared that the Gothic style did not represent modern knowledge. When Barry's plans were publicized, the style was deemed unenlightened in this 'age of railroads'.[84] It demanded a style reflecting modern learning, rather than the Gothic, which invoked the crude 'materialism of the Flemish school'. One commentator in the journal, particularly concerned by the future of architecture which was at stake in the rebuilding of Parliament, was the Devonian architect and engineer, Charles Fowler (1792–1867).[85] Fowler had experienced both disciplines by the 1830s, having constructed Totnes Bridge in Devon between 1826 and 1828. Like Loudon, Fowler believed the new Parliament signified an epoch 'for the development of genius, and the exercise of the arts and science'.[86] Good architecture embraced a wide range of arts and sciences and was the product of 'the profound resources of the philosopher'. Parliament mattered because it would, he predicted, give direction to architecture for years to come. Yet he derided Gothic as a third-rate style, essentially ecclesiastic and collegiate. No Gothic Parliament could be modern 'in this enlightened age'.[87] Outside of the *Architectural Magazine*, Fowler echoed these sentiments, choosing to focus instead on the site, but remaining convinced that Gothic was ignorant and that architecture was inherently connected to a nation's industry.[88] Style then was controversial, not only for politicians, but in wider circles, and Parliament was identified as the vital stylistic battleground. Along with calls for a more mechanistic legislature, style

[83] Ibid., pp. 102–3; for a comparative publication considered suitable for review in a 'scientific journal', see Xylopolist, *A few remarks on the style and execution of the New Houses of Parliament, the insertion of which was refused by a scientific journal for unknown reasons. With some additional observations, occasioned by the debate on the subject in the House of Commons, the 14th February, 1848*, (London, 1848).

[84] Leathlean, p. 149.

[85] Peter Leach, 'Fowler, Charles (1792–1867)', rev. *Oxford Dictionary of National Biography*, Oxford University Press, 2004 [www.oxforddnb.com/view/article/37426, accessed 7 March 2014].

[86] Charles Fowler, *Remarks on the resolutions adopted by the committees of the Houses of Lords and Commons for rebuilding the Houses of Parliament, particularly with reference to their dictating the style to be adopted*, (reprinted from the *Architectural Magazine*, September, 1835), p. 1; copy in Joseph Hume's papers at UCL.

[87] Ibid., p. 34.

[88] Charles Fowler, *On the Proposed Site of the New Houses of Parliament*, (London, 1836), p. 1.

was a subject for radical discussion. Commitments to a mechanistic classical Parliament engendered radical approaches to government.

The Science of Politics and the New Palace

A conviction that science should guide politics informed philosophies beyond those which were politically radical and this had consequences for wider understandings of how Parliament should be rebuilt. Across the political spectrum there was a consensus that the new building should embody enlightened governance. Curiously such interpretations often shared similar intellectual roots. Being scientific provided several political identities with an ideal of what the new Palace should be. To build a structure which embodied scientific learning through practical applications and an appropriate style was a broadly shared value.

In Whig politics science and governance were, in the 1830s, closely connected. Joe Bord has described the strong links between Whig political philosophy and science, as well as between Whig manners and the cultivation of objective knowledge.[89] Four Whig customs demonstrate the ways in which scientific engagement was an expression of identity for Whig statesmen. These Bord termed 'liberality', 'statesmanship', 'cultivation', and 'rational sociability'.[90] Bord used these manners to show how Whigs demarcated themselves from radicals and Tories. For Whigs, good government could be achieved by intellectually equipped statesmen who flaunted knowledge in the execution of their legislative duties.[91] Politicians were also to exude rational sociability, meaning the ability to value and accept all opinions, even if conflicting, in order to work together in coalition for the national good.[92] This paralleled accepting alternate intellectual positions in areas of natural philosophy such as Geology, so as to conduct effective improving investigations, often through learned societies. By liberality Bord has shown that the Whig sentiment of projecting a generosity of spirit towards matters of state concurred with a generosity towards scientific liberality. As a political behaviour, liberality involved opposing war and supporting political reform, while in science, liberality meant a devotion to the finding of truth while ignoring private interests and undue patronage.[93] Finally, the Whig manner of cultivation stemmed from a connection between Whig government and land. An

[89] Bord, *Science and Whig Manners*, p. 2; on Whig political reform and the Royal Society, see Roy M. MacLeod, 'Whigs and savants: reflections on the reform movement in the Royal Society, 1830–48', in Ian Inkster and Jack Morrell (eds.), *Metropolis and Province: Science in British Culture, 1780–1850*, (London, 1983), pp. 55–90.
[90] Bord, *Science and Whig Manners*, p. 3. [91] Ibid., pp. 31–55. [92] Ibid., pp. 56–78.
[93] Ibid., pp. 79–80.

appreciation of agriculture entailed agrarian chemistry and experiments on enhancing produce.[94] This pursuit of improving knowledge extended beyond agriculture to industrial duties. Crucially, it was in this way possible for Whigs to be utilitarian, that is concerned with matters of utility and improvement through enhanced knowledge, without subscribing to the political philosophy of Utilitarianism.[95]

Science shaped more than broad Whig manners, but an approach to government which emphasized an unbiased and objective manner of legislating. The belief that politics should be made a subject comprising of systematic knowledge was an ancient one.[96] In the eighteenth century the Scottish philosopher David Hume (1711–1776) injected this science of politics with an intense vigour which was part of a wider post-Newtonian attempt to apply experimental methodology to moral subjects. Collini, Winch, and Burrow have demonstrated how Dugald Stewart shared this Scottish inheritance with his students, including Henry Brougham (1778–1868), Francis Horner, Sydney Smith, and Francis Jeffrey.[97] These young philosophic Whigs founded the *Edinburgh Review* in 1802; Stewart played a shaping role in forming these men's persistence with politics as a science. Building on David Hume and Adam Smith's (1723–1790) works that proposed a link between advances in commerce and manufacturing, and good government and liberty, Stewart posited a scientific approach to politics.[98] Stewart actually opposed David Hume's philosophy of scepticism, which asserted that facts and assumptions were always open to uncertainty, and instead promoted a common-sense philosophy that held that observable qualities belonging to external objects constituted true knowledge. Stewart taught his students that the enlightened legislator would be directed by an impulse to improve the happiness of society and a consideration of 'general utility'.[99] He coveted a government of general principles, not private interests: this replicated natural philosophy's apparent objectivity. The active study of science was vital in this framework, cultivating improved intellect in legislators. Stewart envisaged 'a moral-cum-intellectual fusion of the purposes of science with the art of legislation'.[100]

If there was a personification of this Whig philosophy, then it was Stewart's pupil, Lord Brougham, who of all Whigs had the most to say about the new Parliament (Figure 1.3).[101] At the University of

[94] Ibid., pp. 102–34. [95] Ibid., p. 111. [96] Collini, Winch, and Burrow, p. 13.
[97] Ibid., p. 25. [98] Ibid., pp. 27, 36. [99] Ibid., pp. 37–38.
[100] Ibid., p. 42; Bord provides a study of the 1806–07 Whig ministry showing how Dugald Stewart's philosophy shaped government approaches to administration, in Joe Bord, 'Whiggery, science and administration: Grenville and Lord Henry Petty in the Ministry of All the Talents, 1806–7', *Historical Research*, Vol. 76, No. 191 (February, 2003), pp. 108–27.
[101] See Henry Brougham, *The Life and Times of Henry Lord Brougham Written by Himself*, Vol. 3 of 3, (Edinburgh, 1871); J. Harwood, *Memoirs of the Right Honourable Henry, Lord*

Figure 1.3 Henry Brougham in training: The embodiment of science and politics in 1830s Britain

Edinburgh academia was a highly political concern; academic offices were the patronage of the Town Council. In 1834 one critic felt that the University's recent decline was attributable to 'political intrigue', while

Brougham, (London, 1840); Robert Stewart, *Henry Brougham, 1778–1868: His Public Career*, (London, 1985).

Tory and Whig interests shaped the city's societies which endeavoured to promote science.[102] Natural philosophy, and particularly medicine, was therefore inseparable from political matters. Between 1790 and 1830 Edinburgh was home to a 'Whig science', in which political economy went hand in hand with chemistry.[103] Brougham had been at the centre of this world, believing that knowledge of chemistry would improve manufacturing and society. In part such beliefs stemmed from the university's inseparability from the city and local industrial area. Students of Edinburgh were often sure that while Oxford University's 'politico-classical dream' was detached from reality, the Scottish university's place in the context of an industrial town shaped an interest in addressing problems of poverty and industrialization.[104] Brougham typified such a perspective, for example, during his inquiry into the health and welfare of Ireland's poor.

Brougham was a prolific author who enjoyed natural philosophy, and was, when arriving in London after his Edinburgh education, the embodiment of a 'modern' Whig.[105] He shared Stewart's conviction that good political science, scientific method, and attention to natural philosophy were all intrinsically connected. Through the 1790s Brougham and Horner loved to visit places of manufacturing around Edinburgh or apply their chemical knowledge to practical problems of agricultural productivity.[106] At the 1812 general election Brougham, when attacking the Tory George Canning (1770–1827), claimed enlightened government displayed a considered use of knowledge and that this was fundamentally a Whig characteristic.[107] Brougham was preoccupied with epistemological matters and engaged frequently in debates over the role of hypothesis in contemporary science and experiment. For Brougham, hypotheses as generalizations based on unobserved phenomena were acceptable, but speculative hypotheses on unobservable phenomena were not.[108] Brougham was both a Whig and a natural philosopher deeply concerned with pursuing what he believed to be good scientific methodology. Maintaining a reliable epistemology in science was important for his political philosophy. As Stewart had taught, government should follow general principles rather than private interests and consider public matters objectively, reflecting on knowledge of observed facts. This was Stewart's definition of being scientific in government and it had ramifications for Brougham's views of Parliament's architecture in the 1830s.

[102] L. S. Jacyna, *Philosophic Whigs: Medicine, Science and Citizenship in Edinburgh, 1789–1848*, (London, 1994), p. 3; Steven Shapin, '"Nibbling at the teats of science": Edinburgh and the diffusion of science in the 1830s', Inkster and Morrell (eds.), *Metropolis and Province*, pp. 151–78, 153.
[103] Jacyna, pp. 6, 3. [104] Ibid., p. 159. [105] Collini, Winch, and Burrow, pp. 49–50.
[106] Bord, *Science and Whig Manners*, p. 119. [107] Ibid., p. 34. [108] Ibid., p. 48.

Collini, Winch and Burrow have demonstrated how Brougham's Whig science of politics shared much common ground with radical Utilitarian concepts of scientific government. This was reflected in the often overlapping approaches to the new Parliament's architecture. In the years running up to the 1832 Reform Act, Utilitarian philosophical radicalism and philosophic Whiggism contested and embraced similar territory. This confrontation climaxed during the 1832 reform.[109] While Utilitarians emphasized radical reform and utility, Whigs refused to allow utility to supersede moral feelings and favoured moderate reform.[110] Whigs envisaged reform which protected talented men in government, such as Brougham, who might not win power through popular support alone. Science in government was thus controversial as Utilitarians and Whigs shared calls for reform and objective government, combined with enlightenment through scientific learning. This was a broad adoption of Stewart's faith that enhanced intellect would secure social progress. These differences and similarities were manifest in the debates surrounding Parliament's rebuilding.

After Barry's appointment in 1836, Brougham raised doubts over how well the proposed Parliament building embodied progressive governance whenever the subject arose in the House of Lords. On one occasion, Brougham warned his fellow Peers that Barry's plans would produce 'a great long low Gothic building, which in a few years ... would become encrusted with smoke, and covered with innumerable Gothic ornaments, until it resembled a large engraving – an eye-sore to every body of taste'. He went on to predict that Barry's 'monument to their [the people of the ninth century] barbarity' would remain a proof of architectural ignorance 'when classical taste shall have overwhelmed ... this Gothic mania'.[111] Though rousing the laughter of fellow Whig, the Marquess of Lansdown, who opposed Brougham's views of the Gothic, these comments demonstrated Brougham's dissatisfaction with the selection of Barry. Lansdown's response, however, reveals the divided Whig opinion concerning the style of the new Palace.

Brougham's clearest evaluation of Parliament's architecture appeared in the *Edinburgh Review*.[112] In April 1837 he reviewed a series of articles assessing how Parliament could best be symbolic of enlightened government. The articles were the work of William Richard Hamilton

[109] Collini, Winch, and Burrow, p. 93. [110] Ibid., pp. 97–98.

[111] House of Lords debate, 17 May 1844, *Hansard*, 3rd Series, Vol. 74, p. 1247.

[112] On Brougham's position in the *Edinburgh Review*, see Joanne Shattock, *Politics and Reviewers: the Edinburgh and the Quarterly in the Early Victorian Age*, (London, 1989), pp. 26–27; on the *Edinburgh Review* and the Tory *Blackwood's Magazine*, see William Christie, *The Edinburgh Review in the Literary Culture of Romantic Britain: Mammoth and Megalonyx*, (London, 2009), pp. 147–66.

(1777–1859) who had attacked Barry's selection and the choice of Gothic. Praising Hamilton's 'good taste' and 'important service' in rejecting Barry's Gothic designs, Brougham noted that Hamilton was a reputable scholar, 'creditably known' and with high political connections.[113] Hamilton had 'manfully' protested 'against the barbaric' Gothic style. Brougham stipulated that the construction of Parliament was no small concern, but an unsurpassed event in the history of art; it was the most monumental work for a free people of the age. The choice to employ Gothic was a 'pain' for Brougham, who agreed entirely with Hamilton's praise for the classical art of 'the most enlightened ages'.[114] Brougham appealed to readers to seek out Hamilton's works and appreciate his insightful observations.[115]

That Hamilton's writings informed Brougham's approach to Parliament's architecture really matters. Brougham was the most outspoken Whig advocating that Parliament's architecture should symbolize enlightened government. He was also the epitome of a Whig statesman who united science and politics. In Hamilton's work, Brougham found the most cogent and sustained analysis of how and why Parliament should mirror scientific learning. Between 1836 and 1837 Hamilton composed three letters addressed to his former Foreign Office colleague Thomas Bruce (1766–1841), seventh Earl of Elgin, arguing for Parliament to represent science and knowledge through the Grecian style of architecture. These letters attracted much attention in the specialist and technical press and had a considerable readership within Parliament, including Hume. Hamilton had won fame in 1801 by foiling a French attempt to transport the Rosetta Stone from Alexandria to France following Napoleon's disastrous Egyptian campaign. He was at the time serving as the attaché to Elgin and the British ambassador to the Ottoman Empire in Constantinople.[116] The son of the Archdeacon of Colchester, Hamilton had attended Harrow and matriculated from St John's College, Cambridge, in 1795. After securing the Rosetta Stone and Parthenon Marbles as British war trophies, Hamilton returned back to London and worked in the Foreign Office until retiring on health grounds in 1824.

[113] Henry Brougham, 'The new Houses of Parliament', *The Edinburgh Review*, Vol. LXV, No. CXXXI (April 1837), pp. 174–79, 174.
[114] Ibid., p. 175. [115] Ibid., p. 178.
[116] William St Clair, 'Bruce, Thomas, seventh earl of Elgin and eleventh earl of Kincardine (1766–1841)', *Oxford Dictionary of National Biography*, Oxford University Press, 2004; online edn, May 2013 [www.oxforddnb.com/view/article/3759, accessed 7 March 2014]; R. E. Anderson, 'Hamilton, William Richard (1777–1859)', rev. R. A. Jones, *Oxford Dictionary of National Biography*, Oxford University Press, 2004; online edn, May 2006 [www.oxforddnb.com/view/article/12147, accessed 7 March 2014].

Aside from his diplomatic career, Hamilton enjoyed antiquarian and scientific pursuits. In addition to translating the Greek on the Rosetta Stone to English, Hamilton was a founding member of the Royal Geographical Society, an ardent supporter of the Royal Institution, and from 1838 a trustee of the British Museum. As president of the Royal Geographical Society, Hamilton oversaw the election of the Cambridge historian and philosopher of science William Whewell (1794–1866) in 1837 and was well acquainted with his writings.[117] Hamilton was also a close friend of the eminent geologist Roderick Murchison (1792–1871). Together they enjoyed days out combining natural history and mechanics. On one occasion Murchison visited Hamilton at his house in Portsmouth. They took breakfast before examining a stretch of coast and then surveying several Royal Navy vessels, including the HMS *Victory*. Following this, they rowed over to the Navy's New Victualing Establishment. There they witnessed the 'most curious thing . . . namely the Baking of Ship's biscuit by machinery'.[118] Hamilton and Murchison shared a mutual interest in the natural and the industrial. Indeed in 1843 Hamilton felt Murchison to be his ideal replacement as President of the Royal Geographical Society: such a man of science would secure the best interests of the society.[119]

In 1836, Hamilton turned his attention to Parliament's architecture. Hamilton's initial work denounced the Gothic style in favour of the Grecian. Gothic, he argued, was monastic in character and reflected a 'barbarous' period of history. Grecian however, Hamilton argued, embodied 'improved knowledge' and learning. He explained that 'Architecture had thus become a mirror of the improvement of science in various periods'.[120] He felt that the post-Reformation abandonment of the Gothic complemented England's 'more wholesome direction' in the arts, literature, and science. Inigo Jones and Christopher Wren produced classical architecture which mirrored the natural philosophy of men like Robert Boyle (1627–1691) and Isaac Newton (1642–1727). The Greek style imitated the 'grandeur of nature' and marked 'progress', while 'Gothic barbarism' was indicative of the ignorance of the Middle Ages.[121] Gothic architecture projected the romance central to the works of Walter Scott, but it also appealed to an age both 'feudal and ancestral',

[117] (Anon.), 'Royal Geographical Society', *The Times*, (London, England), 29 June, 1837; p. 3; Issue 16455.

[118] BL Add. Ms. 46,126, Murchison Papers Vol. II, folio 355, 'Letter from Hamilton to Murchison', (27 October, 1832).

[119] Ibid., folio 367, 'Letter from Hamilton to Murchison', (13 April, 1843).

[120] W. R. Hamilton, *Letter from W. R. Hamilton, to the Earl of Elgin, on the New Houses of Parliament*, (London, 1836), p. 5; UCL holds a copy which Hume annotated.

[121] Ibid., p. 7.

and therefore wholly inappropriate for a reformed legislature. Hamilton argued that architecture should show 'the advancement of national science', rather than fleeting literary fashion.[122]

In his second and third letters, Hamilton specified how Parliament could embody science in two ways. Primarily, architectural construction itself was a science. Hamilton stated that it 'it is idle to discuss whether architecture be a science or an art . . . it is both . . . It is based upon science, and it culminates in art'.[123] Architecture involved 'all the various developments of the properties of nature, of mathematical truths, and of inventive genius . . . it marks the progress of the human race in the powers of composition'. Hamilton described how the Greek style exhibited this employment of nature in a way superior to Gothic. Nature inspired all Greek structures. For example, architectural virtues of mathematics and geometry could be observed in the works of spiders.[124] Greek architecture employed such observations through the 'rude stems of oak or willow placed against each other in parallel lines, [and] the horizontal beams which rest upon them'.[125] To build with Grecian pillars (tree trunks) and in the Corinthian order (the leaves of the acanthus) was to 'copy from Nature'.[126] Such simplicity in weight distribution reflected 'rational faculties, truth and nature'. Hamilton believed Parliament should be an 'exemplification of the simplicity of ancient art applied to modern science'.[127]

According to Hamilton, the second way in which Parliament's architecture should be a work of science was to 'mirror' the science of the age. Architectural works were 'calculated to record the scientific and mechanical discoveries' of a period.[128] The Greek architecture of ancient times paralleled 'the progress of intellectual philosophy, whose real triumph in the person of Socrates was to enquire into, and interpret the phenomena of the moral and physical worlds'.[129] Hamilton believed that modern inductive science was the descendent of this Greek philosophy. Ancient Greece

[122] Ibid., p. 9.

[123] W. R. Hamilton, *Second letter from W. R. Hamilton, esq. to the Earl of Elgin, on the propriety of adopting the Greek style of architecture in the construction of the New Houses of Parliament*, (London, 1836), p. 5.

[124] W. R. Hamilton, *Third letter from W. R. Hamilton, esq. to the Earl of Elgin, on the propriety of adopting the Greek style of architecture in preference to the Gothic, in the construction of the new Houses of Parliament*, (London, 1837), p. 17.

[125] Hamilton, *Second letter*, p. 7.

[126] 'W. E. H.', 'Mr. Barry's design for the new Houses of Parliament', *Westminster Review*, Vol. 3 (25 July, 1836), pp. 409–24, 420; these comments complicate Levine's conception that, in the nineteenth century, 'history finally replaced nature as the sole basis for representation', in Neil Levine, *Modern Architecture: Representation and Reality*, (New Haven, 2009), p. 11.

[127] Hamilton, *Second letter*, p. 15.

[128] Hamilton, *Third letter*, p. 5.

[129] Hamilton, *Second letter*, p. 6.

had lacked the 'commanding necessity' to study and apply the 'physical sciences, such as Mechanics, Astronomy, Optics and Hydrostatics' because they did not have to navigate beyond the Mediterranean.[130] He claimed that in the Mediterranean, no compasses, astronomy, or optical glasses were required. Hamilton observed such techniques were only necessary for the discovery of the new World and passage to India, yet the ancients' application of geometry and mechanics to temples was highly advanced.[131] While Britain did not derive the skills of navigation from the Greeks, Hamilton showed that this was because of differing demands.

Although Hamilton distinguished between the specific subjects of ancient and modern natural philosophy, he felt that enlightened learning was part of a Greek inheritance. Furthermore, as Grecian architecture reflected nature and therefore mirrored the philosophical investigations of the natural world, Parliament should emulate the style to embody a nation which appreciated such enlightenment. Grecian embodied not only the mechanics of architecture, but all contemporary natural philosophy. For Hamilton, this mirroring of modern science was more important for Parliament than any other public building. He asserted that the Palace should reflect 'human intellect' and the reformed political system to project a character of enlightened government. In Parliament, 'politics, trade, justice, religion, property, laws, agriculture, jurisprudence, police, manufactures, roads, enclosures, all our daily wants and interests' were to be 'sifted, debated, and resolved'.[132] Hamilton contended that important discussions had to take place in an atmosphere of scientific enlightenment. Such work should have an architectural style which mirrored Britain's eminence in natural history, the pure and experimental sciences, and navigation. This was an inheritance from the ancients, not the barbaric Middle Ages.[133]

Brougham completely endorsed Hamilton's essays. The close association of science and governance were shared values and Brougham found Hamilton's views compatible with his own Whiggism. Although no Utilitarian, Hamilton's first publication concerning Parliament actually appeared in the *Westminster Review* in July 1836. Writing under the initials 'W. E. H.', Hamilton began by attacking Barry's design. Such a Gothic edifice represented 'the ascendency of the Church, and the triumph of Episcopacy'.[134] Hamilton warned sarcastically that the use of Gothic was part of a 'great ecclesiastical plot' to overturn the newly reformed constitution and restore power to the King and Bishops. He feared that the 'increasing ascendency of the Catholic party over the ministry' was transforming the Commons into 'a great monastic establishment'. Part of this monastic

[130] Hamilton, *Third letter*, p. 40. [131] Ibid., p. 41. [132] Hamilton, *Second letter*, p. 23.
[133] Ibid., p. 25. [134] 'W. E. H.', 'Mr. Barry's design', p. 409.

regime included opposition to allowing women into the gallery of the House. In the context of recent Catholic emancipation and the government-funded church building programme, Gothic perpetuated an increasingly High-Church Anglican, or even Catholic, approach to government. Dissenting non-conformist MPs were the target of such architecture. Barry's design would mean that 'The infidel portion of the Lower House will no longer be enabled to avoid going to church; every committee-room will be a chapel ... [furnished] with Bibles, prayer-books, and useful homilies'.[135] Hamilton believed that the reading of prayers before debates and Parliamentary sermons would soon be accompanied by organ music.[136] With its cathedral and monastic associations, Gothic was not modern, but inappropriate to house a Parliament governing a nation including dissenters, infidels, and Jews. Hamilton demanded the site be changed to one not consecrated, ideally Green Park, and the style be altered to one free of religious connotations.[137]

Hamilton also cited improvements to drainage and air circulation as additional reasons for a change of site. Despite his suspicion that most victims of poor ventilation would be 'elderly astmatical [sic.] gentlemen (who fortunately are all Conservatives)' and therefore harbouring hopes that Westminster's bad air would undermine underhand attempts to unseat worthy liberal representatives, Hamilton believed government should be sheltered from such 'noxious influences' and high mortality rates.[138] As for style, a switch to Grecian promised further practical improvements regarding the durability of stone. London's coal consumption would, Hamilton claimed, damage and discolour any public building of excessive ornament.[139] He felt that Gothic's reliance on intricate mouldings and delicate carvings for aesthetical quality would contrast poorly with Grecian's simplicity in London's deleterious atmosphere. However, Hamilton was not espousing a Utilitarian manifesto.

Hamilton's writings instead seem to fit better within the sentiments of what David Watkin has labelled the 'Cambridge Hellenists'. This group of travelling scholars, which included the prominent designer Thomas Hope (1769–1831) and architect William Wilkins, was active throughout the early nineteenth century in promoting academic interest in Ancient Greece.[140] In particular, Hamilton's thoughts resonate with those of the Hellenist Charles Kelsall (1782–1857) who argued that neoclassical architecture was a central

[135] Ibid., p. 410.
[136] On Parliamentary prayers, see Pasi Ihalainen, 'The sermon, court, and Parliament, 1689–1789', in Keith A. Francis and William Gibson (eds.), *Oxford Handbook of The British Sermon, 1689–1901*, (Oxford, 2012), pp. 229–44, 233.
[137] 'W. E. H.', 'Mr. Barry's design', p. 412. [138] Ibid., p. 414.
[139] Hamilton, *Letter from W. R. Hamilton*, p. 10.
[140] David Watkin, *Thomas Hope, 1769–1831: and the neo-classical idea*, (London, 1968), p. 64.

feature in the liberalization and modernization of the nation.[141] He called for a purifying process of civilization, in which Greek architecture, and specifically the pure Greek Doric order, was central. As early as 1814, Kelsall united architecture and science in a proposed reform of Oxford and Cambridge Universities. He asserted that Cambridge should be made to offer a variety of subjects, taught through six subject-specific colleges, including natural philosophy, mathematics, and agriculture and manufacturing. This scheme for what he referred to as a 'Nurse of Universal Science' envisaged much attention to the architecture of each college, including the mathematics college built in the Doric order. However, this was not shaped by a desire for political reform, as witnessed in the 1830s, but a commitment to intellectual neoclassicism.[142] Hamilton's arguments echo Kelsall's appraisal of the Greek style and science; much more so than Utilitarian notions of function. Neoclassicism was not inseparably linked to republican politics. As Frank Salmon showed, in the 1830s it was classical Roman architecture which was the dominant style of English public architecture.[143]

Hamilton's works are important because between 1836 and 1837 they were a focal point of discussion concerning the use of Gothic at Westminster. The case for the Gothic style was compelling, including appeals to nature and history every bit as elaborate as Hamilton's. Obviously, the style could easily be assimilated with the existing architecture of Westminster Hall and the Abbey. However, the other central arguments were historical and environmental. The idea that architecture was a body of knowledge comparable to science was built on a growing attention to history and nature. This had implications for the use of Gothic for the new palace. Historically, the Gothic was associated with Westminster; the place where it was held that the British constitution and law had been moulded. Importantly, the Gothic was fundamentally Christian in character and could be portrayed as a British style; its selection was an assertion of national confidence. *The Times*, for example, believed the Gothic appropriate because it was England's 'best national style'.[144] Much romantic sentiment had been stirred up

[141] Ibid., p. 71.

[142] Ibid., p. 72–74; David Watkin, 'Kelsall, Charles (1782–1857)', rev. *Oxford Dictionary of National Biography*, Oxford University Press, 2004 [www.oxforddnb.com/view/article/37627, accessed 22 February 2015]; he argued that Windsor Castle should be rebuilt in a Classical rather than Gothic style, in C. Kelsall, *A letter to the Society of the Dilettanti, on the works in progress at Windsor*, (London, 1827), pp. 6–7.

[143] Frank Salmon, *Building on Ruins: The Rediscovery of Rome and English Architecture*, (Aldershot, 2000), pp. 20, 138.

[144] (Anon.), 'Exhibition of designs for new Houses of Parliament, now exhibiting in the National Gallery', *The Times*, (London, England), 29 April, 1836; p. 6; Issue 16090.

by the 1830s thanks to the popularity of Walter Scott's (1771–1832) literary creations, especially his Waverley novels. Such enthusiasm for the medieval was enhanced after the fire of 1834. With Parliament in ruins, the eighteenth- and nineteenth-century alterations were stripped away revealing the staggering beauty of the original medieval walls of St Stephen's Chapel. This rediscovered ancient glory roused popular support for the Gothic.[145] Taken together with the romantic notions of Westminster's role in the formation of the British political system, this presented a powerful basis for those advocating a new Gothic palace.

In terms of nature, the Gothic also seemed appropriate. Promoters of the style noted its suitability for the northern Europe climate, with its sombre appearance fitting of Britain's grey skies and persistent rain. At the same time, its forms were claimed to be taken from nature, including vaulting which was likened to the meeting branches of trees in a forest. More than the neoclassical, the Gothic boasted endless variation in ornament, which was analogous to the infinite variety of botany.[146] As will be shown, this scientific basis for Gothic as a rational style of architecture became increasingly sophisticated during the decades following the 1830s. Such interpretations of the Gothic reached a crescendo in the 1850s with John Ruskin and the architects who followed Barry, including George Gilbert Scott, Alfred Waterhouse, and William Butterfield.

Nevertheless, in the 1830s, Hamilton had a wide readership. One review in the *Architectural Magazine* echoed Hamilton's view of architecture as the 'mirror of the improvement of science' and praised his argument for the superiority of the Grecian style.[147] Others opposed Hamilton's promotion of the Grecian. Colonel Julian R. Jackson (1790–1853), who eventually became secretary to the Royal Geographical Society, felt Hamilton's distinction between the Grecian and Gothic to be superficial.[148] Jackson had contributed several papers to the Royal Geographical Society arguing that geography should be an independent science and in 1834 proposed a systematic terminology for the study of river systems. Within the society, Jackson's work was commonly

[145] Caroline Shenton, *Mr Barry's War: Rebuilding the Houses of Parliament after the Great Fire of 1834*, (Oxford, 2016), p. 33; M. H. Port, 'The Houses of Parliament competition', in M. H. Port (ed.), *The Houses of Parliament*, (New Haven, 1976), pp. 20–52, 30.

[146] Port, The Houses of Parliament competition', p. 31.

[147] (Anon.), 'New Houses of Parliament', *Architectural Magazine*, Vol. IV, No. 37, (March, 1837) pp. 120–32, 121.

[148] J. R. Jackson, *Observations on a letter from W. R. Hamilton, Esq. to the Earl of Elgin, on the New Houses of Parliament*, (London, 1837), p. 10.

regarded to be scientifically rigorous.[149] When it came to architecture, Jackson thought that Hamilton was merely a 'professed amateur of the Greek style'. Jackson argued that true 'genius' was the ability to apply 'skill and science' to any style. The Parliamentary competition, by specifying either Gothic or Elizabethan, was thus effectively saying to architects, 'You have done well in the Grecian style; now show your science by a master-piece of Gothic'.[150] The competition was therefore actively promoting science. Jackson also felt the religious connotations of Gothic were appropriate for a morally upstanding and patriotic legislature. He praised Barry's designs and the Gothic in general. He believed that Gothic vaulting, spires, and buttresses revealed 'such a degree of science in the composition and division of forces . . . as can have resulted only from much mathematical knowledge'.[151]

While Brougham and Hamilton damned the Gothic for its medieval character and unscientific nature, advocates of the style were equally keen to defend its practical and scientific qualities. The *Quarterly Review* noted in 1837 that the philosophies of architecture were 'in a state of war'.[152] A Conservative organ, the journal sympathized with the commissioners of Cust's competition, who were caught in the conflict between Greeks and Goths. It asserted that recent appeals deploying 'technical knowledge' against the Gothic were merely the responses of disgruntled losing entrants to the competition. The journal believed vaulting to be an indication of enlightenment and felt the style to have great 'utility'.[153] Such work was the product of architects who had 'profoundly studied the natural feelings of mankind'. It was a style to conjure up emotion in the ignorant and encourage intellectual cultivation. Authoring the *Quarterly Review's* consideration of Parliament's architecture, the classical scholar John Bacon Sawry (1772[?]–1843) felt that the Gothic clearly displayed 'mechanical skill, and no less intellectual refinement than had been exhibited in the construction of the most finished Grecian temple'.[154]

Even staunch Tories felt compelled to wade into the debates over Parliament's architecture armed with a rhetoric of science and practicality. Sir Archibald Alison (1792–1867) provided a Tory commentary on British architecture in the Conservative *Blackwood's Edinburgh Magazine*. A graduate of the University of Edinburgh and son of an Episcopalian

[149] Elizabeth Baigent, 'Jackson, Julian (1790–1853)', *Oxford Dictionary of National Biography*, Oxford University Press, 2004 [www.oxforddnb.com/view/article/14540, accessed 12 September 2014].

[150] Jackson, *Observations*, p. 12. [151] Ibid., p. 29.

[152] John S. Morritt, 'Review of Hamilton, etc. on architecture', *The Quarterly Review*, Vol. 58 (February, 1837), pp. 61–82, 62.

[153] Ibid., pp. 65, 67. [154] Ibid., pp. 70–71.

cleric, Alison was a frequent contributor to *Blackwood's*.[155] Aside from adopting an anti-Malthusian view of population growth and writing a series of thirteen articles between 1831 and 1832 linking English reform with the French Revolution of 1830, Alison opposed all Whig ideology and argued that the 1832 Reform Act engendered a descent into anarchy. Architecture, Alison asserted, was a noble art which conveyed sentiment through history, defying time and testifying to 'the immortality of man'.[156] He considered Britain's cathedrals and monasteries as evidence of the nation's architectural pre-eminence. However, Alison detested the recent English penchant for building Regency edifices of 'monstrous fragility'.[157] The point of good architecture was to employ quality stone in a style which would defy the decay of time. It was an architect's duty to build works that would last, as the ancients had done. With the wealth of Britain's empire, Alison thought that the ancients would have 'made London the noblest city in Christendom' by using durable stone. As for style, rather than the recent neoclassical piles, a 'more manly' one was called for: one that would glorify the works of the 'Creator'. He felt this was Gothic.

Turning to the new Palace of Westminster, Alison believed a building was called for which honoured God and would endure the ages. He appealed for attention to be paid to the choice of stone and design so that the building might house Britain's government for centuries to come. He advised a construction '*entirely of stone*, fire-proof, and worthy of being the palace of the constitution'.[158] Architects should 'erect on such a scale of durability as may defy alike the war of elements, the decay of time, and the madness of people'. Alison's sentiments reveal a very different conception of what Parliament meant to those held in Utilitarian circles. Rather than a machine for legislation production, or making momentous decisions, Alison's ideal Parliament was a building which would confirm the solidarity of the British constitution.

This Tory call for practical attention to materials was framed with caution. Practical knowledge should be deployed to capture the truth of Anglican government for generations to come and defy time. Yet Alison warned against society's obsession with utility. Although utility was, he

[155] Michael Fry, 'Alison, Sir Archibald, first baronet (1792–1867)', *Oxford Dictionary of National Biography*, Oxford University Press, 2004 [http://www.oxforddnb.com/view/article/349, accessed 7 March 2014].

[156] Archibald Alison, 'The British school of architecture', *Blackwood's Edinburgh Magazine*, Vol. XL, No. CCL (August 1836), pp. 227–38, 231.

[157] Ibid., pp. 227, 232. [158] Ibid., pp. 234, 238.

conceded, vital in bridges, roads, docks, and canals, and was a source of national pride in works such as the Manchester Railway and Thames Tunnel, in architecture it was in danger of overcoming all aesthetics. He feared the nation was becoming a 'mere race of utilitarians', living by dividends. A Parliament determined purely by utility and function would therefore embody not only Utilitarian philosophy, but a more general utilitarian approach to life. Furthermore, it was the 'great convulsion of 1832' that was to blame. Alison argued that great works of durability were rare in 'Democratic societies' where governments lived in fear of expenditure. Reform entailed a stinginess inhibiting the construction of any great and lasting architecture.[159] Alison felt that the consequence of extended franchise was thus ephemeral architecture. Ironically, such an argument was intended to deconstruct utilitarian claims of practicality and reason. To build a Parliament worthy of Britain would entail the use of expensive quality stone; an option that Alison reckoned was severely undermined by recent Whig reforms.[160] Alison demonstrates how all political sides could employ a language of practical knowledge when articulating their ideas. What was contested was not whether the building should be scientific, but what was a suitable manner of being practical and what style best embodied British politics. Interestingly, similar to Hume's Utilitarianism and Brougham's Whiggism, Alison had strong connections with Dugald Stewart. Alison's father, Archibald Alison (1757–1839), was a lifelong friend of Stewart following his time at the University of Glasgow. Hume, Brougham, and Alison's intellectual foundations were not too distant.

Conclusion

Although in the 1830s, science had a broad meaning and curious political value, its prominence in contemporary society means that its adoption as a rhetorical tool in debates over the new Palace should not come as a surprise. In the hands of radicals, science became a rational approach apparently justifying a mechanistic, objective legislature. For others, it was a crucial epistemological resource for improving governance. Indeed, it became something which might help preserve the political system from more extensive reforms, or even dangerous revolutions. Barry's appointment came in

[159] Ibid., pp. 235, 237.

[160] In the late 1820s, Whig and Tory administrations both endeavoured to cut expenditure on public architecture, see Port, 'Parliamentary scrutiny and treasury stringency', pp. 158, 175.

the context of this delicate relationship between science and politics. He was required to produce a Palace emphasizing political stability and traditional continuity, but also one which appeared enlightened, built with attention to the latest scientific learnings. Rebuilding Parliament was an engineering challenge, and a political onc, and Barry intended to provide solutions to both.

2 Architecture and Knowledge: Charles Barry and the World of Mid-Nineteenth-Century Science

> Architecture is a science arising out of many other sciences, and adorned with much and varied learning.
>
> Vitruvius, *De Architectura*

Two years after the fire, in 1836, the competing architectural designs for the new building went on display in the National Gallery at Charing Cross. *The Times* felt that this public display, at such an 'interesting period of our political history', provided an 'unbiased view of the study of architectural skill, taste, and science in England'.[1] This review thus identified that the post-1832 political context mattered for Parliament's architecture, and that building the new Palace raised three questions: who was capable of guiding such an immensely challenging project; what form and style should the new building take; and what was the role of science in this venture? In the years between the fire and this public display, the British government, consisting of MPs in Parliament and officials from the Office of Woods and Forests, endeavoured to resolve these problems. While *The Times*' three questions emphasized the centrality of the political context of rebuilding Parliament, as explored in Chapter 1, its reference to architectural skill and science suggests the importance of the project's scientific context. This chapter addresses this by exploring the intellectual background of Charles Barry's appointment.

Rebuilding the Palace of Westminster was both a political and a technical challenge, and Barry was a solution to both. His attention to aesthetic detail satisfied the architectural demands of the building, while his use of scientific knowledge met the engineering challenge the Palace presented.[2] Barry embraced the latest philosophical learnings, in chemistry and mechanics, employed a rhetoric of science to explain his

[1] (Anon.), 'Designs for the proposed new Houses of Parliament', *The Times*, (London, England), 7 March, 1836; p. 3; Issue 16044; histories of Parliament shaped such nostalgia, for example see Edward Wedlake Brayley and John Britton, *The History of the Ancient Palace and Late Houses of Parliament at Westminster*, (London, 1836).

[2] Mark Girouard, 'Charles Barry: a centenary assessment', *Country Life*, Vol. CXXVIII, No. 3319 (13 October, 1960), pp. 796–97, 797; Mark Girouard, *The Victorian Country House*, (Oxford, 1971), pp. 15–16.

practices, and moved in networks of individuals of scientific learning. However, 'science' in the 1830s was an amazingly slippery term and held mixed, often curious, value in society.[3] The 1830s was a time where science assumed increasing authority, but what that science was, and how it should be made were far from universally agreed. To understand what Barry's work involved, and how he used knowledge, we must first understand the troubled nature of science itself.

This chapter begins by exploring what Barry proposed for Britain's new Parliament, and how his appointment as architect was defended. Barry represented a more professional and accountable approach to public architecture than that seen in the 1820s, combining this with great artistic skill, in part thanks to his collaboration with Augustus Pugin. Barry united the mechanical and the romantic: the technical and the artistic. The emotional and aesthetical experience of Parliament was valued alongside its engineering and technical refinement.[4] Barry brought a more systematic way of doing architecture and was very much part of the scientific world of the 1830s, moving in circles of the nation's premier scientific authorities and looking to the scientific organization of knowledge as something that should be venerated. I then explore what science meant in this community, focusing on the battle between the Utilitarian John Stuart Mill and the Cambridge philosopher of science William Whewell over how to define it. Whewell worked to categorize different types of knowledge, which had moral and political implications. Science could be a stable permanent resource, but it could also be a dangerous, immoral, and politically radical entity. Within this epistemological framework, architecture was a subject of importance. Whewell worked alongside the Cambridge mathematician Robert Willis to show the value of mechanical science for building. What is more, these efforts were actually drawn on in discussions over the new Parliament structure. I conclude by placing Barry in this scientific context. Through his systematic attention to building materials and structural engineering, chemistry and mechanics dominated Barry's notion of science; he was at the very heart of 1830s' scientific culture. By considering the sorts of knowledge Barry used, and the ways in which he used them, it becomes apparent that Barry's work cannot be divorced from this wider scientific scene. He was looking to scientific models of organizing knowledge which were both controversial and political. In later chapters, we will see how science's problematic nature resurfaced throughout Barry's work at Westminster.

[3] James A. Secord, *Visions of Science: Books and Readers at the Dawn of the Victorian Age*, (Oxford, 2014), p. 242.

[4] John Tresch, *The Romantic Machine: Utopian Science and Technology after Napoleon*, (Chicago, 2012), pp. 1, 16.

The relationship between Barry and wider models of scientific knowledge has implications for our understanding of the profession of architect at this time. As Andrew Saint has shown, during the nineteenth century technical expertise was increasingly drawn on in architectural ventures. While the use of new materials and building practices created a growing gulf between architects and engineers, Barry's management of knowledge at Parliament shows how interlinked these positions were.[5] More than any other project before the 1851 Great Exhibition, the Palace of Westminster demonstrated that architects had to direct a wide range of specialist skills and experience. The immense engineering scale of the Palace meant that Barry's ordering of knowledge for the building was radical and represented a new way of approaching architecture. Of course, skill and knowledge of materials had always been important, but it was the systematic, almost industrial, ordering of knowledge which set Barry's work apart.

The Appointment of Barry

When Charles Barry was announced as the architect for the new Parliament building, it provoked immediate hostility within the House of Commons (Figure 2.1). Joseph Hume complained, predictably, that Barry's plans ignored scientific knowledge of light, sound, and ventilation. He lamented the choice of site by the river, and then reiterated his contempt for the competition's restriction on style to Gothic and Elizabethan. In response, he met a united Whig-Tory defence of Barry's entry. While preserving the site was deemed essential to maintaining the 'liberties' of the English people, the Whig Chancellor of the Exchequer and Tories such as Robert Inglis and Frederick Trench rejected Hume's comments.[6] The Commons supported the appointment of Barry by 120 votes to forty-eight.

Although competition entries were anonymous, there was speculation that Barry had been chosen because of his friendship with Edward Cust and his Whig connections. Barry was indeed close to Cust and the Whig government, having friends at the Whig gathering place, Holland House, and in 1837 becoming a member of the Whig dominated Reform Club, which he himself designed.[7] James Pennethorne (1801–1871), an architect

[5] Andrew Saint, *Architect and Engineer: A Study in Sibling Rivalry*, (New Haven, 2007), pp. 486–89.

[6] House of Commons debate, 9 February, 1836, *Hansard*, 3rd Series, Vol. 31, pp. 236–43; House of Commons debate, 17 February, 1836, *Hansard*, 3rd Series, Vol. 31, pp. 503–05.

[7] W. J. Rorabaugh, 'Politics and the architectural competition for the Houses of Parliament, 1834–1837', *Victorian Studies*, Vol. 17, No. 2 (1973), pp. 155–75, 169.

Figure 2.1 Charles Barry as the very model of a professional Victorian architect

who would go on to secure many future government contracts, felt that the Gothic was a natural style for Britain, ideal for the island's 'climate and feelings'. He believed it as grand and imposing as Grecian, while 'in science' it was 'perhaps almost equally correct'.[8] Even so, Pennethorne was sure that 'promises of favor' had been made to Barry, almost guaranteeing his eventual appointment.[9] Unsurprisingly Barry's winning designs came in for much public criticism, particularly from the competition's unsuccessful entrants. The most outspoken of these was the architect Thomas Hopper (1776–1856), who attributed his failure to the exclusion of professional architects from the adjudicating commission.[10] He asserted that as the commissioners lacked experience of complex architectural practices they had chosen Barry for personal, rather than principled, interests.[11] Hopper believed the commission should have consisted of

[8] Geoffrey Tyack, *Sir James Pennethorne and the Making of Victorian London*, (Cambridge, 1992), p. 33.
[9] Ibid., p. 31.
[10] Thomas Hopper, *Designs for the Houses of Parliament*, (London, c.1842), pp. 1–2.
[11] Thomas Hopper, *Hopper versus Cust, on the Subject of Rebuilding the Houses of Parliament*, (London, 1837), pp. 7–9.

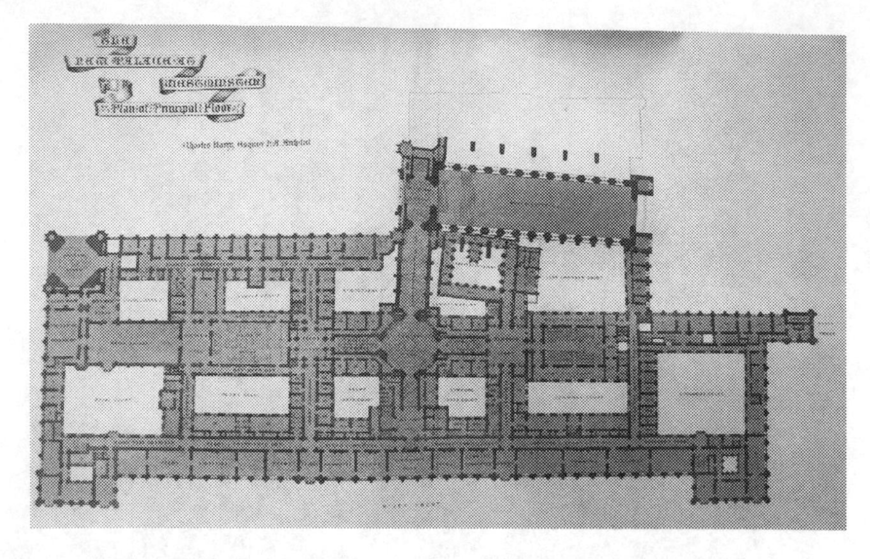

Figure 2.2 Charles Barry's ground plans for the new Palace of Westminster

'men of scientific knowledge'.[12] An absence of architectural knowledge had meant the commission failed to appreciate the complicated relationship between design and stone. Hopper explained how Barry's 'florid design' would not weather well in either Caen or Bath stone and these, he claimed, were the only practical options for such a planned edifice.[13] A man of science would, Hopper was convinced, have known this when adjudicating. Hopper's criticism of Barry was not that his designs lacked artistic merit, but that Barry's knowledge of building was insufficient. Yet despite Hopper's critique, the choice of Barry was surrounded with an assertion that his designs displayed great architectural talent (Figure 2.2).

Despite these criticisms, the commissioners responsible for Barry's selection put up a robust justification for their decision. Importantly they not only argued that Barry's plans displayed artistic quality, but that they also demonstrated technical skill. While they confessed an 'imperfect knowledge ... on the theories of Sound and artificial ventilation', they felt Barry's entry had 'evident marks of genius'.[14] The royal commission's sole MP, Charles Hanbury-Tracy (1778–1858), revealed that ventilation had been considered, but deemed the existing knowledge on the subject to be in such a state of infancy that it could not be relied

[12] Ibid., p. 12. [13] Ibid., pp. 20–21.

[14] *Houses of Parliament Plans. Report of commissioners appointed to consider the plans for building the new Houses of Parliament*, PP. 1836 (66), p. 2.

upon.[15] Hanbury-Tracy declared attention to sound and ventilation to be important, but only after selecting the architectural design: Barry's plans should be adapted to satisfy these practical problems.[16] Hanbury-Tracy spoke with experience, having rebuilt his country seat in Gloucester, Toddington Manor, in a Gothic style in 1819.[17] He had confidence that Barry was capable of addressing these concerns.

Despite this, over the summer of 1836 Hume tried again to reopen the question of Barry's selection, claiming the use of Gothic lacked 'credit on the age'; it was a style for a medieval, rather than an industrial, society.[18] This time Hanbury-Tracy personally quashed Hume's assault by explaining how Barry's drawings 'marked evidence of a great improvement in the science of Gothic architecture'.[19] Barry's designs combined a richness of detail, drawn from the study of medieval buildings, with a Romantic outline (Figure 2.3).[20] Together with the claim that Barry's designs had 'marks of genius', this reference to the 'science of Gothic architecture' really matters. What these comments were intended to show was that Barry's appointment represented a clear break from previous public works, secured through Royal or government favour. In the fallout of John Nash's controversial work at Buckingham Palace this was an important statement. While Robert Smirke's initial appointment to rebuild Parliament seemed typical of the government's old disreputable system of patronage, Barry appeared as a break from the architecture of the 1820s.[21] He had to be competent not only in artistic design, but in the challenge of producing a complex engineering work. Both tasks were crucial for Barry to distinguish himself from the conduct of architects in the 1820s. While Nash at Buckingham Palace had allowed costs to escalate, apparently lacking understanding of the actual practices of building, Barry seemed a professional with accurate knowledge of the materials required to complete the job.[22]

[15] *Report from Select Committee on Houses of Parliament; with the minutes of evidence*, PP. 1836 (245), p. 3.

[16] Ibid., p. 7. [17] Rorabaugh, p. 167.

[18] House of Commons debate, 21 July, 1836, *Hansard*, 3rd Series, Vol. 35, p. 400.

[19] Ibid., p. 404.

[20] Alexandra Wedgwood, 'The new Palace of Westminster', in Christine Riding and Jacqueline Riding (eds.), *The Houses of Parliament: History, Art, Architecture*, (London, 2000), pp. 113–35, 116.

[21] M. H. Port, 'The Houses of Parliament competition', in M. H. Port (ed.), *The Houses of Parliament*, (New Haven, 1976), pp. 20–52, 28.

[22] Rorabaugh, p. 158; see J. Mordaunt Crook, 'The pre-Victorian architect: professionalism & patronage', *Architectural History*, Vol. 12 (1969), pp. 62–78.

Figure 2.3 Charles Barry's Palace of Westminster, at the heart of London's commerce

The British Architectural Profession in the 1830s

Finding a suitable architect for the new Houses of Parliament was difficult because of the complex range of tasks required of the appointee. Not only would the architect have to design the building, but they would also have to oversee its construction, manage the engineering works required to embank land for the palace's northern elevations and erect two colossal towers, perform before select committees, appease irksome MPs, direct contractors, collaborate with technical authorities brought in for their specialist knowledge, keep track of costs and deadlines, take responsibility for building materials, and ensure premier standards of decoration, especially in terms of artwork and ornamental carvings. Finding an architect who could do all this was made harder because the profession of architecture was itself unclearly defined in the 1830s. In order to understand the implications of Barry's appointment, it is worth establishing exactly what it meant to be a 'professional' architect in 1836. The knowledge and skill that could be expected from an architect was not standardized or inevitably provided through any formal training.

From the 1830s until the end of the century the architectural profession expanded, both in scale and definition, and the place of the architect in Victorian social hierarchy changed significantly. By the end of Victoria's reign the position of an architect closely resembled other recognizable professions, complete with standard training, examinations, and an established body of knowledge.[23] The 1830s saw gradual efforts to professionalize the practice of architecture. The founding of the Institute of British Architects in 1834 aimed to facilitate 'the acquirement of architectural knowledge, for the promotion of the different branches of science connected with it, and for establishing uniformity and respectability of practice in the profession'.[24] The institute promoted the idea of architects as increasingly professionalized individuals, receiving training which would distinguish them from builders and engineers.[25] It was initially stipulated that the institute should establish a library of all works connected with the subject, a 'Museum of Antiquities, Models, and Casts', and regular meetings. If architects were to secure a professional status, it was considered essential that the institute provide instruction for students.[26] It hoped to improve and unify the conduct of architects. In 1837 the organization received a Royal Charter making it the Royal Institute of British Architects (RIBA).

Around this, a formal system of education slowly emerged. In 1840, King's College London ran classes which united civil engineering and architecture, and in 1841 University College London appointed a professor of architecture.[27] As much as science, architecture was being treated as a discipline which could be formalized through institutions, reformed education, and dissemination through university teaching. At the root of these efforts to professionalize architecture was the fact that, unlike painters, architects required vast sums of money to realize their projects. This financial urgency encouraged the likes of Barry to adopt a more professional tone. Professionalism became a way for such architects to distinguish themselves from those they perceived to be amateurs.

[23] Frank Jenkins, 'The Victorian architectural profession', in Peter Ferriday (ed.), *Victorian Architecture*, (London, 1963), pp. 39–49, 39; also see John Wilton-Ely, 'The rise of the professional architect in England', in Spiro Kostof (ed.), *The Architect: Chapters in the History of the Profession*, (New York, 1977) pp. 180–208, 193–95; for a history of the architectural profession, see Andrew Saint, *The Image of the Architect*, (New Haven, 1983); also see, Harold Perkin, *The Rise of Professional Society: England since 1880*, (London, 1989).

[24] Jenkins, 'The Victorian architectural profession', p. 41.

[25] M. H. Port, 'Founders of the Royal Institute of British Architects (*act.* 1834–1835)', *Oxford Dictionary of National Biography*, Oxford University Press, May 2013 [www.oxforddnb.com/view/theme/97265, accessed 7 December 2013]; for an example of the problem of training architects, consider Thomas Graham Jackson's (1835–1924) difficulties in planning out space, in William Whyte, *Oxford Jackson: architecture, education, status, and style, 1835–1924*, (Oxford, 2006), pp. 15, 18.

[26] Howard Colvin, *A Biographical Dictionary of British Architects, 1600–1840*, 3rd ed., (New Haven, 1995), p. 42.

[27] K. Theodore Hoppen, *The Mid-Victorian Generation, 1846–1886*, (Oxford, 1998), p. 425.

Despite these efforts, the impact of such measures to raise professional standards was limited. In 1835 the RIBA included only eighty-two members. The majority of architects remained amateurish and gentlemanly. By 1850 Britain was home to about 500 recognized architects, but only half were members of the institute.[28] Voluntary examinations were not introduced until 1863 and they were not made compulsory for associate membership until 1882. Perhaps the one thing that was standardized within the practice of architecture were the wages, with 5 per cent of the overall cost of the building, plus expenses, generally agreed on by the late eighteenth century.[29]

Central to the programmes of professionalizing architecture was a problem: was the architect an artist, comparable to a painter, or a professional? Often these alternate views of architecture carried ideological commitments. As Katherine Wheeler has persuasively argued, notions of what the social status of architects should be were informed by different historical models, often carrying stylistic implications. Some, particularly those who promoted Renaissance architecture, believed an architect should be a professional, equal to lawyers and physicians, and clearly superior to the builder and artisan. An architect in this mould would have complete control over a project and be able to draw on humanist and intellectual aspects of design, acquired through an academic education. In contrast, the traditional architect, which John Ruskin later promoted as a model of the medieval architect, focused on the manual craft of construction, having cut his teeth through an apprenticeship to an established architect.[30] In terms of education, architects in the 1830s could attend occasional lectures at the Royal Academy and the Royal Institution, but in general they learned their trade through apprenticeships.[31] Perhaps the most significant aspect of an architect's training was the traditional Grand Tour across the Continent, providing him with experience of the architecture of France, Italy, Greece, and occasionally Egypt. This was of course, if the architect could afford to travel. While Barry travelled before he was twenty, architect George Gilbert Scott was unable to fund a tour until he was thirty.[32] This traditional combination of travelling and apprenticing was, in some respects, the antithesis of the model professional architect. In the 1830s these two ideologies (architecture as profession or architecture as vocation)

[28] Ibid., p. 416–17. [29] Ibid., p. 419.

[30] Katherine Wheeler, *Victorian Perceptions of Renaissance Architecture*, (Farnham, 2014), pp. 3–4.

[31] An apprenticeship to an established architect remained the primary educational process long after the 1830s. For example, consider George Bodley's education in George Gilbert Scott's office in the late 1840s, in Michael Hall, *George Frederick Bodley: and the Later Gothic Revival in Britain and America*, (New Haven, 2014), pp. 15–29.

[32] Hoppen, p. 425.

were not yet clearly formulated, but there was still tension within the practice of architecture over the proper conduct of an architect.

Charles Barry was very much part of this culture in which architectural professionalism had to be constructed and defined. Rather than see Barry as drawn from a recognizable profession, it would be advantageous to see his work as part of a broader move to professionalize the practice of architecture. Just as those possessing scientific knowledge laboured to stylize themselves as 'experimentalists' and 'natural philosophers', so too did Barry have to work to appear 'professional'. In Shenton's *Mr Barry's War*, Barry appears as the archetypal professional. After training with surveyors Middleton and Bailey of Lambeth from the age of fifteen, learning architectural geometry, financial estimating, and surveying, and then touring the Continent, Barry is portrayed as conducting himself with business-like respectability.[33] However, instead of seeing this as conforming to any kind of agreed notion of professional conduct, we should see his performance in the building of the new palace as itself contributing to the construction of an architectural profession.

Barry and Pugin

Born in Westminster, Barry was the son of a stationer and bookbinder. Before gaining experience with a surveyor in Lambeth, he received little formal education, but displayed a talent for drawing and architectural design. In his youth, Barry toured Italy, Egypt, and the Eastern Mediterranean, developing a considerable knowledge of ancient architecture and a particular enthusiasm for the Italian architecture of Venice, Rome, and Florence.[34] On returning to England in 1820 he began a career as an architect, earning fame through commissions for the Traveller's Club in Pall Mall (1829), King Edward's School Birmingham (1833), and the Reform Club (1837). While Parliament would be the work he would best be remembered for, Barry had little enthusiasm for Gothic architecture and indeed Parliament was in many ways a classical structure with Gothic embellishments. Barry's designs for the new Palace promised a Parliament of dignity and tradition, to inspire national pride from an observing

[33] Caroline Shenton, *Mr Barry's War: Rebuilding the Houses of Parliament after the Great Fire of 1834*, (Oxford, 2016), pp. 19, 49; on the lack of an established profession, see p. 48.

[34] M. H. Port, 'Barry, Sir Charles (1795–1860)', *Oxford Dictionary of National Biography*, Oxford University Press, 2004; online edn, October 2008 [www.oxforddnb.com/view/article/1550, accessed 4 December 2013]; Kathleen Adkins (ed.), *Travel Diaries (1817–1820) of Sir Charles Barry (1795–1860) (Personal Extracts)*, (Privately Published, 1986); also see Peter Fleetwood-Hesketh, 'Sir Charles Barry', in Peter Ferriday (ed.), *Victorian Architecture*, (London, 1963), pp. 125–35.

Figure 2.4 A detail of the north-eastern end of Barry's new Parliament building

public (Figure 2.4).[35] His claims to architectural authority and Gothic competency were enhanced by his close association with Augustus Welby Pugin (1812–1852). Both men had witnessed the burning of

[35] Wedgwood, 'The new Palace of Westminster', p. 116.

the original Palace in 1834, and toured the cities of Belgium together in 1835, exploring the country's town halls which provided examples of secular government buildings in a Gothic style.[36] Pugin assisted Barry in submitting his initial designs and, in the 1840s, endeavoured to perfect the artistic details of the Palace. While Barry was a trained classicist and perfectly capable of planning the broad lines of the Palace, Pugin contributed knowledge and drawings to make the Gothic detail of Barry's work convincing. For example, from Basel, Pugin reported to Barry observations of 'The best modern architecture . . . The [railway] stations are beautiful – all constructive principle'.[37] He supplied drawings of brass and iron works, and clocks in leaden turrets. The only problem with such observations was how to profit from acquired architectural knowledge. As Pugin warned, 'by the time one is very knowing indeed, we shall be almost past profiting by the knowledge'.[38] Being authoritative on Gothic architecture involved more than collecting evidence, but being able to interpret and apply such knowledge. Although their descendants would later disagree over the amount of time Pugin served under Barry, Pugin made a valuable aesthetic contribution to Parliament's construction.[39] This artistic collaboration enhanced claims that Barry understood the 'science' of Gothic architecture.

While Pugin brought artistic knowledge, it is also clear that he was promoting an approach to architecture consistent with wider concerns that architects should act more professionally and that architecture

[36] Phoebe B. Stanton, 'Barry and Pugin: a collaboration', in Port (ed.), *The Houses of Parliament*, pp. 53–72, 60; Alexandra Wedgwood, 'Pugin, Augustus Welby Northmore (1812–1852)', *Oxford Dictionary of National Biography*, Oxford University Press, 2004; online edn, January 2008 [www.oxforddnb.com/view/article/22869, accessed 4 December 2013]; Alexandra Wedgwood, 'Pugin, Auguste Charles (1768/9–1832)', *Oxford Dictionary of National Biography*, Oxford University Press, 2004; online edn, January 2008 [www.oxforddnb.com/view/article/22868, accessed 4 December 2013]; Rosemary Hill, *God's Architect: Pugin and the Building of Romantic Britain*, (London, 2008); H. E. G. Rope, *Pugin*, (Hassocks, 1935); Phoebe Stanton, *Pugin*, (London, 1971); Paul Atterbury and Clive Wainwright (eds.), *Pugin: A Gothic Passion*, (New Haven, 1994).

[37] 'Letter from A. Pugin to C. Barry', (1 August, 1845), in Margaret Belcher (ed.), *The Collected Letters of A. W. N. Pugin*, Vol. 2, 1843–1845, (Oxford, 2003), p. 425.

[38] Ibid., p. 425; on the context of Pugin's selection, see Edward John Gillin, 'Gothic fantastic: Parliament, Pugin, and the architecture of science', *True Principles*, Vol. 4, No. 5 (Winter, 2015), pp. 382–89.

[39] Pugin's son claimed only his father had the 'knowledge' of the Gothic style to produce the Palace, in E. Welby Pugin, *Who Was the Art Architect of the Houses of Parliament. A Statement of Facts, Founded on the Letters of Sir Charles Barry and the Diaries of Augustus Welby Pugin*, (London, 1867), p. 3; on responses, see Alfred Barry, *The Architect of the New Palace at Westminster. A reply to a pamphlet by E. W. Pugin, Esq., entitled 'who was the Art-Architect of the Houses of Parliament?'*, (London, 1868); E. Welby Pugin, *Notes on the reply of the Rev. Alfred Barry D. D. principal of Cheltenham College, to the 'Infatuated Statements' made by E. W. Pugin, on the Houses of Parliament*, (London, 1868).

should involve questions of function and practicality, as much as artistic richness. In a debate with the architect A. W. Hakewill in 1835, Pugin demonstrated this commitment. Hakewill's father had disobeyed the competition rules and entered an unsuccessful classical design, which inspired Hakewill to attack the choice of Gothic.[40] When Hakewill projected classical as the style of 'the arts and sciences', believing it reflected progressive civilisation, he aroused Pugin's animosity.[41] It was Hakewill's comparison of classical as 'a flower' to Gothic as 'a weed' which incurred the greatest irritation. Pugin accused Hakewill of endorsing a style which had not been in vogue for 2,000 years. He cited ecclesiastical architecture as evidence that Gothic masons required great skill and argued that modern architects should employ cathedrals for instruction.[42] The style was shaped by a constantly growing body of knowledge, which had progressed architecture to an enlightened state.[43] Pugin told his readers to view the prints of the German engraver and mathematician Albrecht Dürer (1471–1528) if they doubted that the style combined art with intellectual attainment.[44] As Rosemary Hill put it, Pugin had declared that 'Gothic was best and Gothic was best learned, as Pugin had learned it, empirically', which meant paying great attention to existing architectural works to acquire an understanding of the decorative and constructive principles involved.[45] This argument for the practical superiority of Gothic was not new, however, and did not arise purely through the debates surrounding Parliament. The previous two decades had seen an intensive government church-building programme which had witnessed disagreements over the practical advantages of different styles. Controversies over cost were prominent, with various parties claiming both Grecian and Gothic to be cheapest, but with little consensus achieved.[46]

Pugin's response to Hakewill premeditated his later works, deploying the same arguments and sentiments.[47] He promoted the Gothic, or Christian art as he put it, as a disciplined body of knowledge and practices

[40] Hill, *God's Architect*, p. 147.

[41] (Anon.), 'Thoughts upon the style of Architecture to be adopted in rebuilding the Houses of Parliament', *Architectural Magazine*, Vol. II (November, 1835) pp. 506–07, 506.

[42] A. Welby Pugin, *A letter to A. W. Hakewill, architect, in answer to his reflections on the style for rebuilding the Houses of Parliament*, (Salisbury, 1835), p. 8.

[43] Ibid., p. 9.

[44] Ibid., p. 12; for a review, see (Anon.), 'A letter to A. W. Hakewill, architect, in answer to his reflections on the style for rebuilding the Houses of Parliament', *Architectural Magazine*, Vol. II (November, 1835) pp. 507–09.

[45] Hill, *God's Architect*, p. 148; for example, see 'Letter from A. Pugin to C. Barry, 1 August, 1845', in Belcher (ed.), *The Collected Letters of A. W. N. Pugin*, p. 425.

[46] M. H. Port, *600 New Churches: The Church Building Commission, 1818–1856*, (Reading, 2006), pp. 97–98.

[47] Hill, *God's Architect*, p. 148.

which promised improvement. Christian art was not exempt 'from rule . . . of philosophical and scientific principles'.[48] Pugin was eager to show that aspects of Gothic building, usually considered purely decorative, required 'mechanical skill'.[49] He argued that true Gothic architecture was 'useful'. Pinnacles defied weathering and threw off rain; a quality shared with pointed roofs.[50] Pugin saw his work as '*beautifying* articles of utility', rather than '*disguising*' practical objects.[51] While critics might damn Barry's choice of style as unenlightened, the Gothic could be defended by demonstrating its grounding in scientific principles.

Barry was also keen to show off a more scientific conduct. As with the aesthetic Gothic detail, Barry had to manage huge volumes of technical knowledge in order to build the Palace. This was fundamental to his perception of what it was to be a modern architect. In his 1867 biography of his father, Alfred Barry was eager to emphasize how Charles Barry laboured to make architecture more scientific, adopting a more professional manner. Due to its immense scale, the new Palace presented several engineering challenges, including the task of constructing an embankment by damming off the River Thames around the site and employing fireproof cast iron for construction.[52] The roofs of the Lords and Commons chambers involved spans of immense scale while the Victoria Tower, the world's highest secular square tower, was a work of engineering skill. In these matters Barry demonstrated a mind 'singularly fertile in all kinds of mechanical contrivance', despite disliking the 'systematic study of theory'.[53]

Barry increasingly looked to utilize scientific knowledge for architecture. Alfred Barry explained that his father believed that to be scientific was to employ 'practical knowledge', which moved away from pure 'philosophic knowledge' and could be applied to the material world.[54] This was recognition that architecture was not simply art, but involved

[48] A. Welby Pugin, *Glossary of Ecclesiastical Ornament and Costume, Compiled and Illustrated from Antient Authorities and Examples*, (London, 1844), p. iii; also see A. Welby Pugin, *Ornaments of the 15th and 16th Centuries*, (London, 1836).

[49] A. Welby Pugin, *The True Principles of Pointed or Christian Architecture: Set Forth in Two Lectures Delivered at St. Marie's, Oscott*, (London, 1841), pp. 5, 10.

[50] Ibid., pp. 9–11. [51] Ibid., p. 23.

[52] Denis Smith, 'The techniques of the building', in Port (ed.), *The Houses of Parliament*, pp. 195–217, 197–98; on embanking work, see Dale H. Porter, *The Thames Embankment: Environment, Technology, and Society in Victorian London*, (Akron, 1998); Bennett's description of Christopher Wren shows the errors of divorcing histories of science from histories of architecture, in J. A. Bennett, *The Mathematical Science of Christopher Wren*, (Cambridge, 1982), p 1; for mathematics and architecture, see Anthony Gerbino and Stephen Johnston, *Compass and Rule: Architecture as Mathematical Practice in England, 1500–1700*, (New Haven, 2009).

[53] Alfred Barry, *The Life and Works of Sir Charles Barry*, (London, 1867), p. 9.

[54] Ibid., pp. 9, 11.

knowledge acquired through observation and experience of building. Of course this in itself was not particularly radical, but what Barry envisaged was a move away from inherited traditional practices of building, towards the use of new knowledge created through laboratory experimentation. At Parliament, Barry claimed to apply such knowledge. His labour there was 'not purely artistic', but instead 'taxed heavily scientific knowledge and ingenuity'.[55] Architecture touched 'on one side the domain of science, and on the other the domain of art', while the architect was to take 'advice from both disciplines while retaining overall power'.[56] Years later when Barry died, *The Times* put his efforts in context. It recorded that 1859 to 1860 had been a tragic twelve months for the nation, witnessing the deaths of Robert Stephenson, Lord Macaulay, and Charles Barry: Britain's 'foremost men in science, in literature, and in art'.[57] It observed that the 'chief professor' of improvement in architecture had, fittingly, been buried at the foot of the graves of railway builder Robert Stephenson (1803–1859) and fellow engineer Thomas Telford (1757–1834).

Barry had always moved in networks of men of science. He regularly attended the Friday meetings of the Royal Institution in Albemarle Street and when Faraday lectured it was unusual for Barry to be absent.[58] He was a Fellow of the Royal Society and constant supporter of the British Association for the Advancement of Science (BAAS). In the mid-1830s, 'when it was the fashion to sneer at it', Barry was a keen listener to the transactions of the association.[59] When *The Times* announced the significant figures who declared that they would be attending the 1851 BAAS meeting, it not only identified Prince Albert, William Thomson, and Charles Darwin, but also Faraday, Cust, and Barry.[60] As Barry's son explained, 'The general type of mind found in our great scientific men at all times delighted him; for its union of simplicity of aim and enthusiasm for science with profoundness and originality of thought, was the type of genius with which he felt the warmest sympathy'.[61] As will be shown in subsequent chapters, Barry took what he learned from these circles into his architectural projects.

Although at BAAS meetings Barry listened rather than contributed, he was convinced of the advantages of establishing architecture as a science. Barry was certain that architecture's future should be grounded in scientific learning. As an early advocate of the RIBA, being present at its first meeting in 1834 and one of its earliest vice-presidents, he hoped the

[55] Ibid., p. 199. [56] Ibid., p. 165.
[57] (Anon.), 'Deaths', *The Times*, (London, England), 23 May, 1860; p. 9; Issue 23627.
[58] Barry, *The Life and Works of Sir Charles Barry*, p. 313. [59] Ibid., p. 314.
[60] (Anon.), 'The British Association', *The Times*, (London, England), 2 July, 1851; p. 4; Issue 20843.
[61] Barry, *The Life and Works of Sir Charles Barry*, pp. 313–14.

institution would emphasize that the profession of architect involved 'scientific principles and inventions'.[62] Barry looked to men of science and their institutions, such as the BAAS, for ideas over how to organize architectural knowledge and his son was eager to emphasise this. However, it is not overly clear what exactly this meant. While Alfred Barry was writing in the 1860s, at a time when science carried considerable authority and a much more precise definition, in the 1830s science was a vague term. Clearly then, Alfred Barry's meaning in the 1860s was different to his father's in the 1830s. It is not that Alfred's judgement was incorrect, because Barry was enthusiastic about science and did manage an awful lot of technical information at Parliament, but we need to be clear what precisely this meant in the 1830s. To do this, we need to explore the wider scientific context in which Barry moved. Barry was part of an extensive scientific community but within these circles were concerns over what science actually was.

The Philosophy of Science in the 1830s

To try and be scientific in architecture was a difficult proposition in the 1830s. It is hard to determine what Barry thought about the relationship between science and architecture, but he began his work at Parliament at a moment when two protagonists dominated intellectual debates over science. John Stuart Mill and William Whewell both contested the nature of science, how it should be produced, what its political implications were, and for Whewell, what this meant for architecture. Both protagonists wanted to reform philosophy, believing that a new philosophic system would induce social and political change. To improve science, they believed, would improve morality, politics, and society.[63] As seen in the previous chapter, the study of science had profound ramifications for radical politics. For Mill, the way scientific knowledge was produced mattered to its value as a political resource.

In Science, Mill believed that inductions were to be accumulated and then used to make deductive propositions. In his 1843 *System of Logic* he would go on to argue that an emphasis on deduction after a substantial collection of inductive observations was critical for applying scientific methods to the science of society and the role of science in governance.[64]

[62] Ibid., pp. 311–13.

[63] Laura J. Snyder, *Reforming Philosophy: A Victorian Debate on Science and Society*, (Chicago, 2006), pp. 7–8.

[64] John Stuart Mill, *System of Logic Ratiocinative and Inductive: Being a Connected View of the Principles of Evidence and the Methods of Scientific Investigation*, (London, 1886), pp. 318–23; and see pp. 552–55 on the science of human nature; Jose Harris, 'Mill, John Stuart

Mill's methodology was grounded in a Baconian emphasis on empiricism and was intrinsically linked with utility; Mill's science was a means of producing knowledge which was of material value. The empiricism, rather than idealism, of the mathematical and physical sciences provided Mill with a powerful model of knowledge for politics to follow.[65] Removing intuitionism from science and moral philosophy, and defeating the school of a priori reasoning, was about rejected an epistemology that Mill believed could justify existing inequalities such as slavery and forced marriage.[66]

Such notions of science as a means to material ends and inducer of political reform left William Whewell a worried man (Figure 2.5). A Peelite Conservative opposed to radicalism but open to gradual political reform, Whewell had graduated second Wrangler in the Cambridge Mathematics Tripos exams in January 1816, eventually becoming Master of Trinity College Cambridge in 1841.[67] Whewell feared that Utilitarian moral philosophy had a negative impact on the study of science. His notion of scientific knowledge was that it should not be purely sense based, obsessed with practical applications, but instead that inductive reasoning led to truth and a belief in a Supreme Deity. The pursuit of knowledge confirmed faith and was therefore morally improving. Science could be pursued for its own sake, idealistically.[68] This interest in science brought Whewell into confrontation with Utilitarian politics. While Mill promoted extensive inductive research for practical ends, Whewell stressed the creative role of the mind in science. Great discoveries came from imagination; they rested on ideas, rather than endless sensual impressions. Experiences were important, but for Whewell science began with an act of the mind and then could be supported by inductive investigations.[69] Whewell's science was idealistic, consisting of inspiration and improvement rather than endless data collecting and practical ends. It was something that should not just be measured in terms of utilitarian pleasure and pain, but as a way of creating morally excellent characters.[70]

(1806–1873)', *Oxford Dictionary of National Biography*, Oxford University Press, 2004; online edn, January 2012 [www.oxforddnb.com/view/article/18711, accessed 7 March 2014].

[65] Richard Yeo, *Defining Science: William Whewell, Natural Knowledge, and Public Debate in Early Victorian Britain*, (Cambridge, 1993), pp. 183–85.

[66] Snyder, *Reforming Philosophy*, pp. 204–05.

[67] Jack Morrell and Arnold Thackray, *Gentlemen of Science: Early Years of the British Association for the Advancement of Science*, (Oxford, 1981), p. 25; Richard Yeo, 'Whewell, William (1794–1866)', *Oxford Dictionary of National Biography*, Oxford University Press, 2004; online edn, May 2009 [www.oxforddnb.com/view/article/29200, accessed 7 March 2014]; also see Menachem Fisch, *William Whewell: Philosopher of Science*, (Oxford, 1991).

[68] Snyder, *Reforming Philosophy*, pp. 23–25; Yeo, *Defining Science*, p. 4.

[69] Yeo, *Defining Science*, pp. 13–15.

[70] Snyder, *Reforming Philosophy*, p. 206.

Figure 2.5 William Whewell, the polymathic Master of Trinity College Cambridge and definer of Victorian science

Within Whewell's epistemology there were two types of knowledge. First there was that which was known to be true, well established and beyond contention; what Whewell called 'permanent' knowledge. Then there was that which was still in the process of being made and debated; or 'progressive' knowledge. This division was at the heart of Whewell's notion of the historical development of ideas. Now while both forms were important, progressive could potentially be dangerous. Universities, for example, should only teach students permanent knowledge and encourage them to recognize truths. This ability was important before producing new progressive knowledge, which Whewell rejected as a university subject matter. Teaching progressive knowledge to students would confuse them, making their knowledge production erratic. Whewell worried that such radical, unstable knowledge encouraged dangerous forms of political thought among young men.[71] If they were taught to challenge academic authority too soon, students might go on to question the authority of political and religious institutions.

[71] Ibid., pp. 220–21.

To learn progressive knowledge too soon made critics, not students, who without first knowing truths contested everything. Whewell was sure that this instability created young minds hostile to their nation's institutions, such as in France and Germany, where revolutionary thought ran rife. Progressive knowledge was a threat to the 'conservative reform' which, combined with a love of tradition, characterized Whewell's own politics. It was the basis of an epistemology which aligned Whewell closely with Peel in fearing radical change and threats to the Established Church. It was, after all, Peel who recommended Whewell as Master for Trinity College in 1841.[72] Within this understanding the language of science was essential. Making scientific language fit for public audiences was important, but it was also essential to help the transition of new knowledge into permanent knowledge. For theories to be built from stable bodies of science, they required linguistic reform.[73] Until the mid-1840s Whewell was supportive of the experimental research of Michael Faraday and in early 1834 discussed the terminology of his work. While Faraday's knowledge was new, it was for Whewell a great triumph of rapid progress, which deserved an established terminology to help it become permanent knowledge.[74] Although new and therefore potentially unstable, Faraday's work had proceeded from existing truths and produced in a way Whewell could approve of.

Although not coining the terms 'permanent' and 'progressive' until 1845, Whewell's dichotomy between 'speculative' and 'practical' teaching defined excessive knowledge production as politically radical in much the same way.[75] In contrast to progressive knowledge, with its emphasis on observations and speculations, permanent knowledge provided truth from which ideas could emanate. When excessive observations guided ideas, this was risky. Importantly, under the subjects which Whewell considered typical of permanent knowledge, those based on ideas such as velocity and force, geometry, arithmetic, algebra, and mechanics were prominent. This included architecture, which had been evolving since the Middle Ages.[76] While Whewell was first and foremost a philosopher of science, from an early age he possessed a solid understanding of building. As a child his father had wanted him to become an architect, sending his drawings to a London architect.[77] Growing up with his father, a carpenter, left Whewell

[72] Ibid., pp. 222–23.
[73] Simon Schaffer, 'The history and geography of the intellectual world: Whewell's politics of language', in Menachem Fisch and Simon Schaffer (eds.), *William Whewell: A Composite Portrait*, (Oxford, 1991), pp. 201–31, 209.
[74] Ibid., pp. 226–28. [75] Yeo, *Defining Science*, p. 215. [76] Ibid., pp. 216, 218.
[77] Harvey W. Becher, 'William Whewell's odyssey: from mathematics to moral philosophy', in Fisch and Schaffer (eds.), *William Whewell*, pp. 1–29, 4.

with a practical experience of friction and building techniques. Architecture played a central role in Whewell's work on science. In his 1837 *History of the inductive sciences* the architecture of the Middle Ages was of particular relevance. The 'scientific character' of practical architecture at this time was an important 'prelude to the period of discovery'.[78] Whewell described how the beautiful subdivisions of weight in Gothic architecture showed how 'men possessed and applied ... the ideas of mechanical pressure and support', paving the way for the mechanical sciences of the future.[79] Whewell was quick to stress that science was not a particular subject matter, but a kind of progressively organized knowledge, and architecture fit comfortably into this definition.[80] Architecture provided rich evidence into how forms of knowledge progressed. After touring German churches in 1830, Whewell explained how by the development of the 'pointed arch', architecture could be seen to progress from the classical to more varied forms. Attention to past architecture had been important, but for Whewell the notion of the original 'idea', in this case the arch, in stimulating change was completely relevant to his narrative of progress in science.[81] In this account of progressive architecture Whewell went as far as to proclaim the Gothic as a realization of Christian spirit and in this way approximated a similar understanding to those reached in Pugin's Catholic evaluations of Gothic architecture as a powerful expression of faith.[82] They shared the view that enlightened Gothic architecture embodied Christian faith.

As with science, Whewell taught that to accumulate masses of uninterrupted architectural information produced confusion; building was about ideas and theory. Being scientific in architecture involved clear theory and imaginative ideas rather than a large amount of research.[83] The scientific model of knowledge, of organizing data, was equally valid for architectural projects. Whewell did more than evaluate architecture but, along with Robert Willis (1800–1875), worked to provide a coherent course of mechanical philosophy at Cambridge in during the 1830s and 1840s, including architectural theory. Both mathematicians had a keen interest

[78] William Whewell, *History of the Inductive Sciences, from the Earliest to the Present Times*, Vol. I (1837), p. 343.

[79] Ibid., p. 345. [80] Yeo, *Defining Science*, p. 10.

[81] Ibid., pp. 154–55; see Carla Yanni, 'Development and display: progressive evolution in British Victorian architecture and architectural theory', in Bernard Lightman and Bennett Zon (eds.), *Evolution and Victorian Culture*, (Cambridge, 2014), pp. 227–60; Carla Yanni, 'Nature and nomenclature: William Whewell and the production of architectural knowledge in early Victorian Britain', *Architectural History*, Vol. 40 (1997), pp. 204–21.

[82] Schaffer, 'The history and geography of the intellectual world', p. 216.

[83] Becher, 'William Whewell's odyssey: from mathematics to moral philosophy', pp. 6–7.

in architecture; they wanted to establish a programme of applying mathematics to problems of machinery, securing influence over civil engineering for the University of Cambridge.[84] This project of establishing a 'science of mechanics', or rather the study of geometrical movement and actions which reduced machines to problems of motion, crystallized in Willis' 1841 *Principles of Mechanism*, which Whewell eagerly promoted in his own writings.[85]

An engineer and architectural historian, Willis descended from the physician who had treated King George III during his sickness in 1788. From 1821 Willis attended Gonville and Caius College, Cambridge, before graduating BA as ninth Wrangler in 1826.[86] Willis would go on to be elected Jacksonian professor of natural and experimental philosophy in January 1837 after a fellowship at Downing College.[87] Apart from being active in the Cambridge Philosophical Society alongside his friend Whewell, Willis became an honorary member of the RIBA in 1835, a fellow of the Geological Society from May 1831, and in the same year began lecturing at the Royal Institution. He moved in similar circles to those of Charles Barry. Willis' primary focus was mechanical philosophy and his main interest was machinery. From the 1820s he worked on constructing a programme of mechanics, which he endeavoured to implement in Cambridge throughout the 1830s and 1840s. Ben Marsden has explored how, alongside Whewell, Willis worked to establish a pedagogical culture of the application of mathematical science to engineering.[88] What Willis envisaged was an approach to machinery that focused on the various pieces making up mechanisms which transmitted motion, before then considering forces working on such mechanisms.[89] This science of mechanism was an attempt to remove invention as the privilege of the artisan and place it within the domain of mathematics.[90]

Willis' approach to architecture was, as Alexandrina Buchanan has noted, analogous to his work on machinery. Buildings appeared as systems of weight distribution and structures sustaining forces. In the autumn of

[84] Ben Marsden, '"The progeny of these two "fellows": Robert Willis, William Whewell and the sciences of mechanism, mechanics and machinery in early Victorian Britain', *British Journal for the History of Science*, Vol. 37, No. 4 (December, 2004), pp. 401–34, 422.

[85] Robert Willis, *Principles of mechanism, Designed for the Use of Students in the Universities, and for Engineering Students Generally*, (London, 1841), p. v.

[86] Ben Marsden, 'Willis, Robert (1800–1875)', *Oxford Dictionary of National Biography*, Oxford University Press, 2004; online edn, October 2009. [www.oxforddnb.com/view/article/29584, accessed 7 March 2014]; Alexandrina Buchanan, *Robert Willis (1800–1875) and the Foundation of Architectural History*, (Woodbridge, 2013), pp. 11–29; also see Nikolaus Pevsner, *Robert Willis*, (Northampton, Massachusetts, 1970).

[87] Buchanan, *Robert Willis*, pp. 36–37.

[88] Marsden, '"The progeny of these two "fellows"', p. 403.

[89] Ibid., p. 406.

[90] Ibid., p. 434.

1819 Willis witnessed Baron von Kempelen's (1734–1804) chess-playing Turk automaton. This machine appeared to transcend the limits of machinery and reproduce, mechanically, human thinking.[91] Through what he claimed was inductive reasoning (demonstrating a theory grounded in examples) Willis suggested that a human operated the automaton from inside: it was not a mechanism, but a 'work of art'. The machine suggested it played and thought through 'mere mechanism', but Willis was sure mechanism could not exercise 'faculties of mind'.[92] Rather the mechanism on view was not authentic, but a decorative display. However, Willis reckoned this was still the deployment of 'skill and ingenuity'. Von Kempelen's Turk employed a 'very ingenious mode' for the 'concealment of a living agent'.[93] For Willis it mattered not 'by what means the phenomena' was achieved, either through the 'real' or the 'apparent'.[94] This differentiation was the basis of Willis' architectural work, because he believed that all buildings involved a balance between appearance and structure.[95]

In practice this difference between the real and the apparent manifest itself in the scientific analysis of architecture as mechanical and decorative construction.[96] Willis viewed buildings as constructional systems and in 1835 developed this approach concerning the Gothic. He promoted a 'mode of analysis' which, in order to understand the 'connection and distribution' of weight, differed between the 'Decorative and Mechanical structures of buildings'.[97] Willis believed that the human eye could not be satisfied with architecture, 'unless the weights appear to be duly supported'. Hence 'in all complete styles, part of the decoration is made to represent some kind of construction, and the more completely this is effected, the more satisfactory becomes the result'.[98] The difference between decorative and mechanical construction was that of how a structure's weight was 'really supported' and how it seemed to be supported. As with the automaton Turk, the decorative might conceal the mechanical to the eye, but the mechanical aspect was the most vital construction.

In Greek architecture, the balance between decorative and mechanical construction could, Willis believed, be seen clearly. He explained that 'the actual stones of which the building is composed' were observable, 'and so

[91] Buchanan, *Robert Willis*, pp. 19–21.
[92] Robert Willis, *An Attempt to Analyse the Automaton Chess Player, of Mr. De Kempelen*, (London, 1821), pp. 9, 11.
[93] Ibid., p. 32. [94] Ibid., p. 33. [95] Buchanan, *Robert Willis*, p. 25.
[96] Ibid., p. 87; also see Neil Levine, *Modern Architecture: Representation andR*, (New Haven, 2009), pp. 11, 116.
[97] Robert Willis, *Remarks on the Architecture of the Middle Ages, Especially of Italy*, (Cambridge, 1835), pp. v–vi.
[98] Ibid., p. 15.

decorated as to display their connection, while the whole style of architecture is accommodated to the principle of horizontal weights resting in perpendicular props ... the Decorative construction of the building is identical with its Mechanical construction'.[99] With later post-Roman works, 'barbarous' architects used the decorative to conceal the actual construction of their buildings, reflecting an ignorance of mechanical knowledge.[100] However, in principle the use of buttresses, vaulting, and arches in Gothic architecture reflected a great skill of managing 'perpendicular pressures'. The late Gothic trait of displaying ornaments to support pressure was a great display of 'practical knowledge'.[101]

Whewell and Willis' work provided an important background to debates over Parliament's architecture. William Hamilton's writing against the Gothic, explored in Chapter 1, was actually informed by Willis' mechanics and Whewell's epistemology. Hamilton was well acquainted with Willis' distinction between the decorative and the mechanical, applauding Willis' work as 'a beautiful development of the theory' of mechanical and decorative construction.[102] Hamilton cited Willis' theory as a credible approach to constructing architecture, which had pretensions to be a scientific manner of building. Yet Hamilton interpreted Willis' notion that in Gothic the decorative was often used to conceal the mechanical as evidence that the style was dishonest. Though Willis made no such judgement, believing the decorative to be admirable in both the Turk automaton and Gothic structures, Hamilton felt the use of the decorative, if not essential to the mechanical structure, was to 'substitute falsehood for truth'.[103] Greek architecture, in contrast, Hamilton believed, employed columns to take the weight of a building. In referencing Willis, Hamilton aligned himself with an approach to architecture which saw buildings as mechanical works. The Gothic tendency to make mouldings appear to sustain weights was a deceit which demonstrated that in such structures, the 'scientific faculty was not at work'.[104]

Hamilton also drew on Whewell's praise for Willis' distinction between the decorative and mechanical. He referenced passages from Whewell's *History of the inductive sciences* which supported Willis' reading of Gothic architecture.[105] He directed his readers to pages 344 and 345 which had been published just months before Hamilton's final pamphlet regarding

[99] Ibid., p. 16. [100] Ibid., p. 19. [101] Ibid., p. 21.

[102] W. R. Hamilton, *Second letter from W. R. Hamilton, esq. to the Earl of Elgin, on the propriety of adopting the Greek style of architecture in the construction of the New Houses of Parliament*, (London, 1836), p. 56; on contrasting views over the balance between decoration and construction, see Whyte, *Oxford Jackson*, p. 18.

[103] Hamilton, *Second letter*, p. 56. [104] Ibid., p. 42.

[105] He cites Whewell, *History of the Inductive Sciences*, pp. 9, 344–45, in W. R. Hamilton, *Third letter from W. R. Hamilton, esq. to the Earl of Elgin, on the propriety of adopting the*

Parliament. Whewell echoed Willis when he analysed Greek and Gothic architecture in terms of weight distribution, noting Greek's vertical columns and horizontal masses. Whewell recorded how, under the Romans, the 'struggle between the mechanical and the decorative construction, ended in the complete disorganisation of the classical style'.[106] He then narrated how the restoration of 'true mechanical relations' in construction accompanied the 'progress of science' after the twelfth century.[107] Science in architecture was the concept of 'support and stability in the decorative construction'. Whewell was in accord with Willis completely in associating mechanical and decorative harmony with enlightened building. Whewell continued Willis' mechanical analysis of architecture by explaining how 'by the multiplicity of props assisting each other, and the consequent subdivision of weight, the eye was satisfied of the stability of the structure'.[108]

Hamilton agreed with Whewell's instruction to treat architecture mechanically. Intriguingly though, he also shared Whewell's views of science. Along with citing Whewell's commentary on Willis, Hamilton also referenced page nine of Whewell's *History of the inductive sciences.* Here Whewell explained that the 'long-continued advance' of the history of science was simply 'clear ideas applied to distinct facts'.[109] Wherever there was 'progress', or a 'philosophical discovery', it was through applying ideas to observed facts. For Hamilton, Parliament should reflect Whewell's notion of science. Aligning with Whewell's approach, Hamilton explained that

> in natural philosophy and in the inductive sciences the Greeks failed to attain that eminence to which their native genius entitled them, because they did not pay sufficient attention to the importance of studying facts ... the Greek school of Philosophy must therefore be considered as a great part based upon mistaken hypotheses.[110]

Hamilton's reading of Whewell was politically consistent too, displaying a strong dislike of Utilitarian notions of science and the arts. He bemoaned that 'utilitarianism is the order of the day' and that all arts and sciences were gauged by results. He rejected its emphasis on utility in knowledge production, and drive to measure beauty by utility and science by wealth creation. Like Whewell he argued that great natural philosophy came from inspiration and curiosity, rather than endeavours to be practical. As he put it,

Greek style of architecture in preference to the Gothic, in the construction of the new Houses of Parliament, (London, 1837), p. 42.

[106] Whewell, *History of the Inductive Sciences*, p. 343. [107] Ibid., p. 344.

[108] Ibid., p. 345. [109] Ibid., p. 9. [110] Hamilton, *Third letter*, p. 40.

navigation has been improved by astronomy, watchmaking by the profound investigation of the laws of mechanics ... [and] we search with less loss of time and labour for coal ... since we have become geologists ... But Kepler and Galileo and Newton spent their laborious days, and their nightly oil, without thinking of these results.[111]

Rather than working towards utilitarian ends, either political or simply just prioritizing utility, such great natural philosophers 'were inspired with the pure love of science – with an ardent curiosity to learn'. So when Hamilton called for Parliament to embody science, this was a very different concept to that of Hume. Utilitarian science emphasized the practical, but Hamilton's 'knowledge was power' which once created would be applied to arts if allowed time.[112] Practical science was not immediately created therefore, but was rather the application of existing philosophical knowledge to problems. That Hamilton's writings about Parliament were so much shaped by his reading of Willis and Whewell's works both on architecture and scientific method reveals how closely thoughts about architecture were integrated with questions of science. It was in the context of these links that Barry worked. Understanding how Barry fit in this intellectual context is crucial to understanding how he conceived of rebuilding Parliament.

Chemistry and Mechanics: Vitruvius at Westminster

Working as the building's architect involved dealing with the huge volumes of information. Although Alfred Barry, in defining his father's commitment to science, explained that Charles Barry, despite having a mind 'singularly fertile in all kinds of mechanical contrivance', disliked mathematical mechanics and the 'systematic study of theory', preferring experience and practical knowledge, this is not quite accurate.[113] Actually Barry's earliest conception of architecture was that it was a subject dominated by mechanics and chemistry. Reflecting on Christopher Wren (1632–1723) and Inigo Jones (1573–1652), Barry lamented that neither had applied knowledge of 'the science of mechanics' to their architectural projects. He felt that they both had overlooked ventilation and the comfort of those inhabiting their buildings.[114] Barry believed that in addition to mechanics, architects should be educated in the science of chemistry. Claiming to himself have studied chemistry as a youth, Barry observed

[111] Hamilton, *Second letter*, p. 46. [112] Ibid., p. 47.
[113] Barry, *The Life and Works of Sir Charles Barry*, pp. 8–9.
[114] *Second Report from the Select Committee on Ventilation and Lighting of the House; together with the proceedings of the committee minutes of evidence, appendix and index*, PP. 1852 (402), p. 218.

that recent discoveries of cements had been the achievements of engineers rather than architects. Chemistry transcended cement and embraced geology, the use of iron, lighting, ventilation, and heating; in short the practices of building. Barry explained how 'According to the maxims of Vitruvius it ought to be a part of the instruction of every architect to make himself acquainted with chemistry, as well as other sciences; and I think it would be a very great advantage if all architectural pupils were to take the advice of Vitruvius on that subject'.[115]

Not only was such a focus on chemistry and mechanics typical of wider discourse on architecture, but by invoking the 'maxims of Vitruvius', Barry was promoting a very particular approach to architecture. In his *De Architectura* the Roman architect and writer Marcos Vitruvius Pollio (c.90-c.20 BC) had outlined a specific course of education for architects who had to be 'skilful in drawing, knowledgeable about geometry and familiar with a great number of historical works'. They should employ models for designing, wield knowledge of natural philosophy and medicine, be competent at arithmetic to calculate building costs, understand optics for drawing in light, and give attention to acoustics.[116] Vitruvius taught that philosophy, which included the study of nature was essential to a well cultivated architectural mind, while learning of the laws of nature from studying philosophy was essential to avoid structural failures.[117] Throughout Vitruvius' treatise he paid particular attention to the make-up of cements, the qualities of stone, and to the construction of machines and instruments. Architects had to have a knowledge of elements for colouring and decoration, keep up-to-date with philosophical teachings about how light operated, as well as be able to see how all 'mechanisms owe their origins to nature and are made following the guidance and instruction of the rotation of the universe'.[118] So when Barry declared the relationship between science and architecture to be best instructed by Vitruvius, this really mattered. While the study of mechanics was instrumental to architectural theory in the 1830s, chemistry was an incredibly well established and fashionable science, and Vitruvius' teachings appeared compatible with both. It is perhaps significant that Pugin shared this interest in Vitruvius, owning a 1575 German translation of *De Architectura* and a French edition from 1673. He lamented that so few modern architects really understood Vitruvius' teachings.[119]

[115] Ibid., p. 218.
[116] Marcos Vitruvius Pollio, *De Architectura*, (trans.) Richard Schofield, (London, 2009), p. 5.
[117] Ibid., p. 9. [118] Ibid., pp. 213, 171, and 279.
[119] 'Letter from A. Pugin to John Weale, 6 March, 1843', in Belcher (ed.), *The Collected Letters of A. W. N. Pugin*, p. 21; A. Welby Pugin, *An Apology for the Revival of Christian Architecture in England*, (London, 1843), p. 5.

Accordingly, Barry's attention to chemistry and mechanics at Westminster was intense. Obeying Vitruvius, he worked hard to keep track of building costs. The project was of an immense scale involving a huge office of staff, and the production of somewhere between 8,000 and 9,000 models and drawings.[120] Barry invested huge amounts of time into managing the materials of the work. Taking a recipe straight out of Vitruvius, Barry specified that the Palace's mortar, along with sand and water, should be made with a measure of 'grand premium Italian Pozzolano'.[121] This 'Roman Cement' had been described in *De Architectura* as the volcanic dust of Mt Vesuvius, which when mixed with lime and rubble ensured durability.[122] Vitruvius explained how this Pompeian pumice was unique due to its exposure to hot subterranean sulphur and was vastly superior to any alternatives. The very foundations of Parliament's building were to be made of dust from Pompeii.

Similar attention to the materials of construction characterized the entire rebuilding. Barry selected iron tiles for the building's roof, which were both cheap and relatively fireproof. To protect them from corrosion, he chose to have them galvanised with zinc, imitating the treatment applied to the Paris Opera House, completed in 1821.[123] The French engineer Stanislas Sorel (1803–1871) patented this practice of applying zinc to iron through galvanization in 1837 and the British Galvanization of Metals Company adopted the process soon after. Barry commissioned this company for Parliament's roofing.[124] Similarly he conducted extensive research into ensuring the building's stained glass was of the highest quality.[125] As for the practices of building, Barry provided precise instruction. He demanded all stone carving be done 'under the exclusive control & management of the Architect', and be carved of a fine quality determined by a sample which he approved.[126] Carpenters were to avoid wood with imperfections and use either the best Baltic timber or English oak, while plumbers were only to manufacture pipes from lead cast 'in the presence of the Architect' or someone who he appointed.[127]

Most radical of all was Barry's focus on the use of iron. In his handling of this material he adopted an approach every bit as rigorous as Vitruvius'

[120] M. H. Port, 'Problems of building in the 1840s', in Port (ed.), *The Houses of Parliament*, pp. 97–121, 101, and 105; Port, 'Barry, Sir Charles (1795–1860)', *Oxford Dictionary of National Biography*.

[121] Parliamentary Archives (PA) BLY/58, 'Sketch book of Henry Bailey, 1840–52', p. 17.

[122] Ibid., p. 6; Vitruvius, *De Architectura*, pp. 46–47.

[123] Robert Thorne, 'House of Lords roof', (27 November, 1991); thanks to Robert Thorne for this, which the Parliamentary Works Directorate commissioned privately.

[124] Robert Thorne, 'The galvanising of the iron roof plates at the Palace of Westminster', (1991).

[125] See PA 8/L/10/2(10), 'Barry/Hardman Correspondence, 1845–1860'.

[126] PA BLY/58, pp. 19, 24. [127] Ibid., pp. 9–10, 31.

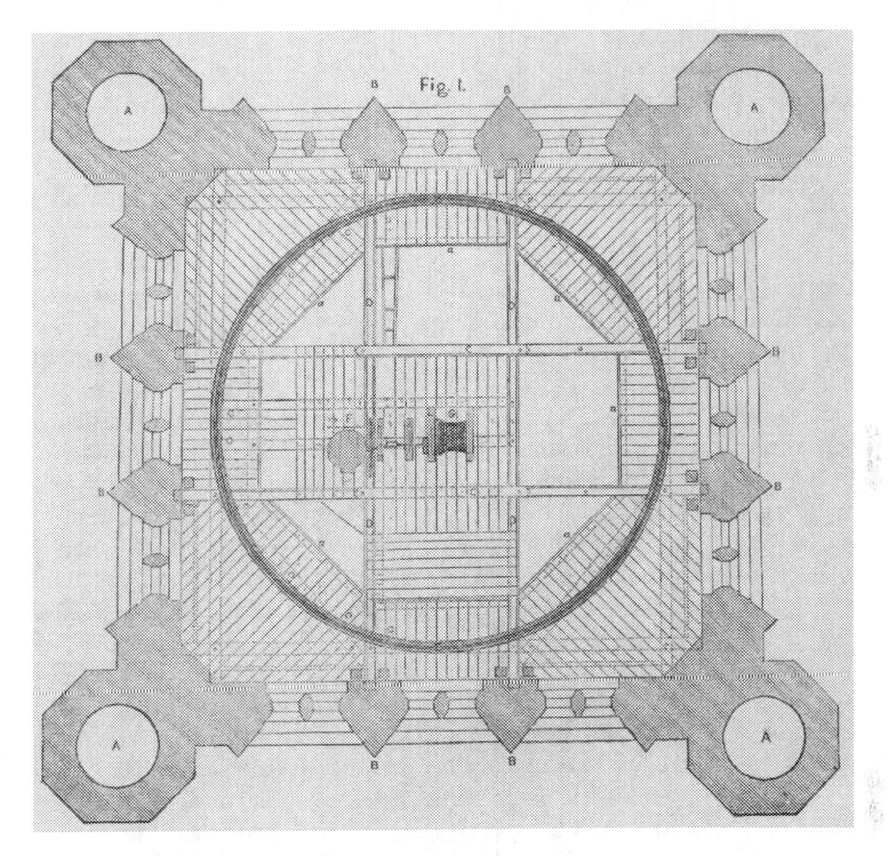

Figure 2.6 A cross-section architectural drawing of the Victoria Tower

recommendations on cement and stone. Although in the 1830s the use of iron on the scale of Parliament was unprecedented, Barry had experience of using cast-iron having adopted it in his first church commissions during the 1820s.[128] At Parliament Barry used vast quantities of iron. In the Lords the benches were supported by a network of cast-iron beams. In the Victoria Tower, to reduce the weight of the floors, iron was again adopted along with an inventive iron spiral staircase (Figure 2.6). To erect the tower a huge scaffold was constructed, capable of holding eighty tons of weight and supporting an enormous steam-powered winch which could lift four tons.[129] Although iron had been in use during the eighteenth century, the

[128] Port, 'Barry, Sir Charles (1795–1860)', *Oxford Dictionary of National Biography*.
[129] Port, 'Problems of building in the 1840s', pp. 199–200, 211.

early nineteenth century witnessed a rapid increase in the employment of the material. Nash notably used iron for Brighton Pavilion between 1815 and 1822, before employing it on a larger scale at Buckingham Palace in the late 1820s, including 35 ft. cast-iron beams for the building's floors and roofs.[130] Indeed Nash was well acquainted with iron construction, with a good eye for judging the quality of the metal having probably experienced casting at the Coalbrookdale and Bersham foundries.[131] Likewise Smirke used iron at the British Museum, in particular in the roof of the King's Library.[132] In the industrial context of 1830s Britain, and with a growing body of literature on the material, it was little wonder so much was used at Westminster. As Andrew Saint concluded, the Palace was the 'apogee' of hidden iron construction.[133]

However, what was different about Barry's approach to iron was both his enthusiasm for the material and his rigorous system of quality control. Iron was a material which, he believed, required chemical knowledge and mathematical skill. His use of the material represented a huge break from the works of Nash and Smirke. While they had relied on practical experience of the metal, Barry established a systematic regime of quality control and testing. Not only would Barry keep track of costs, but to further distinguish himself from the architects of the 1820s, he would avoid any doubts over the strength of his iron; indeed Nash had been criticized for his use of iron at Buckingham Palace while his first iron bridge, over the Teme at Stanford, had collapsed in 1795.[134] Barry ordered all cast-iron work to be 'perfectly sound and smooth', conducted in 'some Foundry in London approved by the Architect'. Once cast, he directed that the castings would be stored in a heated room and rubbed with linseed oil to prevent rusting. As for wrought iron, it was to be the 'best English iron ... submitted in the presence of the Architect or some person appointed by him to a full proof'. Before coating in lithic paint all cast-iron girders would undergo extensive trials.[135] Barry set out mathematical formula for calculating the strength of a cast-iron girder. To ascertain what weight would break a solid cast-iron column he multiplied the cube of its diameter in inches by fifteen, before then dividing the product by its length in feet, giving an answer in tons. So, for a column ten ft. long by three inches in diameter, Barry calculated it could support 40.5 tons,

[130] Saint, *Architect and Engineer*, p. 92.
[131] Jonathan Clarke, 'Pioneering yet peculiar: John Nash's contribution to late Georgian building technology', in Geoffrey Tyack (ed.), *John Nash: Architect of the Picturesque*, (Swindon, 2013), pp. 153–68, 155.
[132] Saint, *Architect and Engineer*, pp. 93–94. [133] Ibid., p. 95.
[134] Clarke, 'Pioneering yet peculiar', p. 156. [135] PA BLY/58, pp. 32–34.

Figure 2.7 The steam-powered crane used in the construction of the massive Victoria Tower

being 3×3×3×15, divided by 10.[136] He provided similar rules for iron of different sizes and shapes.

While such calculations were obviously important for ensuring the physical integrity of the Palace, this work was about much more than managing materials. It represented a clear break with the traditions of public architecture in the 1820s. While architects such as Nash and Smirke were synonymous with the old system of patronage, Barry's commitment to mechanics and chemistry, in accordance with the maxims of Vitruvius, was a way of appearing professional. Essentially the attention to cement and iron was chemical and mathematical, and in the 1830s aligning with these forms of knowledge fit within wider approaches to science. Although as will be shown in Chapter 3, Barry was happy to produce new science for Parliament's construction, he much preferred to reference existing bodies of knowledge. In technical matters the engineer and architect Alfred Meeson (1808–1885) offered able assistance (Figure 2.7). Barry's son recalled how while his father was a fine problem solver, his 'knowledge was more practical than theoretical' and so he benefited greatly from the 'scientific knowledge, ingenuity,

[136] Ibid., unmarked page.

and power of contrivance of Mr. Meeson'.[137] The other great purveyor of science for Barry was the experimentalist Michael Faraday (1791–1867). Even aspects of artistic detail engendered scientific learning, with Pugin and Barry consulting Faraday concerning fresco-painting techniques.[138] In all matters chemical, Barry 'had the advantage of the advice of Professor Faraday'.[139] Barry contacted him regarding concerns as diverse as the House of Commons' acoustics, its ventilation, and over his fears that coke fires at the base of the Victoria and Clock towers would cause oxidation and decomposition of iron and gold work at the top of each tower.[140] Faraday demonstrated his respect for Barry by presenting a lecture at the Royal Institution on Barry's work at Parliament in March 1847.[141] Indeed Faraday and Barry were close friends, perhaps helped by their common background: while Barry was the son of a Westminster bookbinder, Faraday had initially served an apprenticeship in this craft. The extent of their companionship was summed up on Barry's death in 1860. Writing to John Tyndall, Faraday recalled how he had dined with Barry at the Royal Academy, enjoying a long conversation, just a week before Barry's death. Faraday lamented how 'Barry is gone from us; it strikes me the more because I have had occasion to be at the house of parliament lately'.[142]

Understanding this relationship offers valuable insights into how Barry moved in the world of 1830s' science. As will be shown, few scientific authorities found favour with Barry, usually provoking hostility. Faraday was a rare exception. Alfred Barry's explanation was that Barry believed progress in the arts and science was made by men of 'an unscientific audacity', not those possessing pure 'philosophic knowledge'.[143] Indeed Faraday's attention to practical problems and compliance as a willing informant made him an ideal choice for Barry, but it would also seem that

[137] Barry, *The Life and Works of Sir Charles Barry*, p. 200; L. H. Cust, 'Meeson, Alfred (1808–1885)', rev. Susie Barson, *Oxford Dictionary of National Biography*, Oxford University Press, 2004 [www.oxforddnb.com/view/article/18511, accessed 4 December 2013].

[138] Barry, *The Life and Works of Sir Charles Barry*, p. 314. [139] Ibid., p. 169.

[140] 'Letter 2304: Charles Barry to Michael Faraday', (2 July, 1850), in Frank A. J. L. James (ed.), *The Correspondence of Michael Faraday: Vol. 4, Jan., 1849–Oct., 1855, Letters 2146–3032*, (London, 1996), p. 163; 'Letter 1969: Michael Faraday to Charles Barry', (31 March, 1847), in Frank A. J. L. James (ed.), *The Correspondence of Michael Faraday: Vol. 3, 1841–Dec., 1848: Letters 1334–2145*, (London, 1996), p. 608; 'Letter 3473: Charles Barry to Michael Faraday', (6 July, 1858), in Frank A. J. L. James (ed.), *The Correspondence of Michael Faraday: Vol. 5: Nov., 1855-Oct., 1860, Letters 3033–3873*, (London, 2008), p. 405.

[141] 'Letter 1969: Michael Faraday to Charles Barry', (31 March, 1847) in James (ed.), *The Correspondence of Michael Faraday: Vol. 3*, p.608.

[142] 'Letter 3781: Michael Faraday to John Tyndall', (16 May, 1860), in James (ed.), *The Correspondence of Michael Faraday: Vol. 5*, p. 684.

[143] Barry, *The Life and Works of Sir Charles Barry*, p. 11.

Faraday fit within Barry's understanding of what good science should be. Faraday provided knowledge when required or invited to do so; he did not produce new knowledge and then impose it on Barry's work. What Barry sought for the Palace was what Whewell would label permanent knowledge; which was politically and scientifically safe. Later in 1858, when reviewing the opinion of the Cornish experimentalist Goldsworthy Gurney over the ventilation of Parliament, Barry observed favourably how he 'boasts of not having looked into a chemical book for the last 25 years'.[144] New progressive knowledge, which was radical and unstable, was not what Barry valued. Combined with Barry's confidence in Faraday, such sentiments demonstrate how his commitment to being scientific in architecture cannot be divorced from the scientific, epistemological context in which he worked.

Conclusion

Barry certainly brought a more scientific approach to building Parliament than anything seen in architecture before. The range of different skills and knowledge that he had to manage made it crucial that he organize the work in a way that contemporaries would have understood to be scientific. This approach to architecture was particular to the 1830s, being dominated by chemistry and mechanics. Science itself was an unstable term, but what Barry sought was reliable, stable knowledge. We see these efforts throughout Barry's rhetoric and practices. Attending BAAS meetings, citing Vitruvius, relying on Faraday, treating stone and iron mathematically and chemically: these were ways of adopting, even imitating, broader interpretations of science as a fundamentally rational agent. Barry's approach fits with a growing consensus surrounding architecture. Before 1860 there was a conviction that architecture, and in particular that which drew on architectural history, could be as much an objective science as geology or biology.[145] Throughout the 1840s and 1850s the Gothic style came increasingly to be conceptualized as a style of modernity and historical development. Scientific analogies subsequently became more common in architectural discussions, such as those of the art critic and social reformer John Ruskin (1819–1900) emphasizing the importance of including natural objects in Gothic architecture, and asserting the style was analogous to chemistry, composed of various

[144] 'Letter 3473: Charles Barry to Michael Faraday', (6 July, 1860), in James (ed.), *The Correspondence of Michael Faraday: Vol. 5*, p. 405.

[145] Chris Miele, 'Real antiquity and the ancient object: the science of Gothic architecture and the restoration of medieval buildings', in Vanessa Brand (ed.), *The Study of the Past in the Victorian Age*, (Oxford, 1998), pp. 103–24, 103.

moral elements.[146] Mid-Victorian Gothic went hand-in-hand with a close observation of the natural world.[147] Although Barry did not use nature in this way for inspiration, his use of science was part of a broader interconnection between philosophical knowledge and building. Indeed, Barry's use of scientific models of knowledge organization preceded Ruskin's calls to invoke nature through art.

What the works of Whewell and Mill do, is contextualize Barry's approach. Throughout the following chapters these intellectual problems surrounding science resurface, sometimes with political implications. While Barry coveted stable knowledge like Faraday's and, as Chapter 3 shows, was even prepared to help produce new knowledge, when faced with excessively experimental and overtly empiricist forms of science, as in Chapters 4 and 5, Barry was cautious. If permanent knowledge offered order and stability at Westminster, then progressive knowledge could undermine the very fabric of Parliament. As will be shown, Barry's understanding of science was in many ways consistent with Whewell's teachings.

[146] John Ruskin, *The Stones of Venice*, (London, 2000), pp. 141–42.

[147] Michael Hall, 'What do Victorian churches mean? Symbolism and sacramentalism in Anglican Church architecture, 1850–1870', *Journal of the Society of Architectural Historians*, Vol. 59, No. 1 (March, 2000), pp. 78–95, 80–81; the Oxford University Museum, for example, see Carla Yanni, *Nature's Museums: Victorian Science & the Architecture of Display*, (Baltimore, 1999), pp. 62–89.

3 'The Science of Architecture': Making Geological Knowledge for the Houses of Parliament

> Ah! in what quarries lay the stone
> From which this pillared pile has grown.
>
> Dante Gabriel Rossetti, *The Burden of Nineveh*

> Geology, which is connected with so many other sciences, is also connected with Architecture. – The Builder indeed collects his materials from the Earth and therefore that alone is sufficient to show the connexion. But there are many other circumstances besides digging the Clay, making the Bricks & Mortar, quarrying, dressing, & setting the stone; – all intimately allied to the Earth, – which require the superior skill of an Architect, or chief professional advisor; not only in the construction of a building, but in every thing that relates to it; for there is not a contingency of any kind, either in Nature or Art, but should come under his consideration.[1]

This acknowledgement of Geology's prominence in architecture summarizes what William Smith, the 'Father of English Geology', termed the 'science of architecture'. For an architect to embrace both geology and chemistry in the selection of materials and the practice of building was to make his art modern and professional. Smith's sentiments came in January 1839, just three months after his involvement with Charles Barry in a royal commission to identify a suitable stone with which to build Parliament. This selection was to be made on scientific evidence, which geologists and chemists attained by working alongside the building's architect. Together they were to undertake a geological survey across the nation, examining ancient buildings, in order to find a stone which might resist the deleterious atmosphere of London and ensure the Houses of Parliament's longevity.

This chapter is about the selection of stone for Parliament, but it is also about the construction of knowledge. Choosing stone involved the production of a new body of knowledge to make the selection of building materials scientific. On 16 March 1839, Barry and the three commissioners

[1] Oxford University Museum of Natural History (OUM), Papers of William Smith (SMITH), Box 44, Folder 5, Item 6 (6 January, 1839), p. 1.

submitted a report on the selection of stone for Parliament.[2] This report demonstrated how geology, which included chemical understanding, could be utilized to provide accurate scientific knowledge to shape the construction of Parliament and future public architectural projects. It was a manifesto of how architecture should be grounded in experiment and observation; of how man's temples could be built in accordance with nature. This chapter investigates the practices involved in making such knowledge. Part one explores how during the survey, geological practices of observation and collection were employed, while over the following winter the obtained samples were subjected to chemical analysis. This analysis of work conducted in quarries and laboratories demonstrates how building Parliament stimulated the production of new science. Part two employs William Smith's reflections on the survey to situate it within the wider context of mid-nineteenth-century geology. Smith's commentary placed the Parliamentary stone commission's work within contemporary geological science. Part three investigates the divisions within the survey between geologist Henry Thomas De la Beche (1796–1855) and Barry. They debated the proper place to display the findings of the commission. Resolving this problem was itself a practice of constructing knowledge, as the settlement of this disagreement was about organizing and ordering a new body of knowledge. Finally, part four explains how Parliament's stone and its supporting report were scrutinized. Alternative stones for Parliament were proposed which boasted perceived social, economic, and moral advantages. In the face of such challenges, the choice of the commission had to be defended before Parliament could be built. Altogether these practices transformed a desire to select stone scientifically into a physical building. Parliament stimulated the production of new knowledge, but this involved a complex process of making, situating, organizing, and justifying the science of architecture.

The practices employed were not new but were radical in relation to an architectural project. This work for Parliament fits within a broader context of British geology. Martin Rudwick has explored how geology underwent a 'revolution' during the age of political reform between 1820 and 1845. He defined this as a realization that humans were newcomers to the world and a perception that geology was more than geo-history: it was the study of the 'place of Man in nature'.[3] The rapid development of

[2] Charles Barry, *Report of the Commissioners appointed to visit the Quarries, and to inquire into the Qualities of the Stone to be used in Building the New Houses of Parliament*, PP. 1839 (574); for Barry's handwritten original, see National Archives, Kew (NA), WORK 11/17/5/66.

[3] Martin J. S. Rudwick, *Worlds Before Adam: The Reconstruction of Geohistory in the Age of Reform*, (Chicago, 2008), p. 553; for geology before 1815, see Roy Porter, *The Making of Geology: Earth Science in Britain, 1660–1815*, (Cambridge, 1977).

the Geological Society of London through the 1820s, with its emphasis on open discussion within a cross-bench meeting room, often likened to the shape of the House of Commons, underlined this rising interest in the subject.[4] At the heart of this curiosity was an awareness of the potential utility of the science for the materials of industrialization. Geologists endeavoured to locate areas rich in coal and valuable ores.[5] The brightest example of this claim to predict subterranean geological wealth was Roderick Murchison's calculation, in the mid-1840s, that there were large undiscovered gold deposits in Australia.[6] The Parliamentary stone selection was part of this culture of geological promise.

William Whyte has suggested that an architect's choice of stone could convey specific cultural values. For example, the completion of Alfred Waterhouse's Manchester Town Hall in 1877 was a celebrated local event, yet Whyte suggests that the use of three different granites from England, Scotland, and Ireland for the building's three great staircases encouraged notions that the building's opening was also a national event.[7] In demonstrating how geological values were built into the choice of stone for the Palace of Westminster, I show that the survey raised critical concerns over the remit and definition of science in early-Victorian society. While the commissioners believed science provided validating evidence to guide architecture, dissent surrounded the project both in terms of the practices employed and the proper place of science in contrast to economic and social concerns. Above all, I argue that Barry and the commissioners deemed knowledge acquired through experiment to be of the greatest value and that they shared a very broad conception of what experiment was. They did not limit experiment to controlled laboratory induction, but valued experiment in the form of observable chemical action in ancient structures. While Barry observed rock in quarries, he also examined ancient churches and cathedrals to ascertain how stone survived in architecture through time. This offered what he felt were experimental results, unobtainable without centuries of research. Experiment was effectively a way of seeing. In a laboratory, the action of various chemicals on stone might be

[4] Martin J. S. Rudwick, *The Great Devonian Controversy: The Shaping of Scientific Knowledge among Gentlemanly Specialists*, (Chicago, 1985), pp. 22–24.

[5] James A. Secord, 'The geological survey of Great Britain as a research school, 1839–1855', *History of Science*, 24:3 (September, 1986), pp. 223–75, 223; also see Roy Porter, 'Gentlemen and geology: the emergence of a scientific career, 1660–1920', *The Historical Journal*, Vol. 21, No. 4 (December, 1978), pp. 809–36, 816–17.

[6] Ben Marsden and Crosbie Smith, *Engineering Empires: A Cultural History of Technology in Nineteenth-Century Britain*, (Basingstoke, 2005), p. 38.

[7] William Whyte, 'Building the nation in the town: architecture and national identity in Britain', in William Whyte and Oliver Zimmer (eds.), *Nationalism and the Reshaping of Urban Communities in Europe, 1848–1914*, (Basingstoke, 2011), pp. 204–33, 206.

tested, but such work lacked the crucial ingredient of time. The commissioners were sure that observation of architectural structures, centuries old, remedied this problem. To see the durability of these structures was itself a means of procuring reliable evidence. Geological theory, derived from experimental evidence promised the knowledge through which Barry might build a Parliament to defy decay through time and solve at least one problem raised in the debates surrounding the new building. Geology was to bring permanence to British architecture and this claim was to be heard loudest at Westminster.[8]

Making Knowledge: Collecting, Observing, and Experimenting

Soon after Barry's appointment to build the new Houses of Parliament the question arose of what material it should be built from. Inundated with suggestions of stone from interested landowners, Barry's choice was a crucial consideration. At stake was the reputation of the building as well as its longevity. To capture the beauty of his designs Barry wanted a stone that was hard enough to resist the coal-fuelled London atmosphere, yet soft enough to be precisely carved. When Robert Peel suggested that Barry conduct an investigating commission to select a suitable stone, the idea was quickly embraced.[9] In July 1838 Barry proposed to Lord Duncannon that he tour the nation's quarries and ancient structures 'accompanied by two or three scientific persons eminent for their Geological, topographical, and practical knowledge'.[10] He promised to select a stone for Parliament's 'superstructure' that was 'pleasing in color, good in quality, and capable of resisting the blackening and decomposing effects of a London atmosphere'. The dangers of using 'improper stone', selected without reference to its geological and chemical characteristics, were such that he believed 'too much caution cannot be used in selecting a stone for the New Houses of Parliament'. Barry and two geologists would collect 'data', before selecting stone specimens to return to London for chemical analysis. The report transpiring from this stone survey would not only be of value to Parliament, but was intended to become a reference for future architectural projects. As Barry stipulated, the collected

[8] On the broad implications of the stone survey for British architecture, see Edward J. Gillin, 'Stones of science: Charles Harriot Smith and the importance of geology in architecture, 1834–64', *Architectural History*, Vol. 59 (2016), pp. 281–310.

[9] M. H. Port, 'Problems of building in the 1840s', in M. H. Port (ed.), *The Houses of Parliament*, (New Haven, 1976), pp. 97–141, 97.

[10] NA WORK, 11/17/5/10, 'Letter from C. Barry to Her Majesty's Commissioners of Woods & Forests &c', (5 July, 1838).

evidence 'would be useful not only on the present but on all future occasions in the erection of public works'.[11]

To identify a suitable stone, it was first important to select suitable commissioners. The first appointment to this survey was De la Beche. Barry sounded him out before the proposal to Duncannon, considering De la Beche to be a gentleman 'whose eminence as a mineralogist and Geologist is beyond all question'.[12] The two men had initially discussed the potential of a tour to select stone for Parliament in De la Beche's Museum of Economic Geology, itself a place demonstrating the utility of the science.[13] De la Beche grew up on the coast of Dorset at Lyme Regis, renowned for its wealth of Jurassic fossils.[14] Heir to a Jamaican plantation and a close friend of local fossil collector Mary Anning (1799–1847), he had dedicated much time to the pursuit of geological knowledge. After attending a Church of England school in Devon at Ottery St Mary, where a strict education had left him with a disdain of all religious creeds, he pursued geology as a hobby until the 1830s and from 1830 endeavoured to geologically map Devon.[15] By 1832 De la Beche's loss of revenue from his estate endangered this effort; falling sugar prices and looming slavery reform persuaded him to seek government funding for his Devonian survey. The government accepted De la Beche's offer to geologically colour the eight ordnance survey maps comprising the county for the payment of £300, and between 1832 and 1835 he worked towards this project of 'practical utility'.[16]

To secure this patronage, De la Beche lauded the economic advantages of geology. He drew attention to the relationship between geology and the search for coal, metal ores, materials for road repair, and building stone.[17]

[11] Ibid. [12] Ibid.

[13] Jack Morrell, *John Phillips and the Business of Victorian Science*, (Aldershot, 2005), p. 163; on museums, see Samuel J. M. M. Alberti, 'The status of museums: authority, identity, and material culture', in David N. Livingstone and Charles W. J. Withers (eds.), *Geographies of Nineteenth-Century Science*, (Chicago, 2011), pp. 51–72; on the museum's London context, see Iwan Morus, Simon Schaffer and Jim Secord, 'Scientific London', in Celina Fox (ed.), *London – World City, 1800–1840*, (New Haven, 1992), pp. 129–42, 133.

[14] J. A. Secord, 'Beche, Sir Henry Thomas De la (1796–1855)', *Oxford Dictionary of National Biography*, Oxford University Press, 2004 [http://ezproxy.ouls.ox.ac.uk:2117/view/article/1891, accessed 21 July 2013]; Tom Sharpe, 'New insights into the early life of Henry Thomas De la Beche (1796–1855)', in Richard Morris (ed.), *A Journal of Sir Henry De la Beche: Pioneer Geologist (1796–1855)*, (Royal Institution of South Wales: 2013), pp. 5–21, 8.

[15] Henry Thomas De la Beche, 'Some account of myself', in Morris (ed.), *A Journal of Sir Henry De la Beche*, pp. 22–36, 33.

[16] David G. Bate, 'Sir Henry Thomas De la Beche and the founding of the British geological survey', *Mercian Geologist*, 2010, 17 (3), pp. 149–65, 158.

[17] Ibid., p. 160.

Jack Morrell has argued that these promises of advanced mineralogical knowledge largely failed to meet expectations, with geology maintaining a character more of polite gentlemanly learning than economic utility.[18] In the 1830s, however, De la Beche's work was shrouded in optimism. Oxford geologist William Buckland (1784–1856) and Cambridge's Woodwardian Professor of Geology, Adam Sedgwick (1785–1873), were employed to assess his survey work and recommended his efforts continue. Throughout the mid- to late-1830s De la Beche geologically mapped Cornwall and West Somerset, establishing the British Geological Survey (BGS) through a promotion of the economic value of the science. De la Beche was a frequent correspondent to French government-funded geologists and had a firm admiration of the École des Mines in Paris. He envisaged a similar system of state-funded science for Britain.[19] By publicly funding geology the government could move scientific authority away from traditional aristocratic sources. As Duncannon felt in 1837, De la Beche demonstrated 'the application of Geology to the useful purposes of life ... [through] matters exclusively scientific'.[20] Interestingly Peel, who had initially suggested to Barry the idea of a commission, later secured a budget increase for De la Beche's BGS just months after becoming Prime Minister in late-1841. James Secord has shown how De la Beche laboured through the 1840s to ensure that the BGS became a research school for future generations of government-funded science. In the context of the BGS's subsequent expansion, the Parliamentary stone survey appears as an important early assertion of geology's value.[21]

At the time of Barry's stone commission, De la Beche was inclined to view the project as an unhappy distraction from his surveying of South Wales but, as the government's premier geological authority, he could hardly refuse to participate in a project so demonstrative of utility.[22] On 11 August 1838, he was invited on the 'tour of inspection'. Duncannon felt that Barry and De la Beche should be accompanied by a second man of science and 'one practical master mason'.[23] Barry selected the stonemason

[18] Jack Morrell, 'Economic and ornamental geology: the Geological and Polytechnic Society of the West Riding of Yorkshire, 1837–53', in Ian Inkster and Jack Morrell (eds.), *Metropolis and Province: Science in British Culture, 1780–1850*, (London, 1983), pp. 231–56, 248.

[19] Secord, 'The geological survey of Great Britain as a research school', p. 230.

[20] British Geological Survey Archives, Keyworth (BGS) GSM/DC/A/C/11/4, 'Letter from Lord Duncannon to H. De la Beche', (15 February, 1837).

[21] Secord, 'The geological survey of Great Britain as a research school', p. 232.

[22] Paul J. McCartney, *Henry De la Beche: Observations on an Observer*, (Cardiff, 1977), p. 36.

[23] BGS GSM/DC/A/C/11/5, 'Letter from J. Baring to H. De la Beche', (11 August, 1838), p. A.

Charles Harriot Smith (1792–1864), who was to offer advice on the potential of stone to be sculpted, and a second geologist, William Smith.[24] Although accredited with the theory of stratification which underpinned English geology, William Smith was a somewhat side-lined figure. In 1831, his bid to be made the government's 'Geologist Colourer of the Ordnance Maps' was rejected; in all but title this was the same position De la Beche secured in 1832.[25] Barry and De la Beche's initial selection had been Smith's nephew John Phillips (1800–1874). Phillips had trained as a mineral surveyor under his uncle before working as the first keeper of the Yorkshire Philosophical Society Museum in 1826. Phillips was a respected authority on geology and promoter of the utility of science. In 1831, he had been a leading figure in the administration of the first British Association for the Advancement of Science (BAAS) meeting, held at his museum in York to promote science's value in society.[26] From 1834 he had been professor of geology at King's College London. De la Beche contracted Phillips to work in the BGS from July 1838 and subsequently offered him a role on the stone commission for Parliament.[27] Phillips rejected the position on the grounds that he could not spare the time, yet he was really concerned that his uncle and tutor William Smith was 'pained' to see his own nephew offered the role that he felt he himself could fill. Avoiding the 'plague of going to see stones and buildings', Phillips persuaded Barry to select Smith, who welcomed both the remuneration and the potential the project held for geology.[28]

De la Beche and William Smith were both eager that the tour be conducted as a model of how best to manufacture geological research. In the 1830s, geology was a problematic subject characterized by divisive methodological controversies. In his provocative *Principles of Geology* (1830–1833), the Scottish geologist Charles Lyell (1797–1875) argued that the earth's form was best examined by studying geological activity

[24] See E. I. Carlyle, 'Smith, Charles Harriot (1792–1864)', rev. M. A. Goodall, *Oxford Dictionary of National Biography*, Oxford University Press, 2004 [http://ezproxy.ouls.ox.ac.uk:2204/view/article/25787, accessed 31 August 2014]; Rupert Gunnis, *Dictionary of British Sculptors, 1660–1851*, (London, 1953), pp. 354–55; on sculpture and natural philosophy, see Michael Hatt, Martina Droth, and Jason Edwards, 'Sculpture victorious', in Martina Droth, Jason Edwards, and Michael Hatt (eds.), *Sculpture Victorious: Art in an Age of Invention, 1837–1901*, (Yale, 2014), pp. 15–55, 30–34.

[25] Bate, 'Sir Henry Thomas De la Beche and the founding of the British geological survey', p. 150.

[26] *The Times* popularized the BAAS as the 'Parliament of science', see Roy Macleod, 'Introduction: on the advancement of science', in Roy Macleod and Peter Collins (eds.), *The Parliament of Science: The British Association for the Advancement of Science, 1831–1981*, (Northwood, 1981), pp. 17–42, 17.

[27] Morrell, *John Phillips*, p. 162. [28] Ibid., p. 163.

such as volcanoes, earthquakes, and erosion.[29] For Lyell geology was the study of processes, such as rock formation through heat or pressure, which were still in action.[30] By contrast, De la Beche and William Smith asserted geology to be a subject of collecting and observing rocks and fossils. Considering Lyell's *Principles of Geology* to be too theoretical, and even anti-empirical, De la Beche brought out his own *A Geological Manual* in 1831, emphasizing the importance of sample collecting and observation in geology.[31] De la Beche's manual was consistent with the practices of William Smith who had grounded his theory of stratification on collections of fossils belonging to his friends. De la Beche was equally enthusiastic that geological research be based on sustained fieldwork.[32] They both concurred that geology was primarily an outdoor pursuit; part of its appeal was this practice of collecting and observing nature. De la Beche believed that this practice of geological observation was essential to avoid the 'wilderness of crude hypotheses or unsupported assumptions' and worked to make it the BGS's principle value.[33] The survey for Parliament was a model enquiry of these broader techniques. Of course, Lyell's own work was well-grounded in fieldwork, especially around Mt Etna and Mt Vesuvius.[34] However, his emphasis on observing processes presently at work, like volcanic activity, and then grounding explanations for past activity on these actions was not considered appropriate for Parliament. In this intellectual context, it can be seen how De la Beche and William Smith's participation in the Parliamentary survey was a demonstration of geological utility not only to Parliament, but to geological audiences like the Royal Geological Society.

Travelling by carriage, rail, pony, and foot, the four commissioners toured the quarries and ancient buildings of Britain. Meeting at Newcastle for the close of the 1838 BAAS meeting, the commissioners spent late-August and

[29] Martin Rudwick, 'Lyell, Sir Charles, first baronet (1797–1875)', *Oxford Dictionary of National Biography*, Oxford University Press, 2004; online edn, May 2012 [http://ezproxy.ouls.ox.ac.uk:2204/view/article/17243, accessed 23 August 2014]; Rudwick, *Worlds Before Adam*, p. 553; see Charles Lyell, *Principles of Geology: Being an Inquiry How Far the Former Changes of the Earth's Surface are Referable to Causes Now in Operation*, (London, 1834).

[30] Charles Lyell, *Elements of Geology*, (London, 1838), pp. 2, 78–79.

[31] Tom Sharpe, 'Slavery, sugar, and the survey', *Open University Geological Society Journal*, vol. 29, no. 2 (symposium edition, 2008), pp. 88–94, 92–93.

[32] Simon J. Knell, *The Culture of English Geology, 1815–1851: A Science Revealed through Its Collecting*, (Aldershot, 2000), pp. 22–24.

[33] Henry T. De la Beche, *A Geological Manual*, 2nd ed., (London, 1832), p. v; also see Henry T. De la Beche, *Report on the Geology of Cornwall, Devon, and West Somerset*, (London, 1839).

[34] Martin J. S. Rudwick, 'Travel, travel, travel: geological fieldwork in the 1830s', in Martin J. S. Rudwick (ed.), *The New Science of Geology: Studies in the Earth Sciences in the Age of Reform*, (Aldershot, 2004), pp. 1–10, 8–9.

September on tour. First from Newcastle to Edinburgh, then to Glasgow, Carlisle, York, Tadcaster, Doncaster, Derby, Lincoln, and Birmingham, the four men analysed ancient structures before examining the quarries from which the stone had been obtained.[35] All the while samples of stone were collected from quarries which had produced the rock which had lasted well in those architectural structures. These buildings included the ruined St Mary's Abbey next to York Museum, York Minster, Ripon Minster, parish churches, and great country seats such as Castle Howard. Throughout the tour Charles Smith made detailed notes regarding the potential for stone to be carved. At Newcastle, he noted how the local stone tended to blunt workman's tools, causing the quarry's grindstone to be in constant use.[36] At the Hookstone Quarry, near Harrogate, Smith observed a white sandstone which he reckoned to cost '50 per cent more than Portland' to work.[37] He also took note of the various practices of cutting stone in various quarries. After a three-day break in London, the survey resumed from 26 September until 5 October. This time the commissioners visited the colleges and churches of Oxford, Cheltenham, Gloucester, Bristol, Bath, Glastonbury, Dorchester, the Isle of Portland, and Salisbury.[38]

It was the experience of Oxford's architecture which made the most dramatic impression on the commissioners, with the decayed state of the university's colleges offering a bleak warning of the dangers of poorly chosen stone. The commissioners examined the quadrangle of All Souls College, where they observed a shelly oolite much in decay which peeled off in sheets when tampered with.[39] A similar state was witnessed in the cloisters of New College and the quadrangles of 'Brazen Nose College'. They attributed this decay to 'the use of Heddington [sic.] stone, brought from about 1½ mile distant'.[40] Cheap and soft to carve, most eighteenth-century Oxford colleges employed the stone, but when the commissioners examined its performance in 1839 they found the ultimate vindication of their efforts to produce geological knowledge for architecture. As Charles Smith concluded,

[35] OUM SMITH, Box 44, Ms. Journal, 'Journal of William Smith LL. D., Geologist, Civil Engineer, and Mineral Surveyor, on a tour of observation on the principal Freestone quarries in England & Scotland, in company with Mr. Barry the Architect, Mr. De la Beche, and Mr. C. H. Smith, for the purpose of finding stone best calculated for the construction of the new Houses of Parliament', (18 August to 23 October), pp. 1–9.

[36] Royal Institute of British Architects Library (RIBA), London, SmC/1/1, C. H. Smith, 'Abstracts of notes made by C. H. Smith during his travels with the commissioners in search of stone for the Houses of Parliament in the autumn of the year 1838 and spring of 1839', p. 1.

[37] Ibid., p. 44. [38] OUM SMITH, Box 44, Ms. Journal, pp. 11–15.

[39] RIBA, SmC/1/1, p. 92.

[40] Ibid., p. 95; also see W. J. Arkell, *Oxford Stone*, (London, 1947), p. 26.

> It is in the highest degree lamentable to see such fine feeling for architecture displayed on the most fragile stone I have ever beheld; and must, before 50 years have passed, be completely obliterated, and Oxford present one common ruin.[41]

As nineteenth-century geology was considered the observation of the history of nature, so the tour was perceived to be the observation of how well man's past reordering of nature, through architectural construction, had survived the test of decay over time. Stone quarried from the earth, which had been transformed into works of architecture and then exposed to the atmospheric decay of several centuries, provided evidence of the durability of each rock type. The commissioners understood their work to produce experimental evidence, as what they recorded were results of tests on various stone over such an extended period of time as was unrealizable in a laboratory. The way in which the knowledge produced was experimental was something Charles Smith took time to define. He was confident that the strength of geological and chemical knowledge was that they were secured by observable, and therefore comparable, evidence. To obtain this evidence required the study of how stone performed in buildings over time and this involved work beyond the laboratory. Smith described how

> Time is an important element in nature's operations. What is deficient in power is made up in *time*; and effects are produced during myriads of ages, by powers far too weak to give satisfactory results by any experiments which might be extended over perhaps half a century.[42]

Such conditions could not be produced in a laboratory, so to make reliable knowledge the commissioners compared stone in existing structures with the same stone uncut in quarries. Smith asserted that this comparison secured knowledge which was, similar to laboratory produced knowledge, measurable.[43] Smith and the commissioners thus conceived of the quarries and architecture they investigated as a great laboratory in which the effects of time could be witnessed. The tour was conceptualized as the collection of results from an experiment which had taken centuries to unfold. By geologically identifying the quarries from

[41] RIBA SmC/1/1, pp. 95–96; in contrast, when building Keble College between 1868 and 1886, William Butterfield favoured brick rather than Headington Stone, see Geoffrey Tyack, 'William Butterfield and Oxford: adapted from a lecture by Geoffrey Tyack', in Geoffrey Tyack and Marjory Szurko (eds.), *William Butterfield and Keble College*, (Oxford, 2002).

[42] RIBA, SmC/1/2/No. 24, C. H. Smith, 'Something about a hod of mortar', (1865).

[43] RIBA, SmC/1/2/No. 3, C. H. Smith, 'Experiments and selection of stones'; on science and architectural knowledge, see Carla Yanni, 'Nature and nomenclature: William Whewell and the production of architectural knowledge in early Victorian Britain', *Architectural History*, Vol. 40 (1997), pp. 204–21.

which each stone originated, De la Beche and William Smith provided Barry with examples of the various stones in nature, and then compared them to stones removed and employed in churches, abbeys, colleges, and cathedrals. Geology revealed how alternate stone endured time both in its subterranean natural order and exposed to the atmosphere by man's hand.

Beyond observing stone in quarries and in ancient structures, De la Beche and Barry felt it appropriate to subject their collected samples to chemical analysis. Here it is important to distinguish between two different understandings of experimental knowledge. Charles Smith believed that laboratory experiment could not reveal how well a stone would perform in architecture because the crucial factor, time, was absent.[44] He warned that until time acted on stone, nature had 'means of decay which the most scientific chemist never thought of'.[45] However, laboratory experiment was vital to chemically determine the physical differences between sandstones, limestones, and magnesian limestones. Experimental geological knowledge produced beyond the laboratory could provide comparable evidence, but this was to be informed by experimental chemical knowledge produced within a laboratory. Both types of knowledge were the same, but their manners of production were crucially different.

Barry initially advised De la Beche to select Michael Faraday as one of the 'Experimentalists', adding that if he 'offered to do it gratis . . . accept, as we need the cash in hand for further quarry visits'.[46] De la Beche duly invited Faraday to perform the chemical experiments. He explained that having acquired the specimens, 'certain chemical and physical experiments on which would appear very desirable for a right understanding of the subject'.[47] De la Beche noted that as director of the newly established Museum of Economic Geology he was responsible for obtaining this chemical data. He emphasized that chemical knowledge was a crucial element of the survey, recording how they had on tour

found that the atmosphere acts very differently on buildings and therefore it becomes most desirable to find out, if possible, what difference there might be in the stones employed as regards their chemical composition in the first place.

[44] RIBA, SmC/1/2/No. 13, C. H. Smith, 'On the varieties of stone used for architectural carving', (1856).

[45] RIBA, SmC/1/2/No. 3.

[46] 'Letter 66: C. Barry to H. De la Beche', (27 April, 1839), in T. Sharpe and P. J. McCartney (eds.), *The Papers of H. T. De la Beche (1796–1855) in the National Museum of Wales*, (Cardiff, 1998), p. 19.

[47] 'Letter 1115: Henry Thomas De la Beche to Faraday', (5 November, 1838), in Frank A. J. L. James (ed.), *The Correspondence of Michael Faraday: Volume 2, 1832-December 1840, Letters 525–1333*, (London, 1993), pp. 528–29, 528.

There is no difficulty in distinguishing ordinary mineralogical differences, but more delicate work is required for some materials.[48]

In this context, chemistry was conceived of as a vital aspect in the study of geology. Geology could ascertain which stone was and was not durable, but could not explain the chemical reasons for their physical character. De la Beche observed how variable single rock types could be. For example, both York Minster and Southwell Church were of magnesian limestone, yet the former was in a 'wretched state while the latter is nearly in as good condition as when erected by the Normans'. De la Beche believed variation in atmospheric and stone composition could explain the anomalies witnessed on tour. He wanted to know 'the relative porosity of rocks', or rather the 'relative facility with which they imbibe rain, or permit the atmosphere to enter among their particles'.[49]

De la Beche also requested an explanation for the corrosive effect of London's atmosphere. He felt it 'very important to see if there is any thing in the London atmosphere particularly effecting some rocks . . . which ought not affect others'. Faraday was then to experiment on the stone's resistance to 'artificial cold' and pressure. De la Beche had already requested such work from Professor Charles Wheatstone (1802–1875), but encouraged Faraday that his experimental experience would 'assist in dissipating some of the confounded jobbery and humbug which has very generally been hitherto practised as to the stone employed in our public buildings'.[50] This reference to poor past architectural practices of stone selection was one which the four commissioners consistently made, with Charles Smith describing a 'mania' for soft cheap stone in the years before 1839.[51] What De la Beche's comments to Faraday demonstrate so cogently, is the sense that the selection of stone for Parliament was not just drawing on existing geological knowledge, but was actually initiating exciting new avenues of experiment. The survey was not simply a referencing of existing resources, but a creation of enhanced geological and chemical understanding. It was as much an exercise of knowledge construction, as it was a government referencing of scientific authority. Faraday declined De la Beche's invitation, citing his own poor health, but advised Barry to employ the chemist John Frederic Daniell (1790–1845). He warned that De la Beche might be 'expecting too much from Chemistry', but believed the object to be 'a most excellent one'.[52] The commission thus selected Daniell to collaborate with Wheatstone.

[48] Ibid., p. 528. [49] Ibid., p. 528. [50] Ibid., p. 529. [51] RIBA, SmC/1/2/No. 3.
[52] 'Letter 1120: Faraday to Henry Thomas De la Beche', (16 November, 1838), in James (ed.), *The Correspondence of Michael Faraday: Volume 2*, p. 534.

At this time Wheatstone and Daniell were both employed at the recently established King's College London. Wheatstone held the chair of experimental philosophy from 1834, while from 1831 Daniell had been the college's first professor of chemistry and since 1836 held the chemistry and geology chair of the East India Company's military seminary at Addiscombe.[53] Both were projectors of a heavily experimental programme at King's College. In his *An Introduction to the Study of Chemical Philosophy*, Daniell asserted that 'accurate observation' through experiment was the only path to accurate chemical knowledge.[54] He surmised that only experiment provided true knowledge of natural phenomena. In his work, dedicated to Faraday, he argued that experiment provided knowledge of a human understanding, while 'the *theory* of the universe' was comprehendible only to Omniscience.[55] Wheatstone, Faraday, and Daniell held meetings at each other's laboratories to experiment together throughout the 1830s. Daniell's Anglicanism encouraged him to observe nature as the work of God.[56] It was through experimental trial that Daniell pursued this observation. One commentator noted how Daniell's work on a Hygrometer for predicting thunderstorms advanced 'no extravagant theory ... but reasons soberly and acutely upon Meteorological data and on the observations he has made, hence his inference on the deductions of Science and experiment'.[57] Each experiment he performed, 'the greater cause did he find to admire the Power, Wisdom, and Goodness of God'.[58] Daniell was the ideal candidate to conduct the experiments Barry and De la Beche wanted. While some strove to mathematically expose the laws of electricity and magnetism, Faraday and Daniell sought to 'prove ... *experimentally*' the chemical nature of these phenomena. It was this experimental work which Barry believed could invest his choice of stone with chemical authority. It is worth exploring the form this experimentalism took as it reveals the ways

[53] S. P. Thompson, 'Wheatstone, Sir Charles (1802–1875)', rev. Brian Bowers, *Oxford Dictionary of National Biography*, Oxford University Press, 2004; online edn, January 2011 [http://ezproxy.ouls.ox.ac.uk:2117/view/article/29184, accessed 21 July 2013]; Frank A. J. L. James, 'Daniell, John Frederic (1790–1845)', *Oxford Dictionary of National Biography*, Oxford University Press, 2004 [http://ezproxy.ouls.ox.ac.uk:2117/view/article/7124, accessed 21 July 2013].

[54] John Frederic Daniell, *An introduction to the study of chemical philosophy: being a preparatory view of the forces which concur to the production of chemical phenomena*, 2nd ed., (London, 1843), p. 2.

[55] Ibid., p. 5.

[56] David I. Davies, 'John Frederic Daniell, 1791–1845', *Chemistry in Britain* (October, 1990), pp. 946–60, 947–48.

[57] King's College London Archives (KCL), Papers of J. F. Daniell 5/1, 'John Frederic Daniell – Ms life – lectures', (c.1845–50), pages unnumbered.

[58] Ibid.

in which experiment was perceived to create ordered and reliable knowledge of a geological and chemical problem.

The final report consisted of four tables of data. Table A was an extensive list of all the quarries visited, along with rock types located within them, depth of stone beds and distance to rivers or canals for transport. Table B included a list of buildings constructed from stone from each quarry, while Table C was entitled 'of chemical analyses'. Daniell and Wheatstone produced this table, which detailed the chemical makeup of sandstones, magnesian limestones, oolites, and limestones. This table provided evidence of each stone's 'Absorbent Powers when saturated under the exhausted Receiver of an Air Pump'.[59] After being saturated in water, each stone was placed in an air pump before the receiver was exhausted and the extracted water then measured. Table D included further experimental results taken at King's College. Each stone sample was weighed, then saturated with water and weighed again. Daniell and Wheatstone subjected the samples to experiments on pressure. To attain each stone's cohesive power they employed a six-inch hydrostratic press. To determine the weight each stone could resist before producing its first fracture, the cubed samples were individually placed under the press, and one pound weights were added to the lever pushing down on the stone. Each weight was calculated to produce 71.06lbs of pressure per square inch and was added at minute intervals until the stone fractured. To provide accuracy, the results were recorded in terms of weights applied.[60]

These tables provided Barry with what he felt was 'a sufficiently competent knowledge' of the stone observed on the tour.[61] The report omitted granites, porphyries, and other hard stones on account of their expense to carve. Supported by the evidence from King's College, details were focused on the condition of limestones employed at Oxford, sandstones at Derby and Newcastle, and the magnesian limestones of Minster and York. All the 'striking evidence' suggested magnesian limestone provided a sound balance between durability, economy, and malleability. The report noted that the most desirable feature of a stone was a high ratio of crystalline to its 'cementing' substances. This was the essence of a stone that could resist the 'aëriform products' of London's atmosphere.[62] Such 'chemical action' could transform the 'entire matter' of limestone, which then rendered it liable to decay through 'mechanical action'.

Based on the report's evidence, the commission advocated the magnesian limestone of Bolsover Moor as the most suitable for Parliament. It was reported that the Norman portions of Southwell Minster were of Bolsover

[59] PP. 1839 (574), p. 43. [60] Ibid., p. 47. [61] Ibid., p. 3. [62] Ibid., p. 5.

stone and 'in a perfect state', with the 'mouldings and carved enrichments being as sharp as when first executed'. In comparison, the magnesian limestones employed in York, particularly at York Minster, were substantially decayed. The report asserted the explanation for this variance was the crystalline levels of each stone variety. This conclusion was 'in accordance with the opinion of Professor Daniell' who through experiment had observed that 'the nearer the magnesian limestones approach to equivalent proportions of carbonate of lime and carbonate of magnesia, the more crystalline and better they are in every respect'.[63] Perhaps most surprising was that this choice was made instead of Portland stone, which had been popular with previous architects working in London. Portland was examined in the report for Parliament, but was judged to have performed poorly in comparison with Bolsover. In particular, the commissioner's observations of the Portland in Christopher Wren's St Paul's Cathedral, begun in 1675, indicated that the stone had weathered poorly compared to the same stone which had been cut but not used and left to weather on the Isle of Portland (Figure 3.1).[64] Comparatively it seemed that Portland had not performed well, as many parts of the Cathedral were 'mouldering away'. Instead of this familiar stone, the commissioners placed faith in their new research.

Bolsover was listed as a light yellowish-brown magnesian limestone from Derbyshire, with beds of eight-inches to two-feet.[65] It was described as 3.6 per cent silica, 51.1 per cent carbonate of lime, 40.2 per cent carbonate of magnesia, and 1.8 per cent iron alumia. Daniell and Wheatstone determined Bolsover to have the lowest absorption rate of the magnesian limestones.[66] Bolsover's performance could be evaluated through existing architecture and laboratory trials alongside alternative rock types. Daniell and Wheatstone attested that Bolsover was the 'heaviest, strongest, and absorbs the least water' of all magnesian limestones, while it was 'remarkable for its peculiarly beautiful crystalline structure'.[67] This choice combined laboratory experiment with geological observation. As the report summarized, 'for durability, as instanced in Southwell Church, &c, and the results of experiments ... for crystalline character', Bolsover was the most suitable stone for Parliament.[68]

This report was to be an important architectural and geological reference until at least the 1860s. It was not until Edward Hall's *On the building and ornamental stones of Great Britain and foreign countries* (1870) that there was a work of similar scale.[69] Even as late as 1848 Charles Smith was producing lists of building stones based on data collected during the

[63] Ibid., p. 7. [64] Ibid., p. 6. [65] Ibid., pp. 12–13. [66] Ibid., p. 43. [67] Ibid., p. 48.
[68] Ibid., p. 9.
[69] G. K. Lott, 'The development of the Victorian stone industry', *The English Stone Forum Conference*, (York, 15–17 April, 2005), pp. 44–56, 46.

Figure 3.1 The Grove quarry on the Isle of Portland, from which St Paul's Cathedral's stone was cut

government survey.[70] Nevertheless the selection of Bolsover was problematic. The beds in the Bolsover Moor quarry were insubstantial to provide the size and quantity of stone desired for Parliament. Choosing a stone similar to Bolsover, Barry and the commissioners resorted to the magnesian limestone of Anston in Yorkshire, not far from the Bolsover quarry.[71] While the choice of stone was altered after the report, the evidence published remained the substantiating claim for its selection. A different quarry was employed, but selected on its similarity to the stone thought most proper for Parliament. As will be shown, this decision had unfortunate results.

Situating Knowledge: William Smith and the Geology of Architecture

The 1839 report was manufactured through practices of collecting, observing, and experimenting, but this was only the initial process of

[70] BGS GSM/MG/C/20, C. H. Smith, 'Building stones', (5 January, 1848).
[71] Port, 'Problems of building in the 1840s', p. 97; also see Alec Clifton-Taylor, *The Pattern of English Building*, (London, 1987), pp. 94–95.

securing the knowledge to guide the selection of Parliament's stone. In the 1830s geology engendered problems not only of methodology but, for some, religion and utility. The commissioners produced knowledge during the tour, but sought also to situate it within this broader discipline, and for William Smith this meant enunciating the practical advantages and religious implications of using geology for architecture. Of the four commissioners responsible for selecting the stone for Parliament, he displayed the greatest enthusiasm for geology's role within the survey. For Smith the partnership between science and architecture involved more than just bringing scientific order to the building at Westminster; rather it was the beginning of a new age of scientific architecture, where architects like Barry would work alongside geologists like De la Beche to design and construct decay defying national edifices.

Reflecting on the survey of 1838, Smith declared how 'in Britain there has sprung up *a science entirely new*'.[72] He explained how geology had guided Barry from well-preserved ancient structures to the quarries from which their stone had been obtained. Geology facilitated the identification of rock in buildings that had survived the challenges of decay. It therefore provided the architect with a means of controlling the ravages of time. As Smith put it,

> This science therefore sheds a new light upon the mystic gloom of all those ancient structures and upon the science of architecture which must inspire the architects of the present age of Taste & Science with not only the hope of rivalling the ancients but with the confidence of building their Temples with lasting fame; for by the help of Geology they will find a good foundation & durable materials. It is gratifying to see so many societies formed for the advancement of this and other sciences & their meetings so well attended.[73]

Smith's evidence for this optimism transpired from his 'own observation on a recent tour of three thousand miles'. This of course had been the stone survey, which had been preceded by the Newcastle BAAS meeting. Smith hereby connected his work with Barry to new flourishing scientific meetings, like the BAAS, in order to project a new age of scientific architecture in which science would annihilate the problems of time. Geology was bringing permanence to British 'Temples' which would ensure the age a place in future generations alongside the 'ancient' civilizations.

At sixty-nine, Smith brought tremendous energy to the commission. His 'previous knowledge of nearly all the building-stones and quarries in the kingdom' was valuable.[74] Born in the village of Churchill near Oxford in

[72] OUM SMITH, Box 44, Folder 5, Item 6, (6 January, 1839), p. 2. [73] Ibid., p. 3.
[74] John Phillips, *Memoirs of William Smith, LL.D., Author of the 'Map of the Strata of England and Wales'*, (London, 1844), p. 122.

1769, Smith had received only a basic education in childhood. However, in his late teens he gained experience as a surveyor on his uncle's farm with the help of a few geometry textbooks.[75] Later he became surveyor to the Somerset Coal Canal Company, which entailed much levelling work during the early-1790s. Through this role he acquired an understanding of the layers of various rock types beneath the earth's immediate surface.[76] After being sacked from the Canal Company in 1799, Smith worked with the Rev Benjamin Richardson, Rector of Farleigh Hungerford, near Bath, to put the cleric's fossil collection into an order which reflected the earth's various strata. By observing fossils around Bath, Smith concluded that various strata could be identified by the unique fossils contained within them.[77] Following his *Table of Strata Near Bath*, Smith worked until 1815 in producing a coloured geological map of England's strata. The significance of his strata theory took time to secure a consensus of recognition, but by the time of the stone survey Smith had earned fame as the 'Father of English Geology'. Along with this accolade, Sedgwick presented him with the Wollaston Medal of the Geological Society in 1831 before Smith was granted a government pension of £100 a year at the 1832 BAAS meeting in Oxford.[78] Throughout the 1830s Smith regularly frequented BAAS gatherings, before succumbing to influenza on the way to the 1839 Birmingham meeting.

His confidence in the report's potential for scientifically guiding future architecture was immense. He described how the government freestone survey had consisted of,

> accurate investigations which may be of no trifling importance to all the architectural concerns of the country for by it all the best building stone of the Island will be known and the loses from bad and the advantages of good stone will be duly appreciated; for reasonings from phenomena directed by experience and tested by science may be safely taken as guides.[79]

Not only did Smith believe that Parliament could be made out of a stone to defy decay, but that the commission's report ushered in a new age of scientific architecture.

The report provided both an index to the qualities of various building stones and a general guide to the characteristics of different rock types.

[75] John L. Morton, *Strata: the remarkable life story of William Smith, 'the father of English Geology'*, (Horsham, 2004), pp. 13–15.

[76] Ibid., p. 18. [77] Ibid., pp. 30–31.

[78] Ibid., pp. 126–27; see William Smith, *A Delineation of the Strata of England and Wales: with Part of Scotland; Exhibiting the Collieries and Mines, the Marshes and Fen Land Originally Overflowed by the Sea, and the Varieties of Soil According to the Variations in the Substrata, Illustrated by the Most Descriptive Names*, (London, 1815).

[79] OUM SMITH, Box 44, Folder 1, Item 3, (1838).

Smith outlined how it had determined marble to be the most durable of all 'easily' usable stones, while magnesian limestones and oolites were both of less durability but greater practicality.[80] These conclusions had been confirmed by 'chemical & philosophical experiments'. Chemical experiments had demonstrated the crystalline and carbonate composition of these stones, while philosophical inquiries beyond the laboratory assessed their performance in architecture. The variance in durability was experimentally shown to be in relation to the mineral's crystalline structure and compactness. Indeed, Smith felt that the strength of the evidence produced during the stone survey was that it was grounded in experimental practices: experiments within the laboratory had been correlated to observations on tour. Both chemical experiment in the laboratory and the observation of ancient buildings combined to provide the useful knowledge which would guide the science of architecture. As Smith summarised,

> Knowledge is worth but little until we arrive at some generalizations derived either from phenomena or the facts of experiment – the latter is of course the most valuable, but to save the trouble of repeating a multiplicity of experiments a sort of generalizing theory of these must be formed and when this can be done in accordance with phenomena so much more satisfactory and generally useful will such knowledge become. This of course applies to the knowledge of things in nature and though we judge of such objects commonly by phenomena such a mode of judging is not altogether theoretical for the Eye furnishes the mind with the results of experience.[81]

For Smith then, the tour of quarries for Parliament was a great epistemological venture, which was a masterpiece of how experiment could secure a scientific theory of architecture. It was about producing trustworthy knowledge that would guide the construction of Parliament and provide geological theory for future architectural schemes.

Although the final report was in many ways limited, considering only freestones thought suitable for Parliament, Smith was content that Barry had 'found the objects of the Survey so completely and satisfactorily attained and so fully exemplified by the tables as rendered it unnecessary to incumber the Report with any of its Geological details'.[82] The tables provided sufficient information to make the selection for Parliament, but behind the report lay much geological knowledge, including Smith's work on stratification. Smith was confident similar scientific endeavours would

[80] OUM SMITH, Box 44, Folder 1, Item 4, 'Theoretical view of a befitting stone for the Houses of Parliament', (4 December, 1838).

[81] Ibid.

[82] OUM SMITH, Box 44, Folder 1, Item 5, London, (10 March, 1838, but must be 1839), p. 1.

address wider architectural questions of other freestones, brick earths, 'the best limestones', and roofing slates. From the Parliament stone survey a body of knowledge would be produced to guide an architect's work. Buildings were to be completely ordered through a geological interpretation.

Smith's commentary boasted the relevance of the survey's work for Victorian society. By 'noting all the different kinds of building materials in the order of the strata which produce them', Smith hoped 'to induce a large body of scientific men to turn their attention to Geology'. He could not

> refer to better proofs of the necessity of their doing so than are exhibited in the Report. The public however, and especially those in the streets of London and other great Towns may *feel* an interest in having a good footing upon Flagstones (which Architects may consider beneath their notice) but which to a London Geologist may be of some importance.[83]

This use of geology was effectively the application of Smith's theory of strata to the problem of urban architecture. If the Palace could challenge the decay of time through geological authority, then so too could all human constructions, even those quite literally beneath the foot of the architect. Smith felt the stone survey signified a new age in which science and architecture would provide suitable towns for the rapidly urbanizing society of industrial Britain. To emphasize the survey's utility in this way was to situate it within Smith's earlier work on stratification. Indeed, in his 1817 study of the different fossils in various strata, one of Smith's earliest observations on oolitic rock was that it produced the 'finest Building stone in the Island'.[84] The Parliamentary stone survey was the continuation of work Smith had been conducting for over three decades.

It is significant to his perception of Parliament's architecture that Smith's science carried with it deeply religious sentiments. Between 1800 and 1840, geology was frequently interpreted to fit within the depiction of God's creation of the universe described in Genesis. 'Scriptural geologists' argued that God's work could be observed within rock formations.[85] Buckland, for example, was not only reader in geology at Corpus Christi College, Oxford, but in 1808 was ordained a priest in the Church of England, eventually becoming Dean of Westminster thanks to Peel's patronage. Buckland dominated geology in Oxford during the 1820s and 1830s, with the High-Church

[83] Ibid., p. 2.

[84] William Smith, *Stratigraphical System of Organised Fossils, with Reference to the Specimens of the Original Geological Collection in the British Museum: Explaining Their State of Preservation and Their Use in Identifying the British Strata*, (London, 1817), attached table.

[85] Boyd Hilton, *The Age of Atonement: The Influence of Evangelicalism on Social and Economic Thought, 1785–1865*, (Oxford, 1986), p. 23.

Anglican leaders of the Oxford Movement John Keble, Edward Pusey, and John Newman all attending his lectures on geology at the Old Ashmolean Museum.[86] Buckland, though not interpreting the Bible literally, believed geology revealed evidence of a 'geological deluge' consistent with Biblical depictions of a great flood. Buckland accepted that there had been a time before Adam and that the Earth's formation had taken considerably longer than a few days, but felt that the Old Testament was compatible with geological evidence.[87] In his lectures Buckland assured Church of England audiences of this consistency. He laboured to make the concept of 'deep time' in the formation of the earth acceptable to Anglicans.

The stone survey was conducted within the context of these wider geological questions. Specifying his religious denomination is challenging, but Smith's work was produced within a network including several Church of England ministers. Throughout his life Smith constantly referred to the works of the Almighty in relation to geology, while his closest friends were Anglican clergymen.[88] His index linking of fossil types to strata and claims of correlation through layers had been produced alongside the Rev Benjamin Richardson, and the Rev Joseph Townsend of Pewsey. Together they did the fossil-hunting to test the strata claims, while it was Richardson's fossil collection which Smith studied before producing his *Table of strata near Bath*. Smith drew up this table in Townsend's home in Bath.[89] Furthermore, when he took up the care of his seven-year-old orphaned nephew, John Phillips, he had the boy sent for an education under the roof of the Anglican Richardson.[90] Although a choice shaped by financial constraints, this act underlines that Smith had no objections to the Anglican Church and was happy to see his nephew receive a Church of England education.

It is hardly surprising then that when Smith envisaged the application of geology, in particular knowledge of geological strata, to the Houses of Parliament, this too carried religious connotations. Geological knowledge of strata was knowledge of Creation. As Smith explained just two months after the survey, in observing how stone decayed under atmospheric action after removal from its original strata,

> we seem to begin dolting out a line of distinction between the works of Creation & those of Nature, which seems to show us that, that operations of Nature undisturbed ... never extended much beneath the surface of the Earth. This distinction, duly considered, might be of service to some speculative Geologists in connecting their fanciful theories; especially that of 'strata forming from causes

[86] Neville Haile, 'Buckland, William (1784–1856)', *Oxford Dictionary of National Biography*, Oxford University Press, 2004; online edn, October 2007 [http://ezproxy.ouls.ox.ac.uk:2204/view/article/3859, accessed 23 August 2014].

[87] Ibid. [88] Morton, *Strata*, p. 21. [89] Ibid., pp. 30–31.

[90] Morrell, *John Phillips*, p. 1.

now in operation'. But to confine ourselves to common occurrences we find facts innumerable to prove that, all the great changes effected by Nature, are by a chemical process continually going on at or near the surface of the Earth.[91]

Smith's understanding of subterranean strata was that this was the work of God's own creation rather than nature. This interpretation fits within the broader context of scriptural geology. Hilton has demonstrated that such understandings of geology were broadly evangelical, shared within the Church of England and Non-Conformist churches.[92] Smith's theory of strata was often interpreted as the scientific demonstration of God's sustained role in the process of Creation. This religious potential of geology manifest itself as a solution to the problem of a mechanistic universe, which God had set in motion, but working without His involvement. What strata suggested was God's constant action in the process of Creation: Smith's stratification revealed the unfolding of a progressive divine plan from basic to more complicated life forms.[93] Smith's division between nature and Creation placed him within this outlook and emphasizes the notion that the stone survey was both a study of nature and an observation of the work of God.

Smith's reference to geologists with 'fanciful theories' of 'causes now in operation' was a response to Lyell's call to study active processes. Smith was arguing that geology was not the study of present-day phenomena, but the observation of strata which revealed the work of God's hand. To study stone on the earth's surface, as done on the tour, was not to observe geological processes in action according to Lyell's recommendation, but to assess the effects of chemical processes on God's Creation. Considering the use of stone in Parliament, Smith recognized that the problem of decay was due to 'the Chemistry of Nature'. He explained how in building with stone from beneath the earth's top strata, lower strata rock was exposed to the atmospheric elements which produced decomposition. This was an architect's challenge, 'for as soon as Art displaces any of the Earth's solids the powers of nature begin to act upon them'. Bringing stone above 'the surface of the Earth' and exposing it to 'the action of the air' was a reordering of nature which created problems for which science could provide solutions.[94] Chemistry within geology offered knowledge of the atmosphere and stone composition which together revealed the workings of nature; the 'Chemistry of

[91] OUM SMITH, Box 44, Folder 5, Item 4(ii), (2 January, 1839).
[92] Hilton, *The Age of Atonement*, p. 23.
[93] Michael Hall, 'What do Victorian churches mean? Symbolism and sacramentalism in Anglican Church architecture, 1850–1870', *Journal of the Society of Architectural Historians*, Vol. 59, No. 1 (March, 2000), pp. 78–95, 82.
[94] OUM SMITH, Box 44, Folder 5, Item 4(ii), (2 January, 1839).

Geology' had been 'one of the main pillars' of the government survey.[95] Smith's distinction between chemistry at work in nature and God's Creation through stratification really mattered in relation to the report for Parliament. It was a way of claiming that the geology of Smith and De la Beche could yield valuable knowledge without homage to Lyell's definition of geology. Smith's writings established that the commission's endeavours were compatible with theological understandings of geology. Lyell's approach to geology appeared to remove the necessity of a Creator in the formation of the earth, so in rejecting his methodology, Smith was siding with the geology of Buckland who was an outspoken critic of Lyell's emphasis of active natural phenomena. This is interesting because Peel had first suggested the establishment of a commission on stone for Parliament and it was Peel who was Buckland's greatest patron. Although its precise nature is not recorded, in the years after the commission Charles Smith acknowledged that Buckland had provided valuable assistance to the commissioners.[96] While the theology of geology was not directly important for choosing stone, it was vital in situating the knowledge produced within wider controversies.

Decay on the earth's surface was the chemical action of nature, which although not devoid of divine providence, was a very different type of action to the phenomena Lyell described. So, to remove stone from beneath the surface and therefore reorder God's strata was to expose it to nature. In a sense, it was imitating the work of Creation. Yet to use Smith's strata theory successfully at Parliament and to reorder nature at this site called for chemical knowledge to resolve atmospheric challenges. If geology was the observation of Creation, and architecture was the reordering of strata, then chemistry was a means for replicating the work of the divine. In building edifices of endurance and defying the aging of time, architecture could assume the permanence of Creation. This emphasis had particular significance for Anglican church architecture. Employing a Gothic style suggested medieval tradition and continuity. Yet theologians, like Newman before he converted to Roman Catholicism, were in the 1840s keen to show Christianity's authority did not lie in the past, but in the present.[97] With its commitment to the Gothic, Anglican church architecture struggled to move on from its medieval associations. During the 1840s and 1850s, however, Gothic came to embody notions of progress and modernity through an emphasis on scientifically observing the natural world.[98] Architects such as Thomas Deane (1828–1899) and Benjamin

[95] OUM SMITH, Box 44, Folder 5, Item 4(ii), 'Physical effects on those works of man', (18 April, 1839).
[96] RIBA, SmC/1/2/No. 3, C. H. Smith, 'Experiments and selection of stones'.
[97] Hall, 'What do Victorian churches mean?', p. 80. [98] Ibid., p. 81.

Figure 3.2 Various stone samples employed didactically as column shafts in the Oxford University Museum

Woodward (1816–1861), in Gothic works like the Oxford University Museum (1854) and the Trinity College Museum in Dublin (1852), approached building as presenting 'the Book of Nature' through delicate carvings of botanical specimens and various highly polished stone types. Indeed, it was Buckland who had, between 1808 and 1812, collected the rocks and fossils which would eventually form the nucleus of Oxford University Museum's geological collection (Figure 3.2).

Organizing Knowledge: Institute and Museum

Despite William Smith's lauding of scientific architecture as 'one of the uses of Geology in which our Citizens are largely interested', the stone survey aroused concern.[99] Even on the completion of the enquiry, there were early signs of dissent in contrast to Smith's optimism, with the *London Weekly*

[99] OUM SMITH, Box 44, Folder 5, Item 4(ii), 'Flagstones in use in London', (21 January, 1839).

Dispatch noticing the conclusion of 'Another Whig Job'.[100] Evidently for all the claims of scientific practice, there were suspicions of political patronage. Even among the members of the survey there was disagreement. Smith's understanding of how geology should advance architecture was not shared unanimously between the commissioners. In terms of the authority of geology, chemistry, and architecture, this concept was open to interpretation. While William Smith and De la Beche regarded geology as providing the initiative to making building scientific, Barry was keen that the project be architect led. This divergence of opinion was illustrated by the controversy surrounding the specimens collected on tour. Resolving this problem was an important process in the production of the commission's geological knowledge. Managing where stone samples should be exhibited and who was responsible for them was a crucial practice of organizing the tour's evidence.

De la Beche thought that the appropriate place to exhibit the knowledge collected on tour was at the Museum of Economic Geology on Jermyn Street. The stone samples were, he believed, best employed under his control. He asserted that this would be the most effective way of managing the tour's knowledge. De la Beche established the museum in 1837 to display products of geology deemed of economic value to London audiences.[101] With the political support of Peel, architect James Pennethorne (1801–1871) worked alongside De la Beche to build a museum which included marbles in its entrance-hall and decorative Scottish granites and Irish serpentines performing an educative function.[102] Under the 1845 Geological Survey Act this public museum was brought together with the BGS as the home of British geology.[103] Before 1845, with the chemist Richard Phillips (1778–1851) as curator and De la Beche in overall charge, the stone specimens collected during the Parliamentary stone survey were to form the basis of its geological collections. Between the conclusion of the stone survey and the publication of the report, De la Beche and Phillips worked to create a museum 'where instruction can be received in mining or in the application of various sciences'.[104] This was not simply for students, but also for 'Civil Engineers, Architects, and Agriculturalists'.[105] Phillips designated the first

[100] OUM SMITH, Box 44, Folder 1, Item 5, 'London weekly dispatch, Sunday October 14th 1838, p. 485'.

[101] John Smith Flett, *The First Hundred Years of the Geological Survey of Great Britain*, (London, 1937), p. 34.

[102] Geoffrey Tyack, *Sir James Pennethorne and the Making of Victorian London*, (Cambridge, 1992), pp. 179–91.

[103] Secord, 'The geological survey of Great Britain as a research school', p. 227.

[104] BGS GSM/DC/A/C/1, Henry De la Beche, 'Suggestions respecting the Museum of Economic Geology', (28 January, 1839), pp. 51–56, 52.

[105] Ibid., p. 56.

floor of the museum for useful ores and the second for 'all the varieties of British Coal', while on the third there was to be a laboratory displaying chemical substances. The ground floor, however, was to be a selection of stone specimens 'likely to be advantageously employed, either for architectural or engineering purposes'.[106] The basis of this display was the stone specimens collected for Parliament. The museum was to be both a celebration of science's role in the Palace's construction, and also a projector of how the survey could be a model for future architectural ventures.

This ground floor collection had been amassed during the commission but, on Phillips' recommendation, De la Beche felt that a second set of cubed samples should be installed on the museum's exterior. De la Beche believed this second set 'might be usefully employed by being placed on the top of the Museum for the purpose of ascertaining at future periods the different effects produced upon them by the influence of the atmosphere'.[107] This scheme was a continuation of experiments already performed on the samples. Their potential to resist decay would be made observable on the museum's exterior. The practical value of the museum would have been beyond doubt to the likes of William Smith, whose emphasis on collecting stone and fossil specimens encouraged the formation of public displays of geology; his nephew John Phillips' collecting of 4,500 fossils within two years for the Museum of the Yorkshire Philosophical Society being an industrious example.[108] This public display of practical geological knowledge through museums was a vital facet of the culture of English geology during the 1830s. Yet for Barry, De la Beche's ambitions in this regard undermined his own authority, and with it the perceived importance of the architect in the science of architecture.

In the Office of Woods' initial description of the survey, provision was made for the collected stone specimens to 'be cubed and transmitted to the Economic Geology Collection'.[109] This was due to Duncannon being 'anxious that this inquiry, and the Report which will be made, should be useful to the Public in general'.[110] The responsibility for having these six-inch cubes cut fell to Barry, but getting him to deposit them proved contentious. When De la Beche requested not just one, but two sets of cubes, Trenham W. Philipps, secretary to the Board of Woods, warned

[106] BGS GSM/DC/A/C/1, 'Letter from R. Phillips to Henry De la Beche', (22 October, 1839), pp. 106–08, 107; to compare with Oxford University Museum's layout, see Carla Yanni, *Nature's Museums: Victorian Science and the Architecture of Display*, (Baltimore, 1999), pp. 62–90.

[107] BGS GSM/DC/A/C/1, 'Letter from Charles Barry to T. W. Philipps', (29 July, 1839), pp. 141–42, 142.

[108] Knell, *The Culture of English Geology*, p. 115.

[109] BGS GSM/DC/A/C/11/5, 'Letter from J. Baring to Henry De la Beche', (11 August, 1838), p. A.

[110] Ibid., p. B.

that, 'We shall have difficulty in getting our two sets of cubed specimens, or indeed either, from Barry. He is so obstinate'.[111] Philipps wrote to Barry demanding the samples in July 1839, but received a disparaging reply. Barry recalled Duncannon discussing the cubes with him, but repudiated that they were due for the museum. He complained that he had prepared them at his own expense and that while the museum was entitled to one set, the proper place for the second was not at a geology museum but in the RIBA.[112] He claimed that the museum's collection had 'long since been ready for delivery and have been only awaiting the preparation of proper cases for their reception. The second set of cubes I had intended to present to the Institute of British Architects'. These intentions show how Barry wanted the samples to become part of the RIBA in a way that would contribute to making architecture professional. Such sentiments fit neatly alongside wider ambitions to establish the RIBA as a centre of architectural instruction and training, as outlined in Chapter 2. If Duncannon wished, Barry would give his second set to the museum, but lamented that in doing so he would be giving up 'the only object I had in making the collection which was that of having the pleasure to present them to an establishment where they would be highly appreciated and where they might be practically useful'.[113]

Duncannon rejected Barry's insistence that the RIBA was of greater priority than the museum. He considered that he had,

> no second course open to him in the matter when uses have been suggested to him for both of these sets, in connection with the museum founded by the Government; (at whose expense you are aware every proceeding connected with this enquiry has from the beginning been conducted) when all parties to whom the suggestion has been mentioned have concurred in its' propriety: and when it is obvious that the Members of the Institute of British Architects will have as full and as ready access, at all times, to the Museum, as to the Rooms of their own Building.[114]

Duncannon's response did not just emphasize the importance of geology within the aims of the survey, but offered an early sign of the government's commitment to the museum, and with it, state-funded science.

[111] 'Letter 1262: Trenham W. Philipps to De la Beche', (24 July, 1839), in Sharpe and McCartney, p. 91.

[112] See M. H. Port, 'Founders of the Royal Institute of British Architects (*act.* 1834–1835)', *Oxford Dictionary of National Biography*, Oxford University Press, January 2014 [http://ezproxy.ouls.ox.ac.uk:2204/view/theme/97265, accessed 21 August 2014]; for architecture as a profession, see J. Mordaunt Crook, 'The pre-Victorian architect: professionalism & patronage', *Architectural History*, Vol. 12 (1969), pp. 62–78, 68.

[113] BGS GSM/DC/A/C/1, p. 141–42.

[114] BGS GSM/DC/A/C/1, 'Letter from T. W. Philipps to Charles Barry', (2 August, 1839), pp. 142–43, 143.

Duncannon's rejection of Barry's wishes was significant as it showed that architects were not entitled to privileged access of geological knowledge. In not having a private collection, members of the RIBA were to consult the museum, along with other interested members of the public. In effect Duncannon was announcing that architects should take scientific advice from the museum, rather than attempt to produce such knowledge independently. The museum was to guide such professions into a new scientific age. The architect was to reference the authority of the geologist. To add insult to injury, Duncannon informed Barry that the museum had 'better means of analysis and illustration at their command, than can be possessed by any other Institution'.[115] The government hereby situated scientific architecture firmly within the state-funded museum and dealt a blow to Barry's ambitions for the RIBA.

Unsurprisingly Barry was chagrined at this rebuff. In December 1839, Duncannon informed Barry that he had noticed a statement in *The Morning Post*, 'too gross a perversion of the truth to pass unnoticed'.[116] The article claimed to report accurately the dispute over the stone samples. However, the 'impression sought to be conveyed by that statement ... [was] that Lord Duncannon's decision in respect to these duplicates was adverse to your [Barry's] profession'. Barry was reminded that the specimens were 'Public Property'. Duncannon demanded Barry resolve any complaint he had privately and hoped that the story had not been the product of Barry's mischief. He asked for confirmation that he was correct 'in his impression that the report in the Morning Post ... is the gratuitous misstatement of the Reporter'.[117] Whether or not Barry was behind the allegations that he had been robbed of his specimens, despite funding them, and of the impoliteness of Duncannon's department, the article's appearance is illustrative of the struggle between architect and geologist. The question of authority between Barry and De la Beche was argued through the rightful location of these cubed stone samples.

Through this controversy it is demonstrated how Smith's own geological manifesto for architecture could occasionally divide opinion over how to exhibit this new science of architecture, and who should control it. Locating stone samples was a problem of managing knowledge. In wanting to house the samples at the RIBA, Barry was promoting a very different way of organizing the knowledge made for the survey. Managing the evidence constructed for Parliament involved questions

[115] Ibid., p. 143; also see Sophie Forgan, 'Building the museum: knowledge, conflict and the power of place', *Isis*, 96, 4 (2005), pp. 572–85.

[116] BGS GSM/DC/A/C/1, 'Letter from T. W. Philipps to Charles Barry', (14 December, 1839), pp. 143–45, 144.

[117] Ibid., p. 144–45.

over how best to exhibit the commission's findings. John Pickstone has shown that the organizing of knowledge was a crucial concern in the 1830s, with science formed of an increasingly diverse group of analytical disciplines. These concerns manifest themselves through the formation of new institutions like the RIBA and the BAAS, itself divided into a hierarchy of knowledges, including mathematics, chemistry, and geology.[118] Barry's reluctance to hand over the samples to De la Beche was an assertion that the knowledge would have more impact if it was made directly accessible to architects. While De la Beche and Smith saw architecture as an unscientific vocation requiring geology to assist it into an age of enlightenment, Barry considered the science a tool to be referenced when desired. Barry was not suggesting any reduction in the autonomy of the architect. The controversy of the stone samples was a contest over how to manage knowledge for best effect.

Defending Knowledge: Alternative Stone for Parliament

Both the 1839 report and the specimens deposited in the museum were not only scientifically to shape the superstructure of Parliament, but were the media through which the science of architecture might be projected beyond Westminster. Indeed, there is evidence that during the 1840s the work of the enquiry became a reference for architects. While De la Beche himself selected Anston stone for his museum in Jermyn Street, imitating Parliament's selection, he also had requests from geologists and architects for copies of the report.[119] In 1846 De la Beche was asked to submit Caen stone to 'experiments similar to those of the Government Commissioner', and for the analysis to be made in the Museum of Economic Geology.[120]

Even before 1839, claims were made for stone from various quarries around Britain; in particular, Portland stone which had been Christopher Wren's choice for the metropolis.[121] After 1839 such claims for an alternative stone persisted; the commission's use of geological knowledge failed to put the choice beyond contention. A month after the report

[118] John Pickstone, 'Science in nineteenth-century England: plural configurations and singular politics', in Martin Daunton (ed.), *The Organization of Knowledge in Victorian Britain*, (Oxford, 2005), pp. 29–60, 37, 44.

[119] Graham K. Lott and Christine Richardson, 'Yorkshire stone for building the Houses of Parliament (1839–c.1852)', *Proceedings of the Yorkshire Geological Society*, Vol. 51 (1997), pp. 265–72, 271; 'Letter 528: Joshua Fawcett to De la Beche', (4 November, 1842), in Sharpe and McCartney, p. 47.

[120] 'Letter 610: George Godwin to De la Beche', (30 October, [c.1846]), in Sharpe and McCartney, p. 51.

[121] For example, NA WORK 11/17/5/1, 'Letter from William White to Lord Duncannon', (9 April, 1836).

was published, the Limerick landowner M. J. Staunton applied to the Board of Woods to overturn the selection of English magnesian limestone in favour of black 'Limerick Marble'. Staunton argued that the stone from his land, being as hard as granite, would be preferable 'for a great National Edifice, to adorn the country for many centuries'.[122] Although harder to carve, Staunton felt that if it were cut in Ireland, advantage could be taken of Limerick's cheap labour. Staunton's agent, Robert Adams, testified to the rock's durability not as a 'scientific character', but from 'long experience which has given me the means of knowing the qualities of stone and Marble, without reference to chemical analysis'.[123]

This challenge was thus presented as a distinctly unscientific claim, which attempted to draw attention to economic and moral grounds. If chosen for Parliament, the quarries of Ballysimon would become a valuable source of employment and at a rate much preferable to English quarries.[124] Staunton's lobbying for his Limerick stone was referred to Barry who in turn sought advice from his fellow commissioners. He was surprised to find Staunton had quoted De la Beche 'as an authority', but when he wrote to De la Beche in person, he found this to be an inaccuracy. De la Beche had never supported Staunton's cause. Barry found further support from Charles Smith, who pointed out the expense of carving Staunton's marble.[125] Barry rejected Staunton's claims, complaining of both its cost to carve and its 'dark and sombre' colour.[126] This pretext of colour is interesting in that the survey's particular focus on limestones, sandstones, and magnesian limestones implies that the commissioners tacitly agreed that Parliament should be built in a light-coloured stone.

This rejection did not dissuade Staunton, however. Trenham Philipps wrote to De la Beche explaining how,

> Staunton has got up a regular row concerning Limerick Stone and is printing documents. It is scarcely fair for the Chancellor of the Exchequer to let his son put his name to any document of this kind, as Ireland formed no part of your enquiry.[127]

Enlisting the support of Stephen Edmund Spring Rice, the grandson of the first Earl of Limerick, Edmund Pery, and son of the Chancellor,

[122] NA WORK 11/17/5/30, 'Letter from Robert Adams to M. J. Staunton', (20 April, 1839).
[123] Ibid.
[124] NA WORK 11/17/5/33, 'Letter from William Roche MP to M. J. Staunton', (22 April, 1839).
[125] 'Letter 66: C. Barry to De la Beche', (27 April, 1839), in Sharpe and McCartney, p. 19; NA WORK 11/17/5/32, 'Letter from C. H. Smith to Charles Barry', (20 April, 1839).
[126] NA WORK 11/17/5/40, 'Letter from Charles Barry to Lord Duncannon', (26 April, 1839).
[127] 'Letter 1246: Trenham W. Philipps to De la Beche', (24 April, 1839), in Sharpe and McCartney, p. 90.

Thomas Spring Rice (Limerick's MP between 1820 and 1832), Staunton continued to argue for his Limerick stone. Rather than assert its superiority over magnesian limestone in scientific character, he marshalled a cohort of Irish MPs who highlighted the political, economic, and social virtues of the black rock. Staunton warned that the stone employed in London architecture had been a national 'disgrace', while his stone would '*last for ever*'. His quarry produced stone that was a source of Irish pride.[128] To employ Limerick stone at Westminster would strengthen the union between England and Ireland; Ireland's place in the empire would be recognized in stone at the heart of British government. Staunton's hopes found support from ten Irish MPs, along with a document addressing the scientific questions of the 1838 survey.[129] Within this document it was broadly stipulated that the stone had no power of absorption and an 'incalculable' resistance to pressure. Staunton felt the stone's worth was beyond scientific measurement. His agent, Adams, contributed a second report which argued further for the improved reputation of Westminster by creating employment in Limerick through an order for marble. Here the choice of stone was not a question of being scientific, but being politically judicious. Staunton and Adams maintained that the choice of stone should be made in reference to the 'welfare' of the people Parliament governed, rather than the science of Barry's survey.[130]

To this attempt to undermine the value of science in the selection of Anston, the commissioners reacted with staunch resistance. William Smith bemoaned that after 'our laborious investigations' which showed how unsuitable rocks similar to Limerick were, he 'little thought there could be a question on the propriety of using any beds of the Mountain Limestone rock as building stone'.[131] Smith informed Barry that the Limerick stone was no building stone; indeed, it was not even a marble. It was merely a well-polished black limestone. He felt it was unfit for Parliament: to name it marble was 'a sad misguiding perversion'.

Despite disagreements over how best to manage the survey's knowledge, De la Beche and Barry united in its defence. In June 1840, Staunton attempted yet again to recruit De la Beche in his cause, but found him fully committed to the selection's scientific credentials. All four commissioners had 'arrived at this opinion after long and careful investigations'. Above all, the now deceased William Smith was an authority beyond

[128] NA WORK 11/17/5/45, 'Letter from M. J. Staunton to T. W. Philipps', (3 May, 1839).
[129] NA WORK 11/17/5/43, 'Copy of Certificate, Verifying the Superiority, as to durability and cheapness, of Limerick Marble and Clare Limestone', (3 May, 1839).
[130] NA WORK 11/17/5/50, 'Report of Mr. Robert Adams on the Ballysimon Marble & Stone Quarries, Near Limerick', (29 May, 1839).
[131] NA WORK 11/17/5/46, 'Letter from William Smith to Charles Barry', (4 May, 1839).

question. De la Beche explained to Staunton that Smith was the most 'distinguished geologist', indeed he was 'well known to Scientific men as the "Father of Geology" – his knowledge of the applications of his science to the useful purposes of life was extensive'.[132] Scientific authority did not eliminate alternate choices but it was sufficient to defend the selection of magnesian limestone in the face of what could be portrayed as non-scientific opposition. Most importantly, the choice for Parliament reflected a desire to select a stone on what was understood as scientific evidence, rather than social-economic considerations. Defending this choice was a practice every bit as important as producing experimental evidence and organizing the display of samples. This laborious justification was crucial in establishing the work of the survey as appropriate for Parliament's construction.

Conclusion: Building Palaces and Bodies of Knowledge

Deliveries of Anston stone for Parliament began in 1840 and between 1841 and 1844 about 200,000 cubic feet of stone a year was arriving on the banks of the Thames at Westminster.[133] This was accompanied by specially selected granite, free of staining and large quartz crystals, from Devon, Penryn, and Guernsey to be used for the building's plinths.[134] The use of geology, including chemical knowledge, in the construction of Parliament's superstructure promised to be revolutionary. Knowledge of the composition, durability, and geology of stone, yielded through experiment, suggested a human control over buildings which would assist Barry in constructing an edifice to defy decay for generations. Understanding the nature of architectural materials offered a chance to build a lasting national legislature. Yet as shown, the fulfilment of these grand promises, which William Smith so emphatically projected, was deeply problematic. These difficulties were specific to the project, but also inseparable from a science understood to be new. Disagreement surrounded the form the 'science of architecture' should take. At stake were the roles of geologists and architects. Should the architect himself direct geological research and experimental investigation? Or should the architect submit to the authority of the geologist and chemist in matters of natural philosophy? Should science guide public architecture, or should decisions reflect social concerns, such

[132] BGS GSM/DC/A/C/1, 'Letter on behalf of Henry De la Beche to Mr. Staunton', (17 June, 1840), pp. 157–58.

[133] William Cowper, *Report of the committee on the decay of the stone of the new palace at Westminster*, PP. 1861 (504), p. 19; Lott and Richardson, 'Yorkshire stone', pp. 268–9; on transportation, see Christine Richardson, *Yorkshire Stone to London: To Create the Houses of Parliament*, (Sheffield, 2007).

[134] Parliamentary Archives (PA), BLY/58, 'Sketch book of Henry Bailey, 1840–52', p. 15.

as those of impoverished Limerick? Such questions were fought out during the survey and became embodied in the exterior of the new Palace. Resolving these problems were essential practices of knowledge production. Organizing where the report's evidence belonged and who was responsible for it was as important as manufacturing the report through observations, collections, and experiments. Defending the knowledge in the face of challenges from those promoting alternative stone was equally crucial. Once produced, the knowledge projected in the 1839 report had to be situated in wider discourse, ordered in museums and institutions, and justified in the face of criticism. These were processes of establishing both the knowledge made and the stone it endorsed.

This use of geology to guide the selection of Parliament's stone had ramifications for architecture beyond Westminster. In the years following Parliament's construction, the stone mason Charles Smith laboured to expand geological knowledge as a discipline for architects. He extended the research of the original 1838 commission, producing several reports on Caen stone and publishing regular articles in *The Builder* arguing that architects should secure enhanced knowledge of the materials they employed.[135] While architects increasingly turned to iron and glass for construction, Smith remained an outspoken advocate that the future of architecture lay in a greater incorporation of geological knowledge. Yet this attention to choosing stone was a particularly nineteenth-century phenomenon. Before the expansion of Britain's canal and rail networks, stone which architects usually employed was either local or that which could be transported by sea and river. Increased transport infrastructure meant that the selection of stone available to Victorian architects was increasingly extensive.[136] What is also apparent is that the selection of stone was becoming ever more central to the work of a professional architect. As seen in Chapters 2 and 3, it was hard to define exactly what it was to be a professional architect in the 1830s. The considerable energy which Barry expended on the stone survey for Parliament clearly shows that a more diligent attention to building materials, especially those from the quarry, was certainly within the remit of a professional. Travelling the nation to find a suitable material for Parliament was in itself a way for Barry to fashion himself as a professional. The care he invested

[135] RIBA, SmC/1/2/No. 1, C. H. Smith, 'On the various qualities of Caen Stone', (1849); RIBA, SmC/1/2/No. 2, C. H. Smith, 'Experiments, &c., on Caen Stone', (1849).

[136] Lott, 'The development of the Victorian stone industry', pp. 47–48; canal and rail expansion shaped contrasting ecclesiastical architecture in Berkshire and North Hampshire, see J. R. L. Allen, *Building Late Churches in North Hampshire: A Geological Guide to Their Fabrics and Decoration from the Mid-Eighteenth Century to the First World War*, (Oxford, 2009), pp. 8–9; J. R. L. Allen, *Late Churches and Chapels in Berkshire: A Geological Perspective from the Late Eighteenth Century to the First World War*, (Oxford, 2007), p. 6.

into this work marked him out as a man superior to artisans and builders, and placed him within the social status of those of scientific learning, such as doctors and lecturers.

Considering the attention paid to Parliament's stone in the context of projects beyond Westminster, it is apparent that this increasingly intimate relationship between buildings and geology was central to much Victorian architecture.[137] In the 1840s and 1850s, John Ruskin advised architects to employ geological examples from nature to inspire their work.[138] This advice was realized in buildings such as Deane and Woodward's Oxford University Museum, completed in 1861, which included internal shafts of varying polished stone from around the British Isles, presenting a didactic catalogue of different rock types.[139] Other projects, such as William Butterfield's (1814–1900) Balliol College in Oxford employed polychrome stonework providing a metaphorical display of strata. The exterior of Butterfield's All Saint's Church in Babbacombe, Devon, exhibited polished red and grey sandstones in a display comparable to a stone collection.[140] Attention to geology was also paid through architectural interiors, such as George Gilbert Scott's (1811–1878) Exeter College Chapel in Oxford, which he decorated with various coloured stone.[141] While these buildings displayed the importance of geology more obviously than the use of stone at Westminster, the Houses of Parliament embodied what was perhaps the most extensive body of knowledge produced in the mid-nineteenth century which united architecture and science.

[137] See Carla Yanni, 'Development and display: progressive evolution in British Victorian architecture and architectural theory', in Bernard Lightman and Bennett Zon (eds.), *Evolution and Victorian Culture*, (Cambridge, 2014), pp. 227–60, 229–32.

[138] Hall, 'What do Victorian churches mean?', pp. 82–83; also see Michael Hall, 'G. F. Bodley and the response to Ruskin in ecclesiastical architecture in the 1850s', in Rebecca Daniels and Geoff Brandwood (eds.), *Ruskin and Architecture*, (Reading, 2003), pp. 249–76; on nature and European architecture, see Barry Bergdoll, 'Of crystals, cells, and strata: natural history and debates on the form of a new architecture in the nineteenth century', *Architectural History*, Vol. 50 (2007), pp. 1–29.

[139] Frederick O'Dwyer, *The Architecture of Deane and Woodward*, (Cork, 1997), p. 175; for details see Trevor Garnham, *Oxford Museum: Deane and Woodward*, (London, 1992); Eve Blau, *Ruskinian Gothic: The Architecture of Deane and Woodward, 1845–1861*, (Princeton, 1982), pp. 48–81.

[140] Rosemary Hill, 'Butterfield, William (1814–1900)', *Oxford Dictionary of National Biography*, Oxford University Press, 2004 [www.oxforddnb.com/view/article/4228, accessed 29 November 2014]; Paul Thompson, *William Butterfield*, (London, 1971), pp. 35, 140–2; thanks to David Lewis for making me aware of Butterfield's work at Babbacombe; on polychromy, see Neil Jackson, 'Clarity or camouflage? The development of constructional polychromy in the 1850s and early 1860s', *Architectural History*, Vol. 47 (2004), pp. 201–26.

[141] See Geoffrey Tyack, 'Gilbert Scott and the Chapel of Exeter College, Oxford', *Architectural History*, Vol. 50 (2007), pp. 125–48, 143.

4 Chemistry in the Commons: Edinburgh Science and David Boswell Reid's Ventilating of Parliament, 1834–1854

Then the Lord God formed man of dust from the ground,
And breathed into his nostrils the breath of life.

Genesis 2:7

As Charles Barry's new Houses of Parliament took shape on the banks of the River Thames, the business of government continued in Robert Smirke's temporary accommodation. Surrounded by the building site which would one day provide a permanent residence, the Lords in the old Painted Chamber and the Commons in the old House of Lords, governed the nation. The novelty of this arrangement quickly wore off and the new premises were soon judged inadequate. The Painted Chamber was too small to satisfy the dignity of the Lords and aroused continual complaint, while the problem of ventilation was a source of much discontent.[1] The old House of Commons had incurred criticism over the quality of air, yet the new location provoked increased concern from members. In the summer heat of 1835 conditions in the temporary Parliament became intolerable. To resolve these difficulties a select committee, established to investigate the quality of air in Parliament, appointed David Boswell Reid (1805–1863), a Scottish chemist, to construct a powerful system of ventilation in the Commons. Details of Reid's work in Edinburgh built credibility into claims that he could improve Parliament's air. He promised to perform experiments in the Commons to guide the construction of this system. Between 1835 and 1840 his work ensured the Commons was not only a place of political debate, but a site of chemical experiment and knowledge production.

This chapter explores the choice of Reid as a trustworthy authority to perform experiments at Westminster, but it is also about the difficulties of performing science in different locations. Reid asserted that his work, including ventilation, was chemistry and that the practices which guided the form it took were experiments. Such experiments consisted of finding

[1] M. H. Port, 'The new Houses of Parliament', in J. Mordaunt Crook and M. H. Port (eds.), *The History of the King's Works*, Vol. VI: 1782–1851, (London, 1973), pp. 573–626, 575.

ways to determine the chemical composition of air in buildings, before and after human respiration. To ventilate a room thus involved controlling the chemical composition of this air and keeping it free of what were deemed to be dangerous elements, like carbonic acid. During the nineteenth century, architects and engineers strove to assert control over the interior atmospheres of hospitals, theatres, and prisons. Ventilation was subject to several extended studies, most notably Thomas Tredgold's 1824 *Principles of warming and ventilating public buildings*.[2] The conviction that stagnant air conveyed mysterious agents of disease sustained these efforts. It is telling that in 1830s' British political culture, the same debating chamber in which legislation was constructed and scrutinized, was deemed an appropriate place for the production of experimental knowledge. Henrik Schoenefeldt's detailed architectural analysis has astutely drawn attention to the role of experiment in the construction of Reid's ventilation system. Schoenefeldt correctly asserts that Reid promoted a distinct approach to architecture, prioritizing attention to air supply and the health of a building's inhabitants.[3] Furthermore, he shows how Reid transformed the temporary Commons into a place of experiment, effectively reconceiving the building as a laboratory. Yet Reid's efforts at Westminster were also important in a broader epistemological context, where the question of how to make science, and what constituted a valid experiment were not givens.

As will be shown, the construction of credibility was central to the problem of ventilating Parliament, and credibility was not inherent to experimental practices, but contingent on local cultural contexts.[4] As Ben

[2] Robert Bruegmann, 'Central heating and forced ventilation: origins and effects on architectural design', *Journal of the Society of Architectural Historians*, Vol. 37, No. 3 (October, 1978), pp. 143–60, 149–52; see Thomas Tredgold, *Principles of Warming and Ventilating Public Buildings, Dwelling-Houses, Manufactories, Hospitals, Hot-Houses, Conservatories, & c.*, (London, 1824); also see Neil Sturrock and Peter Lawson-Smith, 'The Grandfather of air-conditioning: the work and influence of David Boswell Reid, physician, chemist, engineer (1805–63)', *Proceedings of the Second International Congress on Construction History*, 3, (2006), pp. 2981–98, 2981; Reyner Banham, *The Architecture of the Well-Tempered Environment*, (London, 1969), pp. 11, 29; John Hix, *The Glass House*, (London, 1974).

[3] Henrik Schoenefeldt, 'The temporary Houses of Parliament and David Boswell Reid's architecture of experimentation', *Architectural History*, 57 (2014), pp. 173–213, 173; compare with Henrik Schoenefeldt, 'The Crystal Palace, environmentally considered', *Architectural Research Quarterly*, Vol. 12, No. 3–4 (December, 2008), pp. 283–94. Schoenefeldt's work places Reid's ventilation schemes within a broader history of design, and is certainly not an attempt to 'rehabilitate his scientific theories', as has been suggested in Caroline Shenton, *Mr Barry's War: Rebuilding the Houses of Parliament after the Great Fire of 1834*, (Oxford University Press: Oxford, 2016), p. 166.

[4] Steven Shapin, *Never Pure: Historical Studies of Science as if It was Produced by People with Bodies, Situated in Time, Space, Culture, and Society, and Struggling for Credibility and Authority*, (Baltimore, 2010), p. 19.

Marsden and Crosbie Smith put it, the success of a technology involves its social, as well as material worth. The way a technology is displayed and marketed raises questions over the trustworthiness of its human promoters. The plausibility of a scheme, or its credibility, has to be built through trusting human actors. In turn an actor's trustworthiness has to be manufactured through their behaviour, such as conducting experiments in an appropriate manner or presenting findings in a way acceptable to society.[5] Trust shaped the credibility of an apparatus, theory, or experiment, while the actions and values of a promoter shaped his or her own trustworthiness. Individual character was crucial to being believed. The question then is how did Reid, an Edinburgh chemist, become trusted to experiment in the Commons?[6]

The production of knowledge through experiment took place in physical and social settings which inscribed meaning on what Reid constructed. Local contexts shaped why experiments were performed, who witnessed them, how the results were interpreted, and to what purpose knowledge was applied. Where science was produced was important to how and why it was produced.[7] My argument is not that an environment determined the formation of science, but that local contexts, including networks of human agents, shaped the character and meaning of a body of knowledge. Such a body remained inscribed with the values of its place of production, and carried them as it travelled. At Westminster, Reid's ventilation embodied values specific to the Edinburgh context where it was conceived. Reid maintained that he performed chemistry rather than engineering or architecture and, as he had in Edinburgh, asserted that the purpose of science was to produce useful knowledge for the improvement of man's state. An experiment produced in Edinburgh was not interpreted in the same way in Westminster, and the appropriateness of conducting research varied between the permanent and temporary Parliaments. Reid's initial ventilation work, produced in Edinburgh, was contingent on local cultural resources, including institutions, ideas,

[5] Ben Marsden and Crosbie Smith, *Engineering Empires: A Cultural History of Technology in Nineteenth-century Britain*, (Basingstoke, 2005), pp. 6–8; for example, the Cunard Steamship Company built trust into their ships by avoiding pride and tempting providence, see Crosbie Smith and Anne Scott, '"Trust in Providence": building confidence into the cunard line of steamers', *Technology and Culture*, Vol. 48, No. 3 (July, 2007), pp. 471–96; for trust and questions of taxation and state spending, see Martin Daunton, *Trusting Leviathan: The Politics of Taxation in Britain, 1799–1914*, (Cambridge, 2001), pp. 10–11.

[6] See Steven Shapin, *A Social History of Truth: Civility and Science in Seventeenth-Century England*, (Chicago, 1994), p. xxvi.

[7] Charles W. J. Withers and David N. Livingstone, 'Thinking geographically about nineteenth-century science', in David N. Livingstone and Charles W. J. Withers (eds.), *Geographies of Nineteenth-Century Science*, (Chicago, 2011), pp. 1–19, 1–3.

and social networks. These shaped Reid's work and were inseparable from his approach to chemistry. Debates within Edinburgh University, Reid's relationship with local industrial and commercial leaders, his audiences in the city including the evangelical Presbyterian Thomas Chalmers (1780–1847), and his work for Edinburgh science associations, all shaped his performance of chemical work, including ventilation. It was in this cultural context that Reid laboured to build trust into his apparatus. In Edinburgh, Reid was involved in 'a quest for credibility', in which he actively displayed his work as consistent with local cultural values.[8]

The practice of experimenting was a controversial part of this construction of credibility. Marsden and Smith distinguish between experiment, which often appeared unreliable and fragile, and experience, which could appear reliable and trustworthy.[9] However, I will show how Reid sought credibility by appearing experienced and displaying experiments as producing consistent results. A big problem he faced was the use of humans in experiments. Who was the subject of an experiment was important in its apparent credibility, with individuals constructed as more reliable reporters of sensations experienced during experiments if they lacked education, freedom, or selfhood.[10] This question of managing human subjects as experimental instruments plagued Reid's transferal of skills from Edinburgh to Westminster. Performing experiments on Edinburgh students, Presbyterian ministers, and foot guards was not the same as performing similar trials on MPs in the Commons. Indeed, while experiments on subservient soldiers in the Commons provided suitable subjects with which to impress delegates from *The Times*, such experiments were not sufficient to satisfy independent-minded elected MPs. The Lower Chamber proved a difficult space in which to control experiments and discipline those who witnessed them.[11] Although experiment could be a powerful resource for building conviction, replication to varying

[8] Bruno Latour and Steve Woolgar, *Laboratory Life: The Construction of Scientific Facts*, (Princeton, 1986), pp. 200–1; Bruno Latour and Steve Woolgar, 'The cycle of credibility', in Barry Barnes and David Edge (eds.), *Science in Context: Readings in the Sociology of Science*, (Milton Keynes, 1982), pp. 35–43; for example, see Crosbie Smith, '"Nowhere but in a great town": William Thomson's spiral of classroom credibility', in Crosbie Smith and Jon Agar (eds.), *Making Space for Science: Territorial Themes in the Shaping of Knowledge*, (Basingstoke, 1998), pp. 118–46.

[9] Marsden and Smith, *Engineering Empires*, p. 10.

[10] Alison Winter, *Mesmerized: Powers of Mind in Victorian Britain*, (Chicago, 1998), pp. 7, 4, and 62.

[11] On the importance of disciplining laboratory spaces, compare with Robert Boyle's experiments as explored in Steven Shapin and Simon Schaffer, *Leviathan and the Air-Pump: Hobbes, Boyle, and the Experimental Life*, (Princeton, 1985), p. 39; Steven Shapin, 'The house of experiment in seventeenth-century England', *Isis*, Vol. 79, No. 3 (September, 1988), pp. 373–404, 373.

audiences was troublesome. To account for the success of experimental work, it is important to avoid explanations which claim experiments self-evidently reveal natural phenomena, and instead explore how experiment was constructed as a practice of representing nature.[12]

This chapter begins by examining Reid's chemistry in its urban Scottish context. 1830s' Edinburgh was a crucible in which industry, academia, and religion collided. Reid's science was conducted amid evangelical Presbyterian calls for useful work and rising concerns over rapid industrial and social change. Part two provides an account of how Reid's ventilation apparatus and chemical experiments came to the attention of Parliament between 1834 and 1835. Part three details Reid's experiments in the Commons and analyses how London audiences interpreted this ventilation system. Reid's science carried weight with audiences in Edinburgh, and he built trust with the 1835 ventilation committee, but winning that same science credibility with MPs and in wider scientific networks remained problematic. I conclude by analysing the conflict between Reid and Barry over ventilating the permanent Parliament. Reid found the change from temporary buildings to Palace an altogether different challenge. Transferring knowledge from Edinburgh to Parliament was about much more than experimental techniques. Reid's persistent investigating and absolute commitment to empirical practices won him favour with Edinburgh experimentalists like David Brewster, but the sort of knowledge he manufactured appeared unstable; what William Whewell might have labelled 'progressive' knowledge. It carried with it all the radical connotations this invoked and brought Reid into conflict with Charles Barry.

Chemistry and the Creator: The Edinburgh Context of Reid's Science

1830s' Edinburgh was the site of a great collision of problems and ideas. While in England, the Anglican academic bastions of Oxford and Cambridge were separated from the rapidly industrializing regions of the North, Edinburgh witnessed dramatic economic growth, rising poverty, and religious controversy. Edinburgh University was at the centre of this radicalizing world (Figure 4.1).[13] New mercantile and commercial

[12] David Gooding, Trevor Pinch, and Simon Schaffer, 'Introduction: some uses of experiment', in David Gooding, Trevor Pinch, and Simon Schaffer (eds.), *The Uses of Experiment: Studies in the Natural Sciences*, (Cambridge, 1989), pp. 1–27, 14.

[13] Crosbie Smith, *The Science of Energy: A Cultural History of Energy Physics in Victorian Britain*, (London, 1998), p. 15; on how local contexts shaped contrasting engineering courses, see Ben Marsden, 'Engineering science in Glasgow: economy, efficiency and measurement as prime movers in the differentiation of an academic discipline', *British Journal for the History of Science*, Vol. 25, No. 3 (September, 1992), pp. 319–46.

Figure 4.1 David Boswell Reid's depiction of Edinburgh's poverty stricken industrial classes

groups were finding increasing cultural and political influence, representing a challenge to the traditionally Whig and Tory dominated scientific culture.[14] Edinburgh's 'petty bourgeoisie', self-employed craftsmen and shopkeepers, came together with local lecturers, like the phrenologist George Combe (1788–1858), to form the Edinburgh Philosophic Association in 1832.[15] This association provided cheap lectures on geology and chemistry, and condemned the high fees of Edinburgh University courses.

It was in this context that from 1828 to 1833 Reid taught practical chemistry at Edinburgh University.[16] Although Thomas Charles Hope (1766–1844) held the University's chemistry chair, Reid was responsible for practical chemistry demonstrations.[17] His father, Peter Reid

[14] Steven Shapin, '"Nibbling at the teats of science": Edinburgh and the diffusion of science in the 1830s', Ian Inkster and Jack Morrell (eds.), *Metropolis and Province: Science in British culture, 1780–1850*, (London, 1983), pp. 151–78, 153.

[15] Ibid., pp. 154–55; on Edinburgh science, see James A. Secord, *Victorian Sensation: The Extraordinary Publication, Reception, and Secret Authorship of Vestiges of the Natural History of Creation*, (Chicago, 2000).

[16] See Centre for Research Collections, Edinburgh University (CRC) P.137.25, Thomas Charles Hope, 'Summary of a memorial to be presented to the Right Honourable the Lord Provost, magistrates, and council, respecting the institution of a professorship of practical chemistry in the University of Edinburgh', p. 5.

[17] Edward J. Gillin, 'Reid, David Boswell (1805–1863)', *Oxford Dictionary of National Biography*, Oxford University Press, April 2016 [http://ezproxy-prd.bodleian.ox.ac.uk:2167/view/article/23327, accessed 17 September 2016]; M. F. Conolly, *Biographical Dictionary of Eminent Men of Fife, of Past and Present Times*, (Edinburgh, 1866), p. 377.

(1777–1838), had been an active campaigner for education reform in the city, including increased teaching of what he considered useful subjects like mathematics and modern languages. David Boswell Reid shared his father's interest in improving working-class education and was active with local medical charities promoting knowledge on health between 1830 and 1833.[18] He reduced the admission price of his chemistry classes at the University to increase accessibility to the subject. Along with several publications, Reid's classes built him a reputation in Edinburgh as an authority on chemistry. When Richard Phillips (1778–1851), the editor of *The Philosophical Magazine*, suggested that Hope had guided Reid's *Elements of Practical Chemistry* and that it contained accounts of inaccurate experiments, Reid replied citing his own experience in chemical experimentation.[19]

For Reid, chemistry was important because it was the study of the world's physical resources for the improvement of mankind's condition. As Reid explained, 'unless the key to the nature of the material world be given by some explanatory lessons on science, and especially on Chemistry, the most awakening and fundamental of all the physical sciences, the power of improvement, even in civilised nations, must remain comparatively a sealed book to a large mass of the community'.[20] Reid used the example of iron to demonstrate chemistry's power in advancing civilization. Rock contained iron, but to turn this ore into a useful material could be done only 'by the powerful aid of chemical action'.[21] Reid believed that with chemistry, iron could be produced from nature and this element was the foundation of all 'civilised community'. Iron provided man with surgeon's knives, steamships, steam-engines, compasses, and all the tools of modern progress. As Reid concluded, these were the fruits of knowledge, applied to 'the properties which the Author of nature has impressed upon

[18] Conolly, *Biographical Dictionary of Eminent Men of Fife*, p. 376; Sturrock and Lawson-Smith, 'The grandfather of air-conditioning', p. 2982.

[19] David Boswell Reid, *Elements of Practical Chemistry*, (Edinburgh, 1830); David Boswell Reid, *An exposure of the misrepresentations in the Philosophical Magazine and Annals, for December, 1830, in its attack upon the author's Elements of Practical Chemistry*, (Edinburgh, 1831); Richard Phillips, *A letter to Dr. David Boswell Reid, experimental assistant to Professor Hope, in answer to his pamphlet, entitled 'An Exposure Of The Misrepresentations In The Philosophical Magazine And Annals,' &c.*, (London, 1831); David Boswell Reid, *An exposure of the continued misrepresentations by Richard Phillips, Esq ... in his attempt to vindicate himself from Dr. Reid's first exposure of his misrepresentation in that journal*, (Edinburgh, 1831).

[20] David Boswell Reid, *Illustrations of the Theory and Practice of Ventilation, with Remarks on Warming, Exclusive Lighting, and the Communication of Sound*, (London, 1844), p. xii.

[21] Ibid., p. 5.

the material creation'.[22] In this way Reid believed knowledge which informed the manipulation of chemical elements was power because it had the potential to secure material improvement.

Reid reckoned that chemistry's greatest potential to advance the human state was through enhancing the understanding of the air man consumed. He was sure that chemical experiment, involving repeated examinations and testing on air both before and after human respiration, would yield knowledge which could be applied to improve health. Reid believed that certain chemicals distributed in the atmosphere spread disease, and that to observe how air changed chemically when passing through the human-frame revealed insights into what constituted healthy air. Reid's work was intended to be transformative in a society fearful of disease and often convinced by miasmic theories of its spread through bad air from decaying vegetation and waste. He calculated that a human, on average respiring twenty times a minute, consumed 10 per cent of the oxygen in the air in-taken by volume. In turn, 7.8 per cent of the air discharged was transformed into carbonic acid.[23] Reid believed this consumption of oxygen demonstrated 'the provisions which the Creator has made for giving it [the human-frame] power and endurance'.[24] Oxygen was 'the great agent which the Author of Nature has created for the more immediate support of animal and vegetable life'.[25] As human respiration was contingent on this chemical conversion of oxygen into carbonic acid, Reid maintained that to build architecture which aided human health called for the application of ventilation apparatus. Chemical knowledge of human 'respiration, combustion, and ventilation', was to be 'made a more especial object of attention in the construction of every kind of building'.[26]

Between 1833 and 1835, when Reid ran a private classroom for chemistry demonstrations, he constructed a system of ventilation in the building to extract fumes from experiments. This apparatus was itself conceived of as an experiment in regulating the chemical composition of a building. According to Reid, homes, ships, and churches should be built on the principle of removing corrupted air, filled with carbonic acid, and replacing it with an oxygen rich atmosphere. He used the example of crowded churches to illustrate the problems of vitiated air, describing how,

[22] Ibid., p. 9. [23] Ibid., pp. 16–17. [24] Ibid., p. 19.

[25] David Boswell Reid, *Brief Outlines Illustrative of the Alterations in the House of Commons, in Reference to the Acoustic and Ventilating Arrangements*, (Edinburgh, 1837), p. 10.

[26] CRC S.B.5404/2*13, David Boswell Reid, 'The study of chemistry: its nature, and influence on the progress of society: importance of introducing it as an early branch of education in all schools and academies', p. 12.

The power of the clergyman is often reduced as well as the attention of the congregation. Too often he does not recognise the darkness of the physical atmosphere that, at times, oppresses all his labours, and counteracts or diminishes his usefulness, as much by the power with which it subdues his own energies, as by the careless indifference which it encourages in his congregation. At the very moment that he may be ... pointing out the purifying power of that moral atmosphere which should surround the heart; how often are his labours shorn of their power by the physical poison that sometimes paralyses the best intentions.[27]

The power of improved air was analogous to the power of improved morality. Ventilation facilitated superior conditions for health and worship. In the space of the church, this use of scientific knowledge appeared particularly righteous, with atmospheric purification assisting spiritual purification. As iron ore was in place in rock for man's progress, so too was pure air in the atmosphere for his existence. Chemistry was the ubiquitous tool to harness both. Reid explained that 'Innumerable resources have been provided by an all-bountiful Providence for ministering, perhaps indefinitely, to the necessities, as well as to the comforts of life'.[28]

For Reid, ventilation was chemistry because it involved the manipulation of atmospheric elements, as well as the understanding of the physical qualities of elements such as oxygen. Yet Reid held specific views of what exactly chemistry was, what it was for, and how it should be conducted. These views place Reid firmly within the Edinburgh context in which he laboured, where the boundaries between the city's rapidly industrializing society and the academic work conducted at the university were hard to distinguish. Reid clearly defined his concept of chemistry during a controversy surrounding a proposed practical chemistry chair for the university between 1833 and 1834. Rather than a simple debate over the establishment of a new chair, Reid's attempt to reform the existing programme of chemistry at the university was a contest between what he perceived was theoretical science, and the practical application of knowledge to the improvement of society. Reid argued that the proper place of chemistry was outside of the laboratory.

Having run practical chemistry classes as Hope's assistant, Reid understandably felt himself to be the most suitable candidate for the new chair. New university appointments were controversial, however, with Edinburgh Whigs and Tories often keen to reject reforms that might be seen to hand influence to the city's commercial groups.[29] Reid's proposed practical chemistry chair threatened to do just this. Reid lobbied the

[27] Reid, *Illustrations*, p. 44. [28] Ibid., p. xi.
[29] Shapin, 'Nibbling at the teats of science', p. 158.

Town Council to call for the university to recognise the importance of practical chemistry in its own right: 'It is necessary for all who study Chemistry for any scientific or practical purpose. Its connection with our arts and manufactures is becoming daily more intimate'.[30] Reid offered, 'in the strictest sense, a Course of Chemistry applied to the Arts (including Medicine) and Manufactures'.[31] Practical chemistry offered such potential for improvement in agriculture, art, and manufacturing that it was inappropriate that such knowledge should remain in the domain of Dr Hope's assistants.[32]

Reid's appeal found support in the city, uniting an impressive combination of academics and manufacturers: a network displaying Reid's prominence in Edinburgh's academic and industrial communities. John Baird of the Shotts Iron-Works, James Hay of the Edinburgh Ropery, and John Macfie, owner of the Edinburgh Sugar Works all provided testimony as to the utility of Reid's science.[33] Thomas Dick Lauder, vice-president of the Society of Arts, heralded Reid's interest 'in the advancement of science'.[34] Civil engineers James Leslie and Robert Stephenson also praised Reid's experimental knowledge as beneficial to their own works, while the physician Neil Arnott (1788–1874) spoke in favour of the appointment.[35] Yet Reid's support from within the University was evident too, with George Joseph Bell, Professor of Scottish Law; James Pillans, Professor of Humanity; and the late John Leslie, Professor of Natural Philosophy, all advocating the creation of the new chair.[36]

However, the university rejected Reid's bid to secure academic employment. Hope felt threatened by Reid's application and opposed Reid's claims on the grounds that theoretical chemistry was inseparable from practical experimentation and that his lectures already embraced practical displays of eight- to nine-hundred chemical processes.[37] He argued that the creation of two chairs for what he considered to be one subject would create a dangerous precedent throughout the university, particularly in Theology, Natural Philosophy, and Scottish Law.[38]

[30] David Boswell Reid, *A Memorial to the Patrons of the University on the Present State of Practical Chemistry*, (Edinburgh, 1833), p. 3.

[31] CRC P.89.16, David Boswell Reid, 'Remarks on Dr Hope's "summary," presented to the patrons of the university', p. 11.

[32] Reid, *A Memorial*, p. 4.

[33] CRC P.89.19, David Boswell Reid, 'Testimonials regarding Dr D. B. Reid's qualifications as a lecturer on chemistry, and as a teacher of practical chemistry', (1833), pp. 15, 10, and 20–21.

[34] Ibid., pp. 13–14. [35] Ibid., pp. 1–2, 54, and 42. [36] Ibid., pp. 5–7, 51, and 65–66.

[37] CRC P.137.25, Hope, 'Summary of a memorial', p. 6; Jack Morrell, 'Hope, Thomas Charles (1766–1844)', *Oxford Dictionary of National Biography*, Oxford University Press, 2004 [http://ezproxy.ouls.ox.ac.uk:2117/view/article/13738, accessed 5 April 2013].

[38] CRC P.137.25, Hope, 'Summary of a memorial', p. 8.

Unsurprisingly these specified subjects were those in which Reid's supporters worked. Practical chemistry belonged to the chemistry chair and could not 'constitute a proper object of a separate Chair'.[39] Hope referenced the experience of forty-thousand experiments and thirty-eight years of teaching to his credit. He undermined Reid's reputation by hinting at Reid's interest in a new chair as mercenary. Reid responded with a vigorous defence of both his own integrity and of the proposed chair. He argued that his classes had run at a financial loss, while his position as Hope's assistant 'took away much from Dr Reid's professional standing', as well as from the importance of the subject.[40] He described Hope's teaching as purely theoretical, arguing that 'in the estimation of the public', especially in Edinburgh, the practical element of chemistry was what really mattered.[41] The proposed chair would therefore be more appropriate to the 'progress of science and the interests of the public'.[42] It was constructed as a relevant position for a university situated in the heart of a rapidly changing city.

Nevertheless, Hope found much support in favour of maintaining the status quo, and in doing so raised important questions over Reid's claims for science's role in society. John Wilson Anderson had run Hope's practical chemistry classes before 1828. Anderson defended Hope, casting doubt on Reid's assertions that practical chemistry could be applied outside of the laboratory. He believed that 'No greater delusion can exist than to suppose that such Practical instruction is to be obtained in the Laboratory, any more than that it can be obtained in the Lecture-room'.[43] Anderson argued that the university laboratory was not a place of acquiring practical knowledge, but creating theoretical evidence of the properties of elements and compounds. The problem, as he saw it, was that a laboratory was a site of theoretical knowledge production but was not big enough to trial new methods and apparatus on an industrial scale. In a laboratory one could observe natural phenomena, but one could not see how chemistry could be applied to practical problems. He argued that places of practical knowledge, like breweries and glasshouses, were built for specific arts and therefore more appropriate places to work out improvements in manufacturing. A laboratory was built to investigate chemical laws while factories were places of industry. Anderson acknowledged that 'Scientific knowledge and Practical Skill ought ever to go hand

[39] Ibid., p. 4. [40] CRC P.89.16, Reid, 'Remarks on Dr Hope's "summary"', p. 6.
[41] Ibid., pp. 9–10. [42] Ibid., p. 11.
[43] John Wilson Anderson, *Letter to the Right Honourable the Lord Provost, magistrates, and Town Council of Edinburgh, as patrons of the University, in reference to the contemplated establishment of a lectureship of practical chemistry*, (Edinburgh, 1834), p. 4.

in hand', but also warned that Reid was deluded to think he offered genuine practical knowledge from his laboratory.[44]

The Town Council made a second attempt to secure a new chair for Reid in 1834, but the university rejected this.[45] On Hope's retirement in 1843, Reid applied for his chemistry chair, considering it to have 'so long exerted a powerful influence on the progress of Science', but this too was turned down.[46] Anderson maintained that Reid risked 'reducing their [scientific chemists'] knowledge to the actual business of life'.[47] Hope's supporters constantly asserted that in a university environment, the role of practical chemistry was to demonstrate general principles rather than shape daily life. William Gregory, an ex-student of Hope's, claimed Reid's proposals risked creating a course which paid insufficient attention to theory.[48] Robert Christison, the holder of the Edinburgh Chair of Materia Medica, joined in the criticism with a strongly worded address to the Town Council. Christison disliked Reid's teaching, believing it to be too much 'in the recruit drill-sergeant fashion'. He felt that Reid's students 'came out of his hands ignorant of the simplest manipulations in practical medico-chemistry'.[49]

Although Reid's proposed practical chemistry chair was rejected and he subsequently left the university's employment, the support he attracted demonstrates that there was an understanding among some of Edinburgh's commercial and academic groups that Reid's work was credible. Within the network of industrialists and academics that supported Reid's appointment was an agreement that he and his work were trustworthy. In Edinburgh, there was a consensus that science should be

[44] Ibid., p. 6.

[45] J. B. Morrell, 'Practical chemistry in the University of Edinburgh, 1799–1843', *Ambix*, Vol. 16, No. 1–2 (1969), pp. 66–80, 75; also see David Boswell Reid, *A letter to the Right Honourable the Lord Provost, magistrates, and council, patrons of the University of Edinburgh, on the present state of practical chemistry: with remarks on some statements in a pamphlet by Dr Anderson, assistant to Dr Hope*, (Edinburgh, 1834).

[46] David Boswell Reid, *Professorship of Chemistry in the University of Edinburgh: Testimonials in Favour of Dr D. B. Reid, F. R. S. E.*, (Edinburgh, 1843), p. v.

[47] John Wilson Anderson, *Postscript of a letter to the Right Honourable the Lord Provost, magistrates, and Town Council of Edinburgh, as patrons of the University, in reply to a pamphlet by Dr. Reid, lecturer on chemistry, in reference to the contemplated establishment of a lectureship of practical chemistry*, (Edinburgh, 1834), p. 2.

[48] See William Gregory, *Observations on the proposed appointment of a teacher of practical chemistry in the University; with remarks on some passages in Dr Reid's letter to the council on the subject*, (Edinburgh, c.1834); David Boswell Reid, *A letter to The Right Honourable the Burgh Commissioners, on the evidence of Dr Christiston, Professor of Materia Medica in the University of Edinburgh*, (Edinburgh, 1835).

[49] Brenda M. White, 'Christison, Sir Robert, first baronet (1797–1882)', *Oxford Dictionary of National Biography*, Oxford University Press, 2004; online edn, October 2009 [http://ezproxy.ouls.ox.ac.uk:2117/view/article/5370, accessed 8 April 2013]; Robert Christison, *The Life of Sir Robert Christison, Bart*, (Edinburgh, 1885), p. 344.

practical, by which it was meant that knowledge was most useful if it was conceived of for social improvement. Reid portrayed his work as consistent with such values. Appropriately, he had been an officer in George Combe's Society for Aiding the General Diffusion of Science. Combe led the organization, seeking to remove Tory-Whig patronage from science and to promote its progressive nature beyond the confines of Edinburgh's social elites.[50] While Combe's emphasis on reforming and secularizing Edinburgh science aroused the disapprobation of Whigs, Tories, and the Presbyterian Church, Reid found favour with some from these circles. As Combe struggled to make his science appear apolitical and avoid allegations of irreverence, Reid's work fit comfortably within Edinburgh's increasingly diverse religious and political culture.[51]

Indeed, this religious context is important to understand how Reid's work earned the credibility which it did. One of Reid's keenest promoters was the charismatic leader of the Evangelical Party of the Church of Scotland, the Rev Thomas Chalmers.[52] This support for Reid was important because in 1830s' Edinburgh, Chalmers was at the centre of evangelical interpretations of science: interpretations which were becoming increasingly prominent in society. Bebbington identified nineteenth-century evangelicalism as having four main characteristics. These included a call to conversion (by repenting, which entailed turning from one's self to God and accepting Christ's message), activism (the imperative of bringing others to Christ), a focus on the Gospels (emphasizing Christ's suffering on the Cross and doctrine of Atonement), and a commitment to the Bible as the absolute truth and word of God, to be adopted as a guiding compass.[53] Hilton has shown that these values intrinsically shaped nineteenth-century thought in politics, philanthropy, and natural philosophy.[54] Nowhere was this more apparent than in the Edinburgh circles where Reid sought to make his name.

Chalmers held the theology chair at Edinburgh University from 1828 and, in the years prior to the 1843 disruption of the Presbyterian Church and formation of the Free Church of Scotland, was a prominent member

[50] Shapin, 'Nibbling at the teats of science', pp. 163–64; on Whigs and natural philosophy in Edinburgh at this time, see L. S. Jacyna, *Philosophic Whigs: Medicine, Science and Citizenship in Edinburgh, 1789–1848*, (London, 1994).

[51] Shapin, 'Nibbling at the teats of science', pp. 159, 163.

[52] Hugo Reid, *Memoir of the Late David Boswell Reid*, (Edinburgh, 1863), pp. 11–12.

[53] D. W. Bebbington, *Evangelicalism in Modern Britain: A History from the 1730s to the 1980s*, (London, 1989), pp. 2–3.

[54] Boyd Hilton, *The Age of Atonement: The Influence of Evangelicalism on Social and Economic Thought, 1785–1865*, (Oxford, 1986); on evangelicalism and science, see David N. Livingstone, D. G. Hart, and Mark A. Noll (eds.), *Evangelicals and Science in Historical Perspective*, (Oxford, 1999).

of the network of Edinburgh academics in which Reid built his reputation.[55] Chalmers taught that suffering was a source of improvement, existing to guide men to virtue. Hilton has argued that Chalmers was the most influential preacher of this distinctive attitude, so much at variance to the more 'extreme' evangelical conviction that suffering was simply a sign of sin and depravity. Instead, the natural world was 'an arena' in which man could bring about redemption.[56] Chalmers' Calvinist interpretation of the universe existing in a state of depravity and doomed to inevitable decay had social ramifications. Chalmers advocated minimal government interference, free-trade, *laissez-faire*, and self-help, because such approaches to politics and society allowed for individual, personal morality.[57] Too much charity would create indolence, while free-trade and enhanced personal liberties nurtured conscience through trials, sufferings, and temptations. Quite unlike Utilitarian philosophy, Chalmers' approach to life was not about generating material happiness, but the moral regeneration of society.[58] Alternately, Chalmers taught that the best use of one's time and wealth was to invest it for the improvement of humanity.[59] Chalmers' Presbyterian understanding of 'giving' was not charity, but the maximizing of useful work and the minimizing of idleness and waste.

In the early-nineteenth century, the Evangelical Party of the Church of Scotland's General Assembly adopted a position as defender of the intellectual freedoms of Edinburgh University academics.[60] By the 1830s, Chalmers had very much become the champion of this Edinburgh evangelicalism.[61] His theology held resonance for science

[55] Stewart J. Brown, 'Chalmers, Thomas (1780–1847)', *Oxford Dictionary of National Biography*, Oxford University Press, 2004; online edn, October 2007 [www.oxforddnb.com/view/article/5033, accessed 26 December 2012]; Conolly, *Biographical Dictionary of Eminent Men of Fife*, pp. 111–13.

[56] Hilton, *The Age of Atonement*, p. 55. [57] Smith, *The Science of Energy*, pp. 19–20.

[58] Ibid., p. 20; Hilton, *The Age of Atonement*, p. 7.

[59] Smith, *The Science of Energy*, p. 22.

[60] This position began in 1805 with John Leslie's bid to secure the chair of mathematics at Edinburgh University. Leslie had in a footnote endorsed David Hume's controversial doctrine of causation which taught that cause and effect were in constant sequence. This implied a shift from seeking 'proofs' to 'inklings' of Divine truth which found support from freethinking members of the Evangelical Party but aroused contempt from the Moderate Party of the Assembly; see Hilton, *The Age of Atonement*, pp. 24–25; Jonathan R. Topham, 'Science, natural theology, and evangelicalism in early nineteenth-century Scotland: Thomas Chalmers and the evidence controversy', in Livingstone, Hart, and Noll (eds.), *Evangelicals and Science*, pp. 142–74, 146–49.

[61] David W. Bebbington, 'Science and evangelical theology in Britain from Wesley to Orr', in Livingstone, Hart, and Noll (eds.), *Evangelicals and Science*, pp. 120–41, 123.

in nineteenth-century urban Scotland.[62] In Edinburgh and Glasgow, rising poverty and industry made Chalmers' teachings appear increasingly urgent. For Chalmers, natural science offered a powerful means of understanding the natural laws of the material world. Yet it was natural theology, combined with the scriptural truths of the Gospels, which provided him with an explanation for nature's working. True religion combined both reason and feeling.[63]

According to Chalmers, the problem of nature was that 'all her elements are impregnated with disease ... Even the mute and inanimate things are subject to the power of a decay'.[64] Science revealed the natural laws of motion and matter but showed the state of the world, from mountains to human life, to be in this state of perpetual decay. Only by God's omnipotent power of renewal could these laws be maintained. Natural laws 'created and sustained by God's will' presented the evangelical Chalmers with the most 'palpable argument' for God's existence; the eternal power of God was revealed by knowledge, yielded through science, of the transitory character of laws of nature.[65] While he was occasionally criticized for his attention to academic evidence, rather than the Gospel message, Chalmers' Natural Theology of Conscience, which taught that the supremacy of the human conscience was strong evidence for a Divine architect, secured him influence in evangelical circles.[66] An evangelical disposition taught that God's presence could be seen in every act of nature, and Man's duty was to work in accord with such divinity. While a theological understanding of natural philosophy was central to Chalmers' conception of scientific knowledge, 'experiment' and 'discovery' were valued as the means to cause 'science in conjecture ... to become science in certainty'. Experiment, particularly in chemistry, could serve to 'clear up many of the most recondite problems in physics, and thereby incalculably to extend the boundaries of our present science'.[67]

Chalmers was a keen supporter of scientific works which he felt to be consistent with his teachings and this included Reid's enthusiastic

[62] Smith places Chalmers in the scientific tradition of William Thomson (later Lord Kelvin) and James Clerk Maxwell, and with Scott shows how his shared values with George Burns and Robert Napier shaped the running of the Cunard Steamship Company; see Crosbie Smith, 'From design to dissolution: Thomas Chalmers' debt to John Robison', *The British Journal for the History of Science*, Vol. 12, No. 1 (March, 1979), pp. 59–70, 67; Smith and Scott, 'Trust in Providence', pp. 471–96.

[63] Evangelicalism emphasized the desire to 'feel' the Atonement, see Hilton, *Age of Atonement*, p. 20.

[64] Chalmers, quoted in, Smith, 'From design to dissolution', p. 62. [65] Ibid., pp. 62–64.

[66] Topham, 'Science, natural theology, and evangelicalism', p. 165.

[67] Chalmers quoted in (Anon.), 'Address to Dr Samuel Brown', *The Scotsman*, (29 April, 1843), p. 2.

promotion of chemistry. Chalmers frequently attended Reid's experimental classes, believing the lectures exhibited the most 'lucid exposition' Edinburgh had to offer.[68] Reid displayed his work as consistent with evangelical readings of nature which considered the practice of science as Divinely sanctioned. He explained how,

It has pleased the Author of nature so to form man, that he is forced to attend to the objects with which he is surrounded, and the most of his senses have been given him for this purpose. This world, then, is the arena on which man is called at present to act; the materials of happiness are placed before him, but his skill, activity, and knowledge, are required to enable him to make a proper use of them. He is called upon to obey this great law by the first instincts and most imperious wants of his nature; and it is to the difficulties he has to contend with under such circumstances, that we owe the development of many of the more ennobling qualities of the human mind.[69]

The acquisition of chemical knowledge was the appropriate application of the human senses to understanding God's natural laws. This knowledge could allow man to use such laws to improve his own condition. Even though materials were provided for man's use, Reid believed that to benefit from them required disciplined study. Reid continued that,

To study the laws of nature, then, is to study the laws which the Author of nature has ordained for the happiness and improvement of the human race; and it is not only to ignorance or neglect of the duties which morality and religion have imposed, but also to our ignorance or neglect of those physical laws, that a large share of most of the evils of life can be traced. The physical sciences, then, which investigate these laws, and apply them to the purposes of life, not only form a most essential part of that knowledge which is most necessary to man, but every step we advance tends more and more to exalt our ideas of the wisdom, power, and beneficence of their Author. It is impossible to contemplate the progress of society without seeing how much man is indebted to the cultivation of physical science.[70]

Through experiment, it was possible to reveal and benefit from God's laws. Reid's interpretation suggested scientific investigation was a trial, not only of diligence, but also of morality. Knowledge of nature was knowledge of the Divine, and the pursuit of this would be rewarded by progress.

Chalmers posited a very similar understanding of the material world in his volume of the *Bridgewater Treatises*, financed from the will of the Earl of Bridgewater for work displaying the 'power, wisdom, and goodness of God' through scientific discovery. Referring specifically to the 'moral

[68] Hugo Reid, *Memoir*, p. 11.
[69] CRC, S.B.5404/2*13, David Boswell Reid, 'The study of chemistry', p. 3.
[70] Ibid., p. 3.

elements' which the Author of Creation provided, Chalmers described the relationship between virtue and the materials of progress: great was 'the capacity of that world in which we are placed for making a virtuous species happy'.[71] The elements of physical and moral happiness were, as Reid proposed, provided. It fell to 'the aptitude of the human understanding, with its various instincts and powers, for the business of physical investigation' to make use of these materials.[72] The 'experimental truth' of material nature was to be collected 'by a diligent observation of facts and phenomena'.[73] Through Chalmers, Reid's understanding of nature can be placed firmly within a Scottish evangelical context.

Reid invited Chalmers' wife and children to attend his classes free of charge, and the two men socialized regularly at dinners and meetings over tea. Chalmers applauded Reid's application and understanding of chemical science. Having attended several of Reid's practical chemistry classes at the university, Chalmers recommended Reid for the practical chemistry chair, expressing the

> pleasure and instruction which I received during my attendance ... on your course of popular lectures, where, besides the utmost expertness and address in all the manipulations of Chemistry, you evinced, and more particularly in your lucid exposition of the Atomic Theory, both how thoroughly you had comprehended, and how successfully you could communicate, the principles of the Science.[74]

Reid valued Chalmers' praise and enjoyed reading the theologian's own works, finding them always to be a 'source of pleasure'.[75] Chalmers' support shows how Reid's work was valued in Edinburgh as consistent with his evangelical teachings.

Between Reid and Chalmers, the common interest in the application of science to the material improvement of society was a regular topic of conversation. In 1832 Reid wrote to Chalmers that

> I have been rambling thro all the principle manufacturing towns in England ... One of the most interesting novelties which I saw was a Hydro-Static Bed, invented a few weeks ago by Dr. Arnott of London. It appeared to me so valuable

[71] Thomas Chalmers, *The Bridgewater Treatises: on the Power Wisdom and Goodness of God as Manifested in the Adaption of External Nature to the Moral and Intellectual Constitution of Man*, Vol. II, (London, 1833), p.117.

[72] Thomas Chalmers, *The Bridgewater Treatises: on the Power Wisdom and Goodness of God as Manifested in the Adaption of External Nature to the Moral and Intellectual Constitution of Man*, Vol. I, (London, 1833), p. 9.

[73] Ibid., p. 11. [74] CRC P.89.19, Reid, 'Testimonials', p. 12.

[75] New College Library, Edinburgh University (NCL), CHA4.213.1, 'Letter from David Boswell Reid to Thomas Chalmers', (12 January, 1833).

an invention, that I lost no time in getting one made here, and I am sure you will be much pleased with it when you see it and *try* it.[76]

This hydrostatic bed, which maintained equal pressure throughout a mattress to reduce the chance of bedridden patients developing sores, promised the sort of material improvement which Chalmers had emphasized throughout his career. Chalmers had, in his years at the Tron Church in Glasgow (1815) and the Parish of St John's (1819), attended not only to the spiritual welfare of his parishioners, but also to their material improvement. He felt that the biggest threat to material progress was a lack of education among the urban working classes, combined with increased population. Among his congregations, he promoted the values of thrift, delayed marriage, and limited childbirth, but also encouraged self-help and communal responsibility as he sought to revive a general community spirit. Although statistical science was his principal answer to the threat of Malthusian population growth, practical improvements, such as the hydrostatic bed, were appreciated in Chalmers' moral philosophy.[77] Reid shared Chalmers' concerns for social improvement through enlightenment. Indeed he 'regretted much that my classes for practical chemistry have always met at the same hours at which you lecture, preventing me from attending you, and studying a subject to which I was anxious to have devoted my whole time'. Chalmers sent papers covering the content of his lectures, of which Reid reported that he was 'busily engaged in studying it ... particularly your views as to a compulsory provision for the poor, and of the Christian education of the people'.[78] Reid fully endorsed Chalmers' prescribed education of the poorer classes.

Reid's work was consistent with, and shaped by, local religious and industrial values. As his brother Hugo concluded, Reid had been destined for the medical profession had it not been for the 'growth of the science of chemistry ... with its numerous and daily increasing applications to the arts and manufactures'. It was this expansion which had convinced Reid of the advantage of 'acquiring practical skill in the art of experimenting'.[79] Hundreds of 'miners, manufactures, engineers, [and] agriculturalists' attended Reid's lectures, which emphasized chemistry's potential to improve society, arts, and manufactures.[80] In Edinburgh, Reid's work secured trust and credibility with significant parts of these social groups.

[76] NCL CHA4 188.60–61, 'Letter from David Boswell Reid to Thomas Chalmers', (11 September, 1832).
[77] Brown, 'Chalmers, Thomas (1780–1847)', *Oxford Dictionary of National Biography*.
[78] NCL CHA4 188.59, 'Letter from David Boswell Reid to Thomas Chalmers', (15 February, 1832).
[79] Hugo Reid, *Memoir*, p. 7. [80] Ibid., p. 8.

When he eventually left for London in 1840, the Edinburgh School of Arts summarized this as it celebrated his departure as an act of spreading Christian 'useful knowledge' beyond the Athens of the North to the capital of empire. It heralded Reid's chemistry as a great work of 'philanthropy'. Before presenting him with a watch to remind him of his Edinburgh heritage, the institution praised Reid for revealing 'the workings of Him with whom there is no shadow of turning', and for basing all his philosophy and philanthropy 'upon a sound Christianity'.[81] This reference invoked the New Testament's Epistle of St James which specified that all good gifts were from an unchanging, permanent God, and therefore Reid's works were, as ways of improving health, Divine gifts. In London, Reid worked to continue this promotion of Christian science as a tool of social improvement. Reid delivered a series of lectures in 1842 at the evangelical bastion of Exeter Hall on the 'Chemistry of Daily Life'. This course, of a 'practical character', was aimed at school masters and teachers.[82] The Privy Council Office issued 1,000 tickets to London teachers and promoted the lectures to MPs. Reid boasted how his lectures preceded a government commission into public health by over a year.[83] He later sat on this commission, examining the state of large towns and populous districts, and investigating the condition of the poorer classes in terms of water supply, drainage, dwellings, and ventilation.[84] Through such activities Reid tried to disseminate his chemical knowledge to audiences who could utilize it for social progress.

Edinburgh Science for National Audiences

Following Edinburgh University's refusal to grant Reid a chair in 1833, he left his post and constructed a classroom just behind the hall of the College of Surgeons. It was here that he ran private practical chemistry classes. Reid intended the building to be an ideal site of experiment, with the pillars of the classroom inscribed with chemical formulae.[85] Having

[81] (Anon.), 'Soiree and testimonial to Dr D. B. Reid', *Caledonian Mercury*, (Edinburgh, Scotland), 18 April, 1840; Issue 18765.

[82] (Anon.), 'Ventilation, lighting, and warming of school-rooms.- Dr. Reid's lectures', *The Morning Chronicle*, (London, England), 26 March, 1842; Issue 22574.

[83] David Boswell Reid, *Narrative of Facts as to the New Houses of Parliament*, (London, 1849), p. 12.

[84] *First Reports of the Commissioners for Inquiring into the State of Large Towns and Populous Districts*, PP. 1844 (572), p. v; Reid's association between air and health was consistent with contemporary miasmic accounts of disease, see Benjamin Ward Richardson, *The Health of Nations: A Review of the Works of Edwin Chadwick, with a Biographical Dissertation*, Vol. II, (London, 1887), p. 58.

[85] Sophie Forgan, 'Context, image and function: a preliminary enquiry into the architecture of scientific societies', *British Journal for the History of Science*, Vol. 19, Iss. 1 (March, 1986), pp. 89–113, 113.

built it through the winter of 1833–34, Reid conducted a series of experiments to various public audiences. In March 1834, he invited prominent members of Edinburgh society, including the Lord Provost and Chalmers to witness the lack of echo and clear sound attained in the room. Reid informed Chalmers that, an 'experiment is to be made in my class room on Friday, the 21st ... there will be both vocal and instrumental music that it may be contrasted with other buildings in this respect. All who have hitherto tried it ... stated that it is better adopted for all these principles than any room they have ever been in'.[86] Reid's guests were treated to St Cecilia's Society's vocal and instrumental performance, revealing 'the beauty of the different musical compositions'.[87] Reid attributed the pure sound to a low ceiling, elevated in the middle with an inclination of the roof and an absence of concave areas in the structure. Most importantly, however, he boasted that the classroom provided absolute atmospheric control. As experiments were conducted, fumes accumulated in the class, but these could be drawn out by a current created by the generation of heat outside the room in a chamber connected to perforations in the roof.

The physical shape of the interior contributed to good acoustics, but Reid believed that sound was dependent on the atmosphere for its communication. It was significant then that any experiment of sound performed was accompanied by an explanation of the extraction of vitiated air. Reid recalled how,

> on several occasions, gentlemen may have come in from a distance, foreigners and strangers, to see the working of my ventilating apparatus ... we have put on the fire with a few pieces of wood, and in the course of five minutes we were able with that to bring it into such a state of activity that fumes produced in showing some experiments were carried with great rapidity by those ventilators, which in the course of three minutes would have filled the room to such an extent that we should have been obliged to go out.[88]

Two events raised the profile of Reid's classroom. These two celebrations brought together men of science and politics, transforming Edinburgh into an arena in which Reid could demonstrate his ventilation apparatus to national audiences. The first of these was the 1834 Edinburgh BAAS meeting. On 13 September 1834 Lord Brougham arrived for the final day of the fourth BAAS meeting, which celebrated Edinburgh's scientific

[86] NCL CHA4.227.68, 'Letter from David Boswell Reid to Thomas Chalmers', (15 March, 1834).

[87] (Anon.), 'Communication of sound in public buildings', *Caledonian Mercury*, (Edinburgh, Scotland), 24 March, 1834; Issue 17583.

[88] *Report from Select Committee on the Ventilation of the Houses of Parliament; with the minutes of evidence*, PP. 1835 (583), pp. 37–38.

prominence.[89] As part of the meeting, a tour of Reid's practical chemistry classroom was organized, where his elaborate system of ventilation was on show. On the concluding evening of the meeting, Reid demonstrated his control over the classroom's atmosphere to an audience including a delegation from Parliament. Earl Grey, Lord John Campbell, the Marquis of Tweeddale, and Brougham all attended.[90] Brougham 'examined all the arrangements of the practical class. He also paid particular attention to the system of ventilation'.[91]

A native of Edinburgh, Brougham had a sustained interest in the natural sciences. Between 1808 and 1809 he had reviewed Humphrey Davy's Bakerian Lectures on electricity, and until the 1830s he authored many articles on chemistry, optics, and astronomy in the *Edinburgh Review*.[92] Having matriculated at Edinburgh University in 1792, Brougham was a keen advocate of anti-hypothetical Baconian science. The legacy of the natural philosopher Francis Bacon (1561–1626) and the meaning of Baconian science was increasingly controversial in the nineteenth century. Bacon was a troublesome figure for some Whigs, given his association with the Royal authoritarianism of James I and his career's termination following allegations of corruption.[93] At the same time Bacon's scientific method came under criticism in the 1820s. For example, David Brewster criticized Bacon's total rejection of hypothesis. Brewster believed hypothesis without the scrutiny of experiment lacked value, but equally he felt science without any hypothesis was so limiting of imagination that it undermined the process of scientific discovery.[94] Facts alone, as Bacon prescribed, lacked originality; Brewster argued that they needed interpretation and this involved man's inventive faculty. Brougham avoided these doubts, with hypothesis appearing to him as mere 'work of fancy'.[95] Hypothesis, he believed, was non-scientific practice because it required constant adaption to suit the results of experiment. Brougham argued that if a hypothesis was correct, it was simply a description of facts.[96] He felt that science was the pursuit of truth and that the best way to achieve this was through experiment. Like Reid,

[89] Jack Morrell and Arnold Thackray, *Gentlemen of Science: Early Years of the British Association for the Advancement of Science*, (Oxford, 1981), p. 103.

[90] Reid, *Illustrations*, p. 312.

[91] (Anon.), 'The Lord Chancellor', *Preston Chronicle*, (Preston, England), 20 September, 1834; Issue 1151.

[92] Joe Bord, *Science and Whig Manners: Science and Political Style in Britain, c.1790–1850*, (Basingstoke, 2009), pp. 45–46; Davy (1778–1829) delivered seven of the Royal Society's prize Bakerian lectures between 1806 and 1826.

[93] Bord, *Science and Whig Manners*, p. 73.

[94] Richard Yeo, 'An idol of the market-place: Baconianism in nineteenth-century Britain', *History of Science*, Vol. 23, No. 3, (September, 1985), pp. 251–98. 278.

[95] Bord, *Science and Whig Manners*, p. 49. [96] Ibid., p. 50.

Brougham was confident that science promised material improvement. Displays of the practical value of knowledge were consistent with the celebrated Baconian concept that science's goal should be 'the relief of man's estate'.[97] While Bacon's methodological legacy was debated, his association with the idea that knowledge was a means to material wealth and comfort found favour with Whigs like Brougham.[98] Reid approved of Brougham's own works to improve education. Following Brougham's association with Birkbeck's London Mechanics' Institute in 1824 and his role in founding the Society for the Diffusion of Useful Knowledge (SDUK) in 1826, Reid praised Brougham as a champion of progressing education.[99] He asserted that few had done more to advance 'the state of the mind in youth'.[100] Considering his visit to the classroom pre-dated the destruction of Parliament by little over a month, it can be seen how, when the problem of ventilating the temporary houses in 1835 arose, Reid's work had already received much attention. As Lord Campbell put it, 'on account of the fame you had acquired from the construction of your Class-room in Edinburgh, you were called in'.[101]

The classroom demonstration of 13 September 1834 was not Reid's only opportunity to impress parliamentary audiences. Two days after the BAAS concluded its proceedings, a banquet was held in honour of Earl Grey. This tribute to Grey's career came at the invitation of the Lord Provost of Edinburgh. It followed closely behind the BAAS meeting so as to ensure that a great many notable figures would still be assembled in the city.[102] 'The Grey Festival' was a very Scottish celebration of 'her Patriot Grey', who although English was proclaimed to be Scotland's closest friend.[103] While the BAAS meeting provided Edinburgh with a spectacle of science, the festival was Edinburgh's homage to Whig reform.[104] Following a huge demand for tickets, it was resolved that a large temporary pavilion would be erected. The festival's organizers appointed Thomas Hamilton (1784–1858) as architect, with Reid to advise on

[97] Yeo, 'An idol of the market-place', p. 257.
[98] Bord, *Science and Whig Manners*, p. 74.
[99] Michael Lobban, 'Brougham, Henry Peter, first Baron Brougham and Vaux (1778–1868)', *Oxford Dictionary of National Biography*, Oxford University Press, 2004; online edn, January 2008 [http://ezproxy.ouls.ox.ac.uk:2117/view/article/3581, accessed 28 March 2013].
[100] Reid, 'The study of chemistry', p. 1.
[101] Reid, *Professorship of Chemistry in the University of Edinburgh*, p. 9.
[102] Henry Brougham, *The Life and Times of Henry Lord Brougham Written by Himself in Three Volumes*, Vol. III, (Edinburgh, 1871), pp. 414–15.
[103] John Black Gracie (ed.), *The Grey Festival; Being a Narrative of the Proceedings Connected with the Dinner Given to Earl Grey, at Edinburgh, on Monday, the 15th September, 1834, and a Corrected Report of the Speeches*, (Edinburgh, 1834), p. 3.
[104] Ibid., p. vi.

ventilation and acoustics. The *Caledonian Mercury* felt sure that Scotland would not be embarrassed by the 'Grey Pavilion' because Reid,

> with whose felicitous Genius in Acoustics his townsmen are so well acquainted, is one of the Committee; whence they may gather the assurance that the interests of the Ear will be as much studied as those of the Eye in the scheme of this vast and magnificent entertainment.[105]

Reid's appointment did not inspire unanimous confidence, however. Although enjoying the air of excitement in the city 'full of visitors', present for the BAAS and the forthcoming banquet, the *Morning Chronicle* expressed concern over the pavilion's utility. It was deemed 'very ill adapted for hearing, and as to ventilation, it seems to have been forgotten that the two thousand persons who will meet together on this great occasion will require an occasional breath of fresh air'.[106] While Reid's work received mixed reviews, his involvement with the Pavilion placed him at the centre of Edinburgh political life at a crucial moment before the formation of the 1835 committee to investigate Parliament's air. Along with several MPs, Brougham once again witnessed Reid practically apply knowledge to space.

The projection of his skills of ventilation and sound communication in 1834 made Reid appear as a trustworthy authority for the 1835 committee. The committee 'examined several witnesses of high scientific reputation' in order to ascertain the 'general principles on which a good system of warming and ventilating Public Buildings depends'.[107] With Benjamin Hawes (1797–1862), MP for Lambeth, acting as chair, the committee was impressed with Reid's projection of a perfectly maintained atmosphere. His system promised to utilise chemical knowledge for the material progress of Parliament, but to secure the committee's support, Reid laboured to build credibility. Reid's perceived success in his classroom and the attention the Grey Pavilion attracted appeared to add weight to his promises of securing healthy air at Parliament. The classroom demonstrated that Reid could deliver effective ventilation while the Pavilion showed that he was prepared to publically trial his work on a grand scale. It was

[105] (Anon.), 'Earl Grey's visit', *Caledonian Mercury*, (Edinburgh, Scotland), 1 September, 1834; Issue 17653; a Scottish architect, Thomas Hamilton had secured a reputation through his designs for the Royal High School and Martyrs Monument, both at Carlton Hill in Edinburgh. He later presented a paper on the 'Grey Pavilion' to the RIBA, see Gavin Stamp, 'Hamilton, Thomas (1784–1858)', *Oxford Dictionary of National Biography*, Oxford University Press, 2004 [http://ezproxy.ouls.ox.ac.uk:2117/view/article/12131, accessed 7 April 2013].

[106] (Anon.), 'Edinburgh – Sept. 13', *The Morning Chronicle*, (London, England), 17 September, 1834; Issue 20298.

[107] PP. 1835 (583), p. 3.

the reputation of these projects which contrasted his evidence so favourably to that of the other men of 'scientific reputation'.

Reid argued that to improve the ventilation of the temporary Parliament, 'a column of heated air' should be introduced that would 'create artificial currents through the buildings'.[108] Clean fresh air could enter the Houses through the floor, and would then exit through the roof, drawn by a powerful source of heat outside the temporary accommodation. Vitiated air was to be removed and replaced by pure air which could be heated or cooled, in relation to external conditions, in a chamber beneath the temporary Houses. To provide 'purity', he recommended taking in air from a height above the damp Westminster atmosphere.[109] In his classroom he had specifically implemented this system to extract the fumes from up to '2,000 experiments performed' each hour. As Reid testified, it had performed well: 'I find it absolutely essential to have a power of carrying off those fumes, a power which is perfectly under control, and which can be made to operate to any extent, according to circumstances'.[110] If Reid's system could ventilate a fume-ridden laboratory, accommodating three-hundred students, then by calculating the ideal amount of air an individual should be provided with, in a given time, and then identifying how many people were in attendance, the space of Parliament could come under the control of chemical knowledge.[111]

Reid explained that his ventilation apparatus could secure an atmosphere rich in oxygen and free of carbonic acid. By commanding the laws of nature inside Parliament, improvement could be attained. Efficiency of ventilation would ensure that,

> the atmosphere would never be of that oppressive character which often increases to such an extent in some buildings, where the respired air is not so easily carried away, as to produce a very powerful sedative effect, often accompanied by severe headache, more especially when it is necessary to maintain a continued and anxious attention to any subject under discussion.[112]

Controlling the atmosphere of Parliament with a powerful ventilation system promised to increase MPs' attention to public business.

What Reid delivered to the committee was a promise of control, precision, and power, but also a demonstration of how chemistry had practical applications beyond the laboratory. He believed that because of this, effectively everyone was a 'practical chemist'.[113] Reid's evidence was a

[108] Ibid., p. 34. [109] Ibid., p. 35. [110] Ibid., p. 34. [111] Ibid., p. 37.
[112] Ibid., p. 39.
[113] David Boswell Reid, *Rudiments of Chemistry; with Illustrations of the Chemical Phenomena of Daily Life*, 4th ed., (London, 1851), p. iii.

claim that the practice of ventilating Parliament should be a chemical one. He asserted that his understanding of how air moved and what it consisted of, was knowledge acquired through chemical experiments. As ventilation created a change in an atmosphere's chemical state, Reid deemed it a chemical action.[114] He concluded that any 'change of properties . . . is the grand and leading character of the operations of Chemistry, not a change of motion, place, or figure, as in Natural Philosophy'.[115] The power to effect efficient air purification was therefore chemical action.[116]

Reid's performance before the committee secured him a reputation as someone who could be trusted with improving Parliament's air. His views were of a 'clear, decided and satisfactory nature', while the committee 'saw at once that he spoke as a man thoroughly conversant with the subject'.[117] His proposed system impressed the inquiry, consisting of Hawes, Charles Hanbury-Tracy, Earl Grey, Lord Granville Somerset, Lord Sandon, George Clerk, and Henry Warburton. While Lord Somerset, First Commissioner for Woods and Forests until April 1835, and Hanbury-Tracy had experience of architecture, Warburton and Hawes added medical and scientific authority to the investigations. Warburton had supported Brougham in the founding of London University, and had chaired a parliamentary committee on the study of anatomy in 1828.[118] However, it was Hawes, the Chairman, who brought the most experience of engineering to the committee. In 1820 Hawes had married Sophia MacNamara Brunel, the daughter of Marc Isambard Brunel. A soap-boiler by trade, Hawes was an enthusiastic supporter of several projects of his brother-in-law, Isambard Kingdom Brunel.[119] Both Hawes and Warburton later worked on the committee for Marc Brunel's Thames Tunnel project, where they pursued a line of questioning regarding the excavation's ventilation.[120] Interestingly, Marc Brunel had been a supporter of Reid's work since visiting the Edinburgh classroom. He believed that Reid's scientific principles displayed at Edinburgh would be ideal for the accommodation of Parliament.[121]

[114] Ibid., p. 6. [115] Ibid., p. 7. [116] Reid, 'The study of chemistry', p. 5.
[117] Reid, *Memoir*, pp. 14–15.
[118] H. C. G. Matthew, 'Warburton, Henry (1784–1858)', *Oxford Dictionary of National Biography*, Oxford University Press, 2004; online edn, May 2009 [http://ezproxy.ouls.ox.ac.uk:2117/view/article/28672, accessed 8 April 2013].
[119] Ged Martin, 'Hawes, Sir Benjamin (1797–1862)', *Oxford Dictionary of National Biography*, Oxford University Press, 2004; online edn, May 2009 [http://ezproxy.ouls.ox.ac.uk:2117/view/article/12643, accessed 5 April 2013].
[120] *Report from the Select Committee on the Thames Tunnel; with the minutes of evidence*, PP. 1837 (499), p. 11.
[121] David Boswell Reid, *Testimonials Regarding Dr D.B. Reid's Qualifications as a Lecturer on Chemistry and Teacher of Practical Chemistry, April 1837*, (London, 1837), p. 10.

The committee included individuals, like Earl Grey, who had witnessed, or had friends who had witnessed, Reid's science. As Hawes explained, although witnesses Faraday and William Thomas Brande (1788–1866), both servants of the Royal Institution, 'combined practical and scientific value', the committee found no one else displayed such understanding of 'the science and the practice of Ventilation in large buildings'.[122] In comparison to Reid's evidence, none of the other witnesses produced such a trustworthy solution to the problem of parliamentary ventilation. The first, George Birkbeck (1776–1841), was a lifelong friend of Brougham, and had inaugurated the London Mechanics' Institute, as well as conducted work for the SDUK.[123] Birkbeck presented the committee with a very damning review of the state of national architecture in terms of ventilation. Due 'to the want of practical knowledge on the subject', he could not recall a single architectural example that employed reliable ventilation. Birkbeck suggested that through 'strict calculation' of how many would sit in a building and 'what impression they will make upon the atmosphere', it might be deduced how to bring about effective air purification.[124] With reference to Lavoisier's experiments on air in Parisian theatres, Birkbeck confirmed Reid's assertion that ventilation was a problem of chemistry, rather than architecture or mechanical engineering. A poorly ventilated crowded room would create a high temperature, a considerable deprivation of oxygen gas, and a drastic expanse of carbonic acid gas. Birkbeck explained how Lavoisier had found that poorly ventilated theatres became 'charged with azotic gas' and carbonic acid: both detrimental to health.[125] Yet Birkbeck lacked Reid's experience of the practical application of such chemical knowledge. He could not recommend a 'good system' to improve the temporary buildings. Considering Parliament's atmosphere, Birkbeck 'for one would not endure it for the service of the public'.[126]

The second witness, John Sylvester, claimed that he had produced a reputable system of ventilation at the Kent Lunatic Asylum. As early as 1819, Sylvester's father, Charles, had published a work considering ventilation in which he argued that natural philosophy, if it were really

[122] Reid, *Professorship of Chemistry in the University of Edinburgh*, p. 6.
[123] Matthew Lee, 'Birkbeck, George (1776–1841)', *Oxford Dictionary of National Biography*, Oxford University Press, 2004 [http://ezproxy.ouls.ox.ac.uk:2117/view/article/2454, accessed 7 April 2013].
[124] PP. 1835 (583), p. 5. [125] Ibid., p. 7.
[126] Ibid., p. 11; Antoine Lavoisier (1743–1794) had, along with producing an early substantial list of elements, identified oxygen and hydrogen.

to improve mankind's state, should be directed to domestic comfort.[127] John Sylvester, continuing his father's work, proposed a system at Parliament based on his asylum near Maidstone, which included a large tunnel under the chambers in which air could be warmed if desired. Nevertheless his system lacked the credibility of Reid's. After a vague explanation of how he controlled the temperature in the asylum, he stated he was not prepared to outline how his system might be adapted for Parliament.[128] After much questioning, Sylvester revealed that his control of the temperature at the asylum required at least fifteen minutes to respond to commands.[129] Later, when Faraday was interviewed, the committee asked about the suitability of Sylvester's scheme. Faraday warned that the whole system would introduce dry 'unpleasant' air into the chambers.[130] Sylvester thus appeared to lack the practical experience which Reid had demonstrated in Edinburgh. Apart from Reid, two other 'eminent' men of science were called to the committee after Sylvester. Brande reported on his experiments on the air in the Covent Garden Theatre, which he recorded as having 3 per cent carbonic acid by volume during performances in front of a full house.[131] Brande's early career had been prodigious, being elected a fellow of the Royal Society aged twenty-one, and in 1813 replacing Humphry Davy as Professor of Chemistry at the Royal Institution, where he lectured until 1848.[132] However, like Birkbeck and Sylvester, he did not offer a convincing system, and advised the committee to 'be guided by practice rather than theory'.[133]

Before interviewing Brande, the committee invited Faraday to present evidence. He reported on the scientific principles on which his Royal Institution lecture-room was based. Yet Faraday's lecture-room contrasted poorly to Reid's classroom. He warned that,

> no arrangement of the present air passages, or the present mode of heating, has enabled us to give that free and proper draught through the room which shall sufficiently move away bad air and give good without violent partial currents.[134]

[127] Charles Sylvester, *The Philosophy of Domestic Economy*, (Nottingham, 1819), pp. x, 1.
[128] PP. 1835 (583), p. 17. [129] Ibid., p. 16. [130] Ibid., p. 23. [131] Ibid., p. 31.
[132] Frank A. J. L. James, 'Brande, William Thomas (1788–1866)', *Oxford Dictionary of National Biography*, Oxford University Press, 2004 [http://ezproxy.ouls.ox.ac.uk:2117/view/article/3258, accessed 7 April 2013].
[133] PP. 1835 (583), p. 27.
[134] Ibid., p. 20; with his brother Robert Faraday, Michael Faraday later worked to ventilate lighthouses, see Frank A. J. L. James, '"The civil-engineer's talent": Michael Faraday, science, engineering and the English lighthouse service, 1836–1865', *Transactions of the Newcomen Society*, 70 (1998–99), pp. 153–60, 155.

Faraday's experience of ventilation was in this way framed with caution. Faraday warmed air beneath the lecture room, allowed it to flow into the room, and then regulated its exit with a valve placed above in the roof. As Faraday observed, this was sufficient for the Institution, but was unsuitable for Britain's Parliament, where members 'ought to have the full command of temperature without depending on the admission of a given quantity of air'.[135] Faraday defined the chemical problem that any ventilation system should overcome to be deemed successful. He calculated that a man 'destroys about a gallon of air per minute', so a supply of ten to twenty times that amount was desirable.[136] As to the question of how to introduce this supply, 'experiments are required . . . we want some numerical data, something more than general impressions or mere opinions'.[137] He reckoned that experiment could provide trustworthy, measurable, evidence.

The committee's witnesses collectively identified the problem of Parliament's atmosphere to be chemical. They agreed that to improve Parliament's air, a system should be built that would ensure its atmosphere was kept oxygen rich, while levels of carbonic acid were minimized. Faraday outlined the importance of finding a trustworthy man of science to direct Parliament's air purification, believing

> that ventilation and warming are so important with regard to an assembly of such a nature, that a full and efficient system ought to be adopted . . . [delivered by] a man of judgement and observation, uniting some knowledge of architecture and construction with an acquaintance with natural principles.[138]

Considering Reid's evidence and work in Edinburgh to be impressive, the committee felt the chemist was just such a man of judgement. They recommended to Parliament that,

> the alterations suggested by Dr Reid should, if practicable, be submitted to the test of actual experiment during the ensuing recess, as the only means of ascertaining with accuracy the soundness of the principles laid down in the Evidence, and their useful application to the future Houses of Parliament.[139]

This conclusion went further than simply declaring Reid's evidence credible, but explicitly demonstrated that he had convinced the committee to sanction experiments to validate his work in Edinburgh. This decision to allow Reid to prove his evidence through experiment paralleled the scientific method he developed in Edinburgh. Following Reid's death in 1863, his brother asserted that his career had been built on the strength of his competence at using experiments as demonstrations. It was not only that

[135] PP. 1835 (583), p. 26. [136] Ibid., p. 22. [137] Ibid., p. 25. [138] Ibid., pp. 23–24.
[139] Ibid., p. 4.

Reid's lectures were lucid, but that his experimental displays usually fulfilled his predications in front of audiences. The extent of Reid's preparation and knowledge meant that he 'was seldom if ever placed in the awkward position of failing in an experiment – of telling his class that such a change would take place, while the inert materials defied him, and refused to exhibit the predicted phenomenon'.[140] Therefore the strength of Reid's public experiments was that the results of his chemical manipulations realized his predictions. His approach was to show he knew an experiment's results before its completion. In asking Reid to validate the evidence he presented to the committee, what was requested was the performance of this experimental method on a large scale. The challenge would be to fulfil promises made from evidence collected in Edinburgh, at Westminster. The process of experimentally demonstrating predicted phenomena was synonymous with Reid's series of classes. Reid was subsequently appointed to improve the atmosphere of Parliament, and to conduct experiments at Westminster which might confirm the 'soundness of the principles' of his earlier endeavours.

Edinburgh Science at Westminster: Parliament as a Laboratory

Although Reid's proposed system impressed Hawes' committee, he was not immediately entrusted with the ventilation of Barry's new permanent building. His appointment to experiment in the temporary accommodation provided an opportunity to deliver his Edinburgh classroom system at Westminster, but securing trust in London was hard. In Smirke's temporary Commons, Reid introduced an upcast shaft system of ventilation. Exiting via a specially built chimney, air was drawn through the House from the floor and out through the roof (Figure 4.2). A twenty-five horsepower steam engine created heat at the base of the chimney; this controlled the current through the House.[141] Incoming air was first filtered, and then heated in a chamber beneath the House.[142] A single valve regulating the exit of air into the roof and down towards the furnace controlled the speed of the current. Reid introduced a lower ceiling, of the same shape as that in his classroom to assist the air flow and acoustics of the room. In October 1836, *The Times* confidently reported that the interior had been completely 'remodelled, and rendered in every particular more convenient and eligible for its purposes. The ventilation of the

[140] Hugo Reid, *Memoir*, p. 12.
[141] Denis Smith, 'The building services', in Port (ed.), *The Houses of Parliament*, pp. 218–31, 220.
[142] Sturrock and Lawson-Smith, 'The grandfather of air-conditioning', pp. 2984–85.

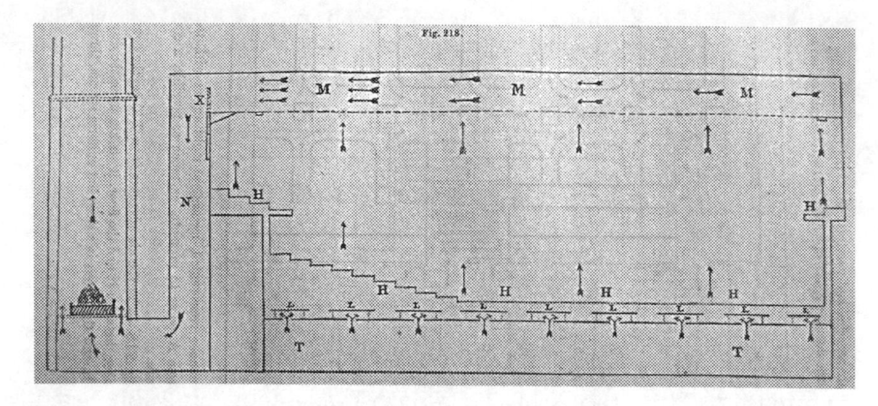

Figure 4.2 Reid's designs for the flow of air through the temporary House of Commons, up through the floor and then extracted through the roof by the action of a furnace

House will be much improved, by preserving an equal temperature without the sudden gusts of cold air'.[143] Yet while Reid's alterations appeared improving, the nature of his work, as the 1835 committee directed, was experimental. The committee expected Reid to advance knowledge in the scientific principles of ventilation. Reid worked to transform the political space into a site of chemical knowledge production.

When the committee examined Smirke about the practicality of Reid's alterations, they proposed experiment, in the form of repeated testing of the composition of the atmosphere both when under Reid's control, and without the aid of ventilation. Reid convinced the committee that the Commons, like his classroom, could be treated as an experimental space. The committee wanted 'to try the experiment, considering that a new House is to be built, and that something in the way of experiment should be tried previously'.[144] Smirke was requested to 'give a fair trial to the principles upon which Dr. Reid has founded his plan'. Smirke did not share this confidence in transforming what he perceived to be a functioning legislature into a chamber of experiment. He agreed to follow Reid's directions, provided that it was the chemist who accepted sole responsibility for the system. Smirke preferred the construction of a smaller room to experiment on first, being 'very much afraid of trying so hazardous an experiment upon such a room' as the House of Commons.[145] To alter a building,

[143] (Anon.), 'Workman have been actively engaged . . .', *The Times*, (London, England), 20 October, 1836; p. 3; Issue 16239.
[144] PP. 1835 (583), p. 52. [145] Ibid., p. 53.

or adapt architectural plans to provide effective ventilation was untried and dangerous. Smirke warned that 'Should the experiment fail … from any difficulties in the application of the principle … the inconvenience … during the session of Parliament, would be very serious'.[146] When Reid was recalled, he suggested that a small laboratory may be erected to obtain information regarding ventilation, by 'actual experiment'.[147] Such 'a model of the new House of Commons' would provide knowledge to guide the construction of the permanent building. Reid offered to do this work in Edinburgh, for the sum of £2,000. He advised that with an additional £1,000, he could convert this experimental model into a church after the experiments were complete.[148]

Using a model chamber, built in his Edinburgh classroom, and the temporary Commons, chemical knowledge would be acquired through experiment, which would be used in reference to the new permanent Parliament.[149] Experiments would consist of air-testing the atmosphere in the Commons when subjected to ventilation. In the Commons, Reid developed a device for testing and regulating the purity of the atmosphere. The 'carbonometer', which he likened to a thermometer of purity, consisted of lime water in a phial through which air was passed to test for carbonic acid levels (Figure 4.3).[150] Such an apparatus would allow for the examination of the atmosphere and provide comparable evidence. Meat also provided an instrument for testing air. Reid found that in the area around Parliament, meat ten to twenty feet off the ground went bad within twenty-four hours, yet meat suspended thirty to forty feet remained fresh for as long as three days.[151]

As in Edinburgh, Reid claimed authority by displaying his work to public audiences. These spectacles were experimental, but they were also about wider issues of credibility. In November 1836, he invited a press delegation, along with Hawes and several gentlemen of 'literature and science', to witness a demonstration of the power of his system. *The Times* reported this display in detail. Reid filled the House with 540 guests, including 412 off-duty foot guards marched down from local barracks. This provided an environment similar to that encountered in the House when full. Reid explained how the air moved into the house via

[146] Ibid., pp. 52–3. [147] Ibid., p. 55.
[148] Ibid., p. 56; on the shortage of Presbyterian Church space and the post-1823 building programme, see Allan Maclean, *Telford's Highland Churches: The Highland Churches and Manses of Thomas Telford*, (Inverness, 1989).
[149] Schoenefeldt, 'The temporary Houses of Parliament', p. 178.
[150] Reid, *Illustrations*, pp. 65–66.
[151] *Report from Select Committee on Ventilation of the New Houses of Parliament; with the minutes of evidence, and appendix*, PP. 1841 (51), p. 13.

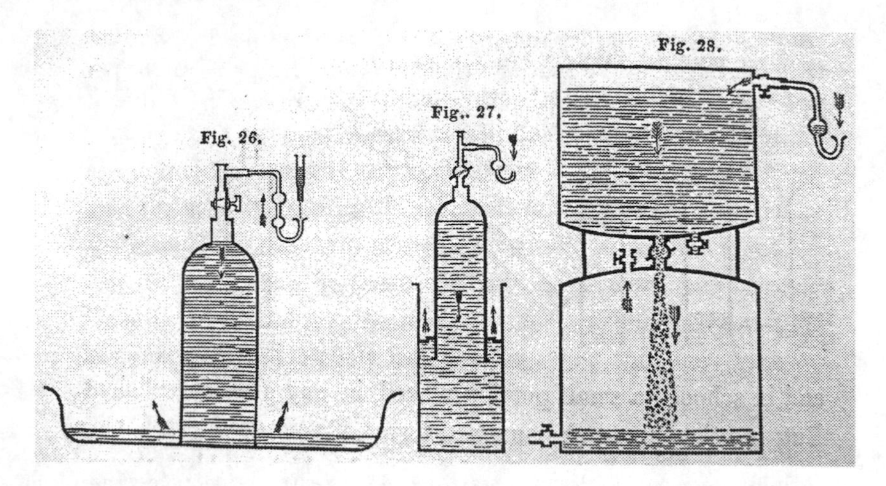

Figure 4.3 Variations of Reid's 'carbonometer', for testing the purity of air inside the Palace of Westminster

one-sixth of an inch perforations in the ancient floor, totalling 'the almost incredible number of 350,000'. To 'pump out' the vitiated air through the ceiling, 'an exceedingly large coal-fire' at the base of the chimney erected in the Cotton Garden provided the desired power. This source of heat affected a 'rapid withdrawal of the air within the house and substitution of fresh air for foul'.[152] Showing the power he had at his command,

> Dr Reid then proceeded to try the following experiments, with a view to show the rapidity of circulation through the house:- He first caused the introduction of a smoke so dense that it was impossible to see five yards forwards. In about one minute and a half, by the action of the shaft, it was entirely expelled. He next introduced the odour of ether, which was strongly perceptible to every person present, and dispersed in an equally short space of time by the active but imperceptible introduction of heated air. In like manner was the scent of oranges raised and dispersed.[153]

This exhibition transformed the locus of government into a spectacle of science. To be able to apparently introduce and evacuate atmospheric elements at will demonstrated precision and control. *The Times* noted the lecture-room-like quality of the House of Commons under Reid's direction. (Figure 4.4) Throughout the demonstration, Reid had exhibited,

[152] (Anon.), 'The alterations in the House of Commons', *The Times*, (London, England), 5 November, 1836; p. 3; Issue 16253.
[153] Ibid.

Figure 4.4 The temporary Houses of Parliament as a laboratory under Reid's chemical control

by means of a glass model on the table of the house, the operation of his plans, in order to render the experiments subsequently made in the House itself more comprehensible to his numerous auditory. During the experiments which followed, the learned gentleman, in an able manner, and at considerable length, delivered what may be termed a lecture on ventilation.[154]

Not only did Reid show off the power of his system, but he detailed how his practice at Westminster advanced scientific learning and established his credibility as a result.

The findings of Reid's experiments were published in his 1844 *Illustrations of the theory and practice of ventilation*. This text, although including much evidence collected beyond the walls of Parliament, was the printed embodiment of Reid's work at Parliament. Its findings were presented to guide public building in accordance with the natural laws of atmospheric composition and respiration. In this publication, Reid called for 'mechanical power' to deliver the amount of air each individual required and for 'each building to be ventilated having been treated as a piece of apparatus . . . absolute power obtained over the ingress and egress of air'.[155] This work was to 'contribute to assist the Architect in designing – the Physician in practising – and others in regulating the atmosphere in which they live, in unison with the principles of ventilation'.[156]

Reid defined his work as that of a 'Professional Chemist', involving 'Chemistry of the Atmosphere; of Respiration and Transpiration; of

[154] Ibid. [155] Reid, *Illustrations*, p. viii. [156] Ibid., p. xv.

Figure 4.5 The treatment and purification of air beneath the House of Commons before entering the chamber

Combustion ... and the regulation and control of an external atmosphere, ever changing in its natural qualities'.[157] Reid outlined the threats of carbonic acid, sulphureted hydrogen, sulphurous acid, hydrochloric acid, ammonia, and other air impurities. These he had tested at Parliament. Local manufacturing had reduced the quality of the Westminster atmosphere; Reid had found arsenic, copper, and lead impurities through his experiments. Reid's evidence was very much an answer to local conditions; he complained of the smell of the local gas works and 'barges laden with manure' passing on the River Thames.[158] (Figure 4.5) Yet the knowledge produced and the solutions provided at Parliament could, Reid asserted, be replicated. Indeed, the same practices which had secured the health of Edinburgh students appeared to have been transferred to Westminster's unhealthy atmosphere.

Building trust at Westminster was an altogether different challenge. Not only did Westminster provide a considerably larger audience than his Edinburgh classroom, but MPs proved difficult subjects to satisfy. This question of subject was one prominent in Victorian society. The question of who it was that was experimented on mattered. Alison Winter notes that when humans became instruments in experiments, the main problem was

[157] Reid, *Professorship of Chemistry in the University of Edinburgh*, p. iv.
[158] Reid, *Illustrations*, p. 298.

how to make their accounts seem reliable. She shows that this was often achieved by portraying human subjects to lack education or apparent freedoms; human accounts seemed most reliable when the subjects appeared like unthinking machines.[159] *The Times* observed that MPs would provide a sterner test than the rows of silent foot guards under 'strict obedience to the word of command', and joked that 'a more orderly house' had never been seen before.[160] While atmospheres and foot guards might be controlled, honourable members did not make such compliant subjects.

Just a year after the experimental displays to *The Times*, Thomas Wakley (1795–1862), the Devonian MP for Finsbury, raised concerns in the Commons over the health of members. He believed that ventilation still demanded 'serious consideration', and asked other members if they too had not 'suffered from irritation of the throat and lungs'.[161] Wakley had examined the House with Arnott, who recommended air descend rather than rise through the house, along with a constantly wet cloth suspended above the seating. This introduction of increased moisture, Arnott and Wakley claimed, would reduce the dust inhaled by members.[162] Despite Reid publicly responding to such criticism, groups within the Commons remained dissatisfied with the system.[163] Members reiterated their objection to Reid's method of drawing air up from the floor. Wakley was particularly concerned following his consultancy with Arnott. Arnott had much praise for Reid, believing the Commons to be 'the only instance in existence of effectual ventilation for such a place'.[164] Nevertheless he expressed concern at the upcast system in place due to the atmospheric dust created. Although raised a Roman-Catholic, Arnott was a prominent member of London Benthamite circles and had introduced Edwin Chadwick to Jeremy Bentham. Arnott was a keen disseminator of the Chadwickian miasmatic theory of disease, maintaining that filth and odour were causes of poor health.[165] In Wakley he found a willing

159 Winter, *Mesmerized*, p. 62.
160 (Anon.), 'The alterations in the House of Commons', *The Times*, (London, England), 5 November, 1836; p. 3; Issue 16253.
161 (Anon.), 'House of Commons', *The Times*, (London, England), 25 December, 1837; p. 3; Issue 16608.
162 Ibid.
163 *Ventilation of the House. Copy of a letter from Dr. Reid to Lord Duncannon, dated February 4th, 1837, relative to the acoustic and ventilating arrangements lately made in the House of Commons*, PP. 1837 (21), p. 1.
164 Neil Arnott, *On Warming and Ventilating: with Directions for Making and Using the Thermometer-Stove, or Self-Regulating Fire, and Other New Apparatus*, (London, 1838), p. 86.
165 Bill Luckin, 'Arnott, Neil (1788–1874)', *Oxford Dictionary of National Biography*, Oxford University Press, 2004 [http://ezproxy.ouls.ox.ac.uk:2117/view/article/694, accessed 7 April 2013].

audience for his concerns over the air of Parliament. Wakley himself had earned a reputation in medicine as the founder and editor of the medical journal *The Lancet* (1823). He campaigned for medical reform, drafting various medical acts throughout the 1830s and 1840s alongside Warburton.[166] Together the medical concerns of Wakley and Arnott represented a serious challenge to Reid's system.

Concerned MPs chose Sir Frederick William Trench (c.1777–1859), MP for Scarborough, to voice this lack of confidence in Reid's system. He argued that Reid's work was 'admirable', but that the House remained devoid of a system that incorporated safe ventilation and sufficient lighting. Trench was a self-proclaimed authority on architecture.[167] He proposed what he felt was a distinctly non-scientific solution to the problem of dust, in which the haircloth carpet covering the floor would be lifted, and the floor oiled, before replacing the covering.[168] This cheap 'experiment' would create 'a constant flow of the improved Wakley and Arnott air, without a particle of dust'. Trench portrayed this plan as common sense, rather than scientific. There was 'nothing of philosophical science to recommend it'.[169]

Reid defended his system by explaining that the dust was not due to the upcast current, but to the incompetence of the 'cleaning department'. The carpet did accumulate dust, but Reid's instructions for it to be lifted and beaten daily had been ignored.[170] He maintained that without 'proportionate care in the cleaning department . . . it is impossible that any system of ventilation can give satisfaction'.[171] Reid nevertheless revealed that he wanted to convert to downcast ventilation following experiments in his Westminster laboratory. This would draw in air through the roof of the Commons before evacuating it through the floor, therefore reducing the chance of dust from the carpet contaminating the room's atmosphere. Trench rejected the possibility of a downcast system, believing it would ruin the chamber's candle lighting. He felt Reid's blaming of the cleaning department was a poor attempt to show the faults were 'not from any error

[166] W. F. Bynum, 'Wakley, Thomas (1795–1862)', *Oxford Dictionary of National Biography*, Oxford University Press, 2004 [http://ezproxy.ouls.ox.ac.uk:2117/view/article/28425, accessed 5 April 2013].

[167] M. H. Port, 'Trench, Sir Frederick William (*c.* 1777–1859)', *Oxford Dictionary of National Biography*, Oxford University Press, 2004 [http://ezproxy.ouls.ox.ac.uk:2117/view/article/27699, accessed 5 April 2013].

[168] *Ventilation of the House. Copy of a letter from Sir Frederick Trench to Lord Viscount Duncannon, on the subject of ventilating the House of Commons, with Lord Duncannon's answer*, PP. 1837–38 (204), p. 2.

[169] Ibid., p. 2.

[170] *Ventilation of the House. Copy of a letter from Dr. Reid to the Viscount Duncannon, in reply to observations addressed to His Lordship by Sir Frederick Trench*, PP. 1837–38 (277), p. 1.

[171] Ibid., p. 5.

in his science'.[172] Even if the 'housemaids' made the carpet 'as clean as Her Majesty's toilet table', dust would still be a problem. As Trench put it, 'common sense (*versus* philosophy and science)' suggested dirt would be continually carried in on the shoes of members.[173] Trench also rejected the suitability of having Reid administer the system from Edinburgh. Someone 'resident in London' would be preferable to the constant expense of having Reid travelling down to monitor the ventilation[174] (see Map A).

Rather than reply, Reid arranged a series of demonstrations to quiet the complaints, continuing to believe that the best way to build credibility was through performance. Several members of the press were invited to witness Reid's new descending system of ventilation, combined with experimental gas lighting. Filling the Commons benches with men armed with buckets of water should a fire break out, the gas lighting was kept separate from the Commons atmosphere. With air exiting via the floor and the combusted gas 'effluvium being prevented from descending', the system had been adapted to meet the concerns of Wakley, Arnott, and Trench.[175] Reid invited Trench at a later date, 'to see his experiment of the effect of the descending current'.[176] Trench admitted that he had 'been mistaken'. He considered Reid's system was an effective way of placing the Commons' atmosphere under the regulation of a furnace. Trench described how he had seen,

> four experiments tried: three different odours, orange, lavender, and cinnamon, were in succession distributed rapidly and effectually over the whole body of the House, and carried through the hair-carpet and the perforated floor into the apartment below, where a gale of wind hurried them off to feed the furnace which created this current. Gunpowder was then exploded between the roofs; the smoke instantly pervaded the whole of the House and was seen descending regularly till it was drawn through the hair-carpet into the regions below.[177]

Through witnessing Reid's power, Trench was 'now convinced of the truth'. He agreed that Reid's solution was sufficient for the parliamentary atmosphere. Successfully managing an experimental spectacle proved a powerful resource for building credibility, even with sceptical audiences.

[172] *Ventilation and lighting of the House. Letters from Sir Frederick Trench to Lord Duncannon, on the subject of ventilation and lighting the House of Commons*, PP. 1837–38 (358), p. 2.

[173] Ibid., p. 2.

[174] Reid's expenses for the Westminster experiments totalled over £512, see *Ventilation of the House. Return of the detailed expenses incurred in experiments for improving the ventilation, &c. of the House of Commons, in the experiment of lighting with gas, also in lighting with candles, ending with the present lustres and shades*, PP. 1837–38 (725), p. 1.

[175] (Anon.), 'Domestic notices', *The Architectural Magazine*, Vol. V, (London, 1838), p. 87.

[176] *Ventilation and lighting of the House. Letters from Sir Frederick Trench to Lord Duncannon, on the subject of ventilation and lighting the House of Commons*, PP. 1837–38 (358), p. 5.

[177] Ibid., p. 5.

Figure 4.6 The three steamships of the Niger Expedition about to commence their ascent up the river

Reid's system of ventilation was but one scheme in a series of projects in which he laboured to secure his chemistry recognition of its usefulness. Reid implemented a complex system of ventilation on the three ships comprising the ill-fated Niger Expedition of 1841. (Figure 4.6) Initiated during an inaugural address by Prince Albert at the 1840 meeting of the Society for the Extinction of the Slave Trade and for the Civilization of Africa, held in Exeter Hall, this project was intended to carry Christianity into Africa and eradicate the evils of slavery.[178] Crucially, an aim of this society was to demonstrate to the African continent that British medical science could tame the tropical miasmas of Africa's coasts and fever-ridden river areas. Christian science was to be exhibited as a great tool of 'civilization'.[179]

[178] (Anon.), 'Address on behalf of Africa', *The Friend of Africa*, Vol. 1, No. 1, (London, England), 1 January, 1841, p. 5; on the expedition, see Howard Temperley, *White Dreams, Black Africa: The Antislavery Expedition to the River Niger, 1841–1842*, (New Haven, 1991).

[179] (Anon.), 'Advantages of medical science to Africa', *The Friend of Africa*, Vol. 1, No. 2, (London, England), 15 January, 1841, p. 17; see Stewart J. Brown, *Providence and*

Appreciating Reid's experimental work at Parliament, the MP Benjamin Smith invited Reid to replicate his methods of ventilation onboard the expedition vessels.[180] He devised a system whereby tubes for the ingress and egress of air ran through each ship, and air was drawn in through filters using the rotation of fans connected to the paddle-wheels of the ships.[181] What this was, as Reid informed the BAAS at the Plymouth meeting of 1841, was a showcase that the practices developed in Parliament could be replicated beyond Westminster and carried over space to bring order, control, and improved health to the most far flung corners of the world.[182] This system of ventilation was to 'illustrate the importance of a knowledge of practical chemistry being acquired generally by those who may have to visit a distant country', and also those at home.[183] At Westminster, Reid worked to perfect techniques that he believed could carry civilization beyond Britain. If buildings could be conceptualized as air pumps, then so too could ships.

Giving an account of Reid's work, Elisha Harris explained how demonstrations at Westminster and on the Niger had ramifications for all society. Reid's promotion of health through ventilation displayed 'the soundest principles of political economy and the precepts of Christian duty'.[184] Ventilating the homes of the poor furnished the 'most reliable indices of the state of intellectual and moral advancement in any community', and its improvement was a means of 'moral elevation'.[185] At Parliament then, Reid claimed to do more than secure the legislature's health, but demonstrate how to induce material improvement. Parliament was foremost a work of architecture, but architecture was itself 'an art upon which the principles of vital chemistry ... have claims'.[186] Chemistry, alongside architecture, was to be applied to Parliament and the 'wants of common life'. Reid explained that ventilation was a 'new power'

Empire: Religion, Politics and Society in the United Kingdom, 1815–1914, (Harlow, 2008), pp. 139–43.

[180] David Boswell Reid, 'Dr Reid on the ventilation of the Niger steam vessels', *The Friend of Africa*, Vol. 1, No. 4, (London, England), 1 February, 1841, p. 43.

[181] Ibid., pp. 44–45.

[182] (Anon.), 'Arts and Sciences. British Association', *The Literary Gazette and Journal of Belles Lettres, Arts, Sciences, &c.*, (London, England), 28 August, 1841, pp. 561–63; for an account of Reid's ventilation system, see James Ormiston M'William, *Medical History of the Expedition to the Niger During the Years 1841–2 Comprising an Account of the Fever Which Led to Its Abrupt Termination*, (London, 1843).

[183] David Boswell Reid, 'Dr Reid on the ventilation of the Niger steam ships', *The Friend of Africa*, Vol. 1, No. 5, (London, England), 24 March, 1841, p. 70.

[184] Elisha Harris, 'An introductory outline of the progress of improvement in ventilation', in David Boswell Reid, *Ventilation in American Dwellings; with a Series of Diagrams, Presenting Examples in Different Classes of Habitations*, (New York, 1858), pp. iii–xxxv, iv.

[185] Ibid., p. iii. [186] Ibid., p. xvii.

and product of 'modern times'. When done with 'proper experimental illustrations', chemically informed ventilation was an advancing science.[187]

Securing support for his ventilation was difficult beyond Westminster, especially with national scientific audiences, such as those of the BAAS. Although missing the inaugural BAAS meeting at York in 1831, Reid shared in the sentiments of the association and hoped 'that a permanent society will be established'.[188] Reid became an active attendant of meetings, exhibiting his Edinburgh laboratory in 1834, and giving papers in Dublin (1835), Newcastle (1838), Birmingham (1839), and Plymouth (1841). At Newcastle, he delivered a paper to the medical section of the association 'On the Amount of Air Required for Respiration', which contained the results of his experiments on respiration performed in the Commons. He explained how 'precise experiments' within Parliament had proven that the minimum supply of air in a crowded public building should be 'thirty cubic feet for each individual' per minute. He also suggested 'methods of filtering the air of its impurities when desired.[189] At Plymouth, however, speaking on the ventilation of ships, Reid aroused not praise, but scepticism over the utility of his apparatus. Reid presented at Devonport alongside papers from Marc Brunel on his Thames Tunnel project, and watchmaker Edward John Dent on recent improvements to chronometers. Despite appearing with such eminent authorities, the Section E audience received Reid poorly, with one reviewer describing how in his paper, Reid had taken up 'so much space in opening the *valves* ... that he left himself no time for its sufficient *winding* up'.[190]

Despite these questions surrounding Reid's work, there was a consensus that he had improved Parliament's atmosphere. Thomas Graham (1805–1869), the Professor of Chemistry at London University College, felt the 'magnificent experiment' to be 'one of the grandest applications of physical science that has lately been attempted'.[191] Physician to the Queen and Prince Albert, James Clark, believed the work would 'do more to improve the public health than any measure

[187] Reid, *Ventilation in American Dwellings*, pp. 4–5.

[188] Bodleian Library, Oxford (BOD) Ms Dep. Papers of the British Association for the Advancement of Science, 1, Correspondence of John Phillips, Folios 134–35, 'Letter from David Boswell Reid to John Robison', (20 September, 1831).

[189] 'Notices and abstracts of communications to the British Association for the Advancement of Science', in *Report of the Eighth Meeting of the British Association for the Advancement of Science; Held at Newcastle in August 1838*, (London, 1839), pp. 131–32; also see David Kennedy, 'Dr. D. B. Reid and the teaching of chemistry', *Studies: An Irish Quarterly Review*, Vol. 31, No. 123 (September, 1942), pp. 343–50.

[190] (Anon.), 'Arts and sciences. British Association', *The Literary Gazette and Journal of Belles Lettres, Arts, Sciences, &c.*, (London, England), 28 August, 1841; p. 563; Issue 1284; also see Morrell and Thackray, *Gentlemen of Science*, p. 265.

[191] Reid, *Professorship of Chemistry in the University of Edinburgh*, p. 1.

with which I am acquainted'.[192] Physician John Forbes believed Reid, 'in the laboratory provided for him by Government in the Houses of Parliament', had progressed practical scientific knowledge of ventilation in public buildings.[193] Crucially Reid also found support from Edinburgh natural philosopher David Brewster, as well as T. Lloyd, the chief engineer of the Royal Dock-Yard at Woolwich, Alexander Milne, Commissioner of Her Majesty's Woods and Forests, the Duke of Sutherland, and the Canon of Westminster, John Jennings.[194] Furthermore, Lord Campbell, of the 1834 BAAS delegation, believed that in the Commons 'the temperature is regulated by the thermometer with the most complete accuracy and steadiness; and the air is at all times as pure as that breathed on Hampstead Heath'.[195]

Hanbury-Tracy, now Lord Sudeley, shared Campbell's enthusiasm for Reid's apparatus. He recalled how the,

> pestilential atmosphere of the House of Commons was notorious; its baneful effects on the healths and energies of its members were painfully felt and admitted ... and the most sanguine never dreamt that it could be cured, much less that the ventilation of the Houses could be brought to such a degree of perfection ... To your skill, zeal, and determination it is owing that the members of the House of Commons can now pursue their senatorial duties without a sacrifice of either health or comfort – to you we owe the solution of our problem.[196]

Practical chemistry had effected a material improvement in government and the administration of its duties. Reid had secured enough credibility that his practices and apparatus could be considered suitable to address the atmospheric challenges of the permanent Parliament.

Knowledge for a Palace

Reid's experiments in the temporary Commons earned him much praise, but the question of ventilating the permanent building was an altogether different challenge. Once Reid had exported his apparatus and practices from Edinburgh to Westminster, he had to transfer his work from the temporary buildings to the new Parliament. This raised two problems. Primarily working on the permanent building involved cooperating with Barry. As shown in Chapters 2 and 3, Barry felt himself to be a scientific authority in his own right. With Reid used to exerting complete control, as he had when experimenting in the temporary buildings, cooperation between the two men proved impossible to sustain. The second problem

[192] Ibid., pp. 4–5. [193] Ibid., pp. 5, 7. [194] Ibid., pp. 29, 25, 10, 12, and 16.
[195] Ibid., p. 9. [196] Ibid., p. 8.

was the nature of Reid's experimental knowledge. Reid's approach in the temporary Commons was to experiment and allow his results to guide the form of ventilation adopted. As his experiments continued, so the system evolved. In temporary accommodation this practice was acceptable, but when Reid attempted to continue this in the permanent building, experimental knowledge appeared inappropriate. Every experiment involved changes to his ventilation plans and as his relationship with Barry broke down, the national press interpreted his experiments as dangerous and unstable. Constant experimenting appeared to produce disordered knowledge. In contrast, Barry referenced men of science, like Faraday, who provided consistent information. In public and private debates, Barry's supporters used this perceived instability to assert that Reid's experiments were inappropriate for a permanent Parliament.

After lobbying Lord Duncannon, Reid secured appointment as the practical engineer to the permanent building in October 1839.[197] Duncannon's choice was initially popular. *The Times* observed that Reid's experiments in the temporary House had been enlightening: he was 'the gentleman who introduced and perfected the system at the Commons'.[198] Duncannon advised Lord Melbourne of the advantage of enticing Reid to leave Edinburgh and work permanently at Westminster, and in November 1839 Melbourne and the Chancellor of the Exchequer agreed to finance Duncannon's request.[199] Ventilating the permanent building required permanent attention, so Reid took accommodation at Duke Street. The ventilation and heating of Parliament was thus placed in Reid's hands, but with a clause that Barry should not agree to anything which would affect the building's architectural character or solidarity. Nevertheless, Duncannon instructed Barry to assist Reid in all matters of a scientific nature.[200]

Reid outlined his proposed ventilation system to an 1841 select committee on ventilating the new building. To create a 'pure atmosphere', Reid explained how his experiments in the temporary House revealed three components to a successful system of ventilation. Firstly, air for Parliament should be taken from a high altitude to avoid impurities. Secondly, it was vital to reserve a large space beneath the building as a

[197] Alexandra Wedgwood, 'The new Palace of Westminster', in Christine Riding and Jacqueline Riding (eds.), *The Houses of Parliament: History, Art, Architecture*, (London, 2000), pp. 113–35, 120.

[198] (Anon.), 'The Houses of Parliament', *The Times*, (London, England), 25 January, 1841; p. 4; Issue 17576.

[199] Reid, *Narrative*, p. 8.

[200] Ibid., p. 8; also see Moritz Gleich, 'Architect and service architect: the quarrel between Charles Barry and David Boswell Reid', *Interdisciplinary Science Reviews*, Vol. 37, No. 4 (December, 2012), pp. 332–44.

'reservoir' for prepared air. In summer, this was to be cooled and in winter heat applied. Finally, to drive this system, a large source of power was desirable.[201] It was proposed that air enter from the Victoria and Clock towers at either end of the Palace, and then be drawn through the building before extraction via a new central tower. This system would sustain a current, constantly driving the air in and out of Parliament. Heat beneath the central tower would assist this 'plenum movement'. The Palace of Westminster was conceived of as an enormous 'tube' through which air would pass, 'modified' along the way 'according to circumstances'.[202] He described how he intended Parliament to operate as a giant air pump; he wanted to establish 'a constant plenum impulse sustained by natural causes'.[203]

Reid stipulated that the central tower over Barry's central hall would create enough pressure to drive the system without mechanical power. It was to be a structure which created a motive 'power' of ventilation.[204] Vernon Smith, Whig MP for Northampton and member of the committee, was sceptical of these assertions of power, but Reid advised the member to visit his 'experimental room' where empirical evidence as to the efficaciousness of a central shaft could be presented.[205] Throughout his explanation of this scheme, Reid consistently emphasized how his method was a product of experiments and observable phenomena. Although he boasted of the power of a central tower, he wanted to have a source of heat available in addition which might be deployed to create additional power during busy sessions in the two debating chambers.[206] Barry agreed to Reid's demand of a tower which would, he believed, improve the architectural character of the building. He recommended a high spire both to aid ventilation and to create 'the most picturesque' appearance alongside the other two towers.[207] This 299-foot octagonal tower, erected over the central lobby between the Commons and Lords, was covered with exquisite external Gothic detail and conceptualized as a giant chimney. (Figure 4.7) Barry estimated that the cost of construction would be around £20,000, but was cautious over Reid's claims for so much space as an air reservoir beneath the building.[208] It was a remarkable architectural structure in that it was almost completely guided by scientific theory.

201 PP. 1841 (51), pp. 7–8. 202 Ibid., p. 8. 203 Ibid., p. 24. 204 Ibid., p. 10.

205 Ibid., pp. 12, 31; W. R. Williams, 'Vernon, Robert , first Baron Lyveden (1800–1873)', rev. H. C. G. Matthew, *Oxford Dictionary of National Biography*, Oxford University Press, 2004; online edn, January 2008 [www.oxforddnb.com/view/article/25898, accessed 25 February 2015].

206 Explained in, Reid, *Illustrations*, pp. 270–309. 207 PP. 1841 (51), p. 17.

208 Ibid., p. 20.

Figure 4.7 The central ventilation tower of the Palace of Westminster as it stands today

Reid's evidence at the select committees of 1842 and 1844, both established to monitor the work on the permanent building, was produced through experiment. The 1842 committee agreed that Reid's experiments should continue.[209] Reid convinced the committee with his experimental evidence based on 'daily observations'.[210] Via experiment, Reid asserted that he was securing ever more power over the atmosphere. His observations confirmed his belief that steam engines should be used to produce heat to draw air through the Palace.[211] At the 1844 select committee he explained that

[209] *Report from the select committee on ventilation of the new houses of Parliament; with the minutes of evidence*, PP. 1842 (536), p. 3.
[210] Ibid., p. 5. [211] Ibid., p. 12.

efficient ventilation was essential to an efficient legislature. He believed that some bills, such as the 1832 Reform Bill, required time and diligence to be made effective. What Parliament demanded was a system of ventilation that could facilitate 'any particular bill'. As Reid put it, 'The Power of individual Control, by closing or increasing the Introduction of Air at the particular Benches, I consider will facilitate considerably the public Business of both Houses'.[212] Reid's ventilation system was about providing a model of chemical practices replicable beyond Westminster, but also enabling effective government.

Reid's early claims to provide scientific ventilation found support both in Parliament, and in the British press. *The Times* agreed that Reid's science was effective and contributed to a modern legislature. Its only regret was that such worthy work was wasted on MPs. While in Parliament Lord John Russell complimented Reid's ventilating 'powers', *The Times* could not 'refrain from saying that the Doctor's abilities seem to have been sadly thrown away upon the Whigs, for truly a worse lighted or a more ill-aired set have never sat in Parliament'.[213] Early on then, the greatest problem facing Parliament was not the standard of its air, but the perceived competence of its members.

Despite such public support, Reid's work increasingly became a subject of discontent with Barry. In June 1844, Reid complained to the select committee assessing the progress of the building of Parliament, of a dispute with Barry regarding the space beneath the two debating chambers. Reid believed the space was best employed as a reservoir in which to condition air before it entered into the Houses of Commons and Lords. Barry was promoting 'different modes' for this space with suggestions of employing it as a store for Parliamentary records, or even as a 'sub-hall' for horses and carriages.[214] Apart from the noise this would create in the Commons, Reid felt that Barry's 'various claims upon the space' would undermine his 'power' of controlling the temperature and humidity of air entering into the Lords and Commons.[215] Reid asserted this space was a 'scientific concern', rather than 'architectural part'.[216]

[212] *Brought from the Lords, 9 August 1844. Second Report from the select committee of the House of Lords appointed to inquire into the progress of the building of the Houses of Parliament, and to report thereon to the house; with the minutes of evidence taken before the committee*, PP. 1844 (629), p. 69.

[213] (Anon.), 'Sir Robert Peel', *The Times*, (London, England), 18 September, 1841; p. 6; Issue 17779.

[214] *Report from the select committee on Houses of Parliament; together with the minutes of evidence taken before them*, PP. 1844 (448), pp. 26–34.

[215] Ibid., pp. 29–30. [216] Ibid., p. 34.

There had initially been accord between Barry and Reid but by 1843 disagreement was rife. Barry was unhappy with Reid's continually changing demands over the form of ventilation, while Reid disliked Barry's persistent architectural alterations.[217] Reid's reliance on experimental evidence to determine the form of his ventilation proposals meant that his demands changed, as the findings of his tests varied. As a result, the knowledge on which Reid was basing his judgement appeared inconsistent and unstable. With obscured lines of authority between architect and chemist, communication disintegrated as each reworked their plans and intentions without reference to each other. Both men made claims over various sections of the building, and proposed different schemes. In the Lords, Reid demanded a high roof for the gallery which he believed would assist the air circulation of the chamber. Barry preferred a low ceiling for aesthetic qualities, and the 'general effect of the building'. To the commissioners of the Office of Woods, Reid promoted this conflict as a question of architecture versus science. They responded by demanding that Barry, in the interests of 'public advantage', respect Reid's 'professional knowledge'.[218]

Another area of dispute erupted over the fireproofing of the building. Barry's mandate included instructions to employ only incombustible materials in the building's superstructure. Yet Reid's ventilation threatened to create a space throughout the building in which a fire might rapidly spread. When Reid responded angrily to Barry's refusal to prioritize ventilation over fireproofing, and stormed into Barry's office, Barry had two gentlemen take notes covertly. These included some unspecified offensive comments on Reid's behalf, which Barry had copied and circulated among the Peers and press.[219] Barry later took legal action for slander over these 'strong' expressions, but the court found in Reid's favour.[220] Reid felt that Barry and his engineer, Meeson, had acted in concert to ruin his reputation and had forged the minutes of the meeting regarding the fireproofing. Reid believed a court was not adequate to value the concerns he had raised, but rather that his efforts should be judged by individuals he considered to be 'men of science'.[221]

Barry's move to discredit Reid was more than a personal attack, but an assertion that in architecture it was the architect who should have authority. Although Duncannon maintained a low opinion of architects, the Office of Woods, now under the Earl of Lincoln, appointed Joseph Gwilt

[217] Reid, *Narrative*, p. 13. [218] Ibid., pp. 13–14. [219] Ibid., p. 15.
[220] Hugo Reid, *Memoir*, p. 22; on the trial, see (Anon.), 'Court of Common Pleas. – July 6', *John Bull*, (London, England), 13 July, 1850; p. 448; Issue 1544.
[221] (Anon.), 'Court of Common Pleas, Saturday, July 6', *The Times*, (London, England), 8 July, 1850; p. 7; Issue 20535.

(1784–1863) in 1845 to arbitrate between Reid and Barry.[222] An architect himself, Gwilt supported Barry's claims against Reid. It was an opinion which carried weight. Gwilt had over forty years of architectural experience, as well as of engineering works with the sewers of Kent and Surrey between 1805 and 1846.[223] His 1842 architectural encyclopaedia had become a seminal handbook. In it he included substantial sections of Barry's 1839 stone report for Parliament and praised the credibility such an experimental approach held with 'every scientific person'.[224] Like Barry, Gwilt had been an early supporter of the RIBA, attending the institute's second meeting in December 1834. He not only shared Barry's conception of architecture as a science, but lauded the architect's own skill in practically applying knowledge to the problems of construction. The Reform Club, where Barry had overseen the installation of a ventilation system in 1839, powered by a five horse-powered steam engine, impressed Gwilt.[225]

Gwilt was sure that Reid's plans were incompatible with fireproofing the Palace and that Barry should take complete control over the project.[226] He conceded that Reid was a man of 'skill and science', but felt that he lacked knowledge of design and construction.[227] The problem, as Gwilt saw it, was that Parliament was a building without parallel of scale within Europe and that it required the practical knowledge of an architect to bring it to completion. When the Tory naval authority, Howard Douglas, asked how long architecture and ventilation had been 'distinct professions', Gwilt felt it to have been about fifteen years, but stipulated that in both domains it was the architect who should maintain authority.[228] Reid's plans appeared 'incomprehensible' and demonstrated that ventilation should be considered a branch of architectural knowledge.[229] Reid rejected Gwilt's judgement on the grounds that he did not have a true scientific understanding of ventilation. The government appointed a select committee, consisting of

[222] M. H. Port, 'Problems of building in the 1840s', in M. H. Port (ed.), *The Houses of Parliament*, (New Haven, 1976), pp. 97–121, 113, and 116.

[223] Roger Bowdler, 'Gwilt, Joseph (1784–1863)', *Oxford Dictionary of National Biography*, Oxford University Press, 2004; online edn, October 2007 [www.oxforddnb.com/view/article/11811, accessed 4 December 2013].

[224] Joseph Gwilt, *An Encyclopaedia of Architecture, Historical, Theoretical, and Practical*, (London, 1842), pp. 457–78.

[225] John Olley, 'The reform club: Charles Barry', in Dan Cruickshank (ed.), *Timeless Architecture: 1*, (London, 1985), pp. 23–46, 38.

[226] *Reports from the Select Committee of the House of Lords appointed to inquire into the progress of the building of the Houses of Parliament, and to report thereon to the house: together with the minutes of evidence taken before the said committee*, PP. 1846 (719), p. 15.

[227] *Third Report from the Select Committee on Westminster Bridge and new palace; together with the minutes of evidence, appendix, and index*, PP. 1846 (574), p. 37.

[228] Ibid., p. 41. [229] Ibid., p. 43.

three individuals of scientific and engineering experience, to investigate further and assess the confrontation.[230] Philip Hardwicke, George Stephenson, and Thomas Graham reported on Reid's system as a work of practical science, but deemed it too complicated.[231] In response, once again, Reid declared the report to be biased and demanded 'a fair and impartial inquiry' which would consider his success in the temporary Commons.[232] At stake here were questions of Reid's professional autonomy.

The Lords subsequently appointed a select committee, at which Barry changed his approach. Rather than question Reid's credentials as a gentleman, or the practicality of his ventilation system, Barry instead laboured to undermine Reid's claims to be scientific. He blamed Reid for the delays in completing the Upper Chamber before establishing himself as an authority on ventilation. He understood 'the theory of the system', which he felt had 'nothing whatever that is new in it'. Reid's was a scheme which appeared similar to those 'adopted by all scientific persons engaged in the practice of warming and ventilating buildings'.[233] Seemingly the only novelty of Reid's work was the scale and importance of the Palace. Barry believed that ventilating Parliament demanded 'great mechanical skill, a thorough knowledge of the arts of construction, sound judgement and decision . . . in all which attainments and qualities of mind Dr. Reid is . . . most certainly deficient'.[234] Barry excluded any reference to experiment or chemistry in this assertion. In short, he attacked Reid's reputation as someone who practiced useful science.

Barry's attack on Reid found support from Goldsworthy Gurney, himself a practitioner of experiment, who presented evidence undermining Reid's boasts of a practical system of ventilation. Gurney spoke with a 'good deal of practice in the way of experiment' when he declared that Reid's system lacked power.[235] Gurney performed experiments which indicated that heating 1,000 cubic feet of air from 60 to 500 degrees Fahrenheit consumed one pound of charcoal. To draw air through Parliament would demand a temperature of 500 degrees in the central tower. Gurney calculated that if Parliament covered ten acres, then this amounted to 2,393,600 cubic feet of air passing through the system every minute which he believed to require 2,000lbs of charcoal a minute to sustain.[236] Barry reiterated that he himself had the knowledge and skills to introduce a scientific system of ventilation.[237] Interestingly though, this

[230] *Report from the Select Committee on Westminster Bridge and new Palace*, PP. 1846 (177); *Second report from the Select Committee on Westminster Bridge and new Palace*, PP. 1846 (349).

[231] PP. 1846 (574), pp. 172–73. [232] Ibid., p. 174. [233] PP. 1846 (719), p. 9.

[234] Ibid., p. 9. [235] Ibid., p. 24. [236] Ibid., p. 25. [237] Ibid., p. 38.

included maintaining the 'great spaces' Reid had claimed above the ceiling and beneath the floor of the debating chambers.[238] Barry preferred Gurney's advice, having consulted Faraday over his credentials and believing the experimentalist to have avoided looking into a chemistry textbook for twenty-five years.[239] Gurney avoided the constant experimenting that made Reid so unreliable.

The press soon caught wind of Reid and Barry's disagreement and the controversy became public. During the dispute, neither Whig nor Tory ministry gave a firm lead to the work, partly because neither side wanted to take responsibility for something which might end in failure. At the same time neither wanted to abandon the project.[240] To lead it might have been embarrassing, but to abandon it risked condemnation for undermining work which might aid MPs' material well-being. By 1845 this political refusal to remove Reid appeared paradoxical in the press. Prime Minister Peel was lampooned for his refusal to abandon the chemist:

> Peel's patronage of Dr. Reid,
> Is very natural indeed,
> For no one need be told
> The worthy scientific man
> Is acting on the Premier's plan
> Of blowing hot and cold.[241]

Reid's work appeared analogous to the state of British politics. Such political comparisons were part of wider concerns over the direction architecture was heading and the role that science had to play in this. From 1845 the press orchestrated an intense campaign against Reid. Reid believed, probably correctly, that it was Barry who had initiated this public attack. (Figure 4.8)

In June 1844, *The Times* noted that no one, except Reid, really cared about the air of Parliament, other than the Commons and Lords.[242] The engineering press raised early concerns over Reid and Barry's relationship, but dismissed these as a simple question of authority.[243] By 1845

[238] Ibid., p. 40.
[239] 'Letter 3473: Charles Barry to Michael Faraday', (6 July, 1860), in Frank A. J. L. James (ed.), *The Correspondence of Michael Faraday: Vol. 5: Nov., 1855–Oct., 1860, Letters 3033–3873*, (London, 2008), p. 405.
[240] M. H. Port, 'The failure of experiment', in Crook and Port (eds.), *The History of the King's Works*, pp. 209–49, 228.
[241] (Anon.), 'Epigram', *The Times*, (London, England), 2 July, 1845; p. 4; Issue 18965.
[242] (Anon.), 'We have had on our table . . .', *The Times*, (London, England), 3 June, 1844; p. 5; Issue 18626.
[243] (Anon.), 'Mr. Barry – The lords, and the Parliament Houses', *The Civil Engineer and Architect's Journal*, Vol. VII, (London: June, 1844), pp. 217–23, 217.

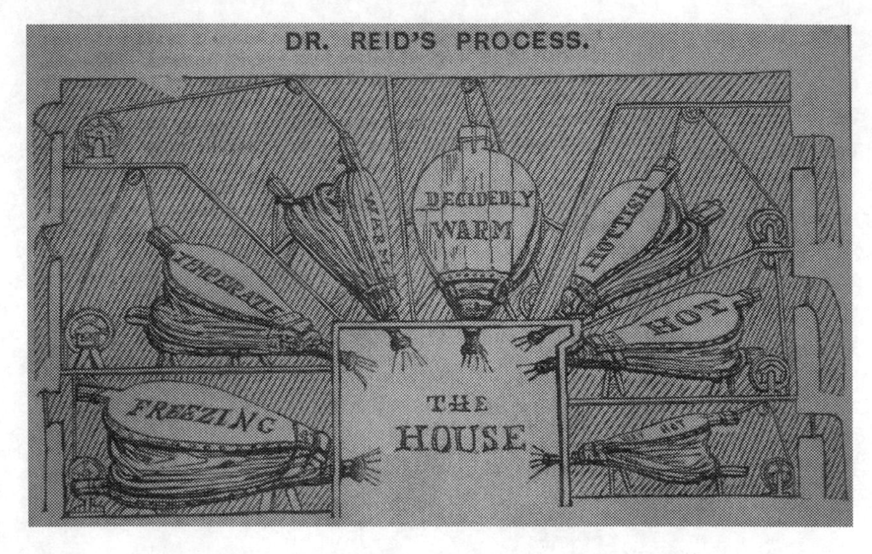

Figure 4.8 *Punch*'s satirical depiction of Reid's ventilation system

this indifference had turned into condemnation. While Reid employed the term 'experiment' as a mark of empiricism, *The Times* deployed the epithet as a derogative. That the Earl of Lincoln sanctioned further experiments, when the building was under pressure to be completed, was construed a monstrosity. The publication believed that a 'more egregious failure than Dr. Reid's "experiments" have hitherto proved' could not be imagined. Correspondents from *The Times*, having sat uncomfortably in the gallery of the temporary Commons, testified to his failings. Reid's experiments were not improvements but 'delusions', and his apparent ignorance of true science demanded scrutiny from 'some men of undoubted science'.[244]

The publication's sudden change from mild praise to utter contempt of Reid's work mirrored Barry's own mounting dislike of the ventilator's schemes. It revealed how the appropriateness of experiment was contingent on the building under investigation; experiment in the permanent was not the same as in the temporary Parliament. Barry and the press believed enough time had been spent producing knowledge and that to be scientific at Parliament involved the referencing of existing knowledge. *The Times* continued this criticism throughout the summer of 1845. After

[244] (Anon.), 'We heard last night . . .', *The Times*, (London, England), 21 March, 1845; p. 5; Issue 18877.

Figure 4.9 Though a hardened veteran of wars in India and the Iberian Peninsula, the hero of Waterloo, the Duke of Wellington was no match for Reid's experiments. *Punch* reported how the Duke was reduced to encasing 'his venerable head in a Welsh Wig' and demolishing enormous bowls of thick gruel to recover

an intensely hot night in the temporary Commons, the publication damned those still 'deluded by his pseudo-scientific pretensions'. With the Commons hotter than 'a chamber at Sierra Leone', Reid's experiments appeared to be endangering the legislature's health.[245] (Figure 4.9)

[245] (Anon.), 'The great ventilator', *The Times*, (London, England), 28 June, 1845; p. 5; Issue 18962.

In select committees Reid cited experiments as evidence, appearing as truthful observations, but beyond the confines of government his methods were characterized as dangerous. This reputation was exacerbated by accounts filtering through into the press of experiments gone wrong. In 1846, while 'Dr. Reid was trying some experiments with wood, as a substitute for other fuel', he managed to set fire to the door of the Commons. It was reported how this mishap almost erupted into a re-enactment of the 1834 fire before immediate action and a fateful supply of water rescued the chamber.[246] Although costing only a door, such experiments damaged Reid's reputation.

Dangerous experimenting was one thing, but when Reid's scientific pretensions began to influence government business, public outcry escalated into a fearsome storm. As with so many careers at Westminster, Reid was to find Irish Catholicism particularly problematic. In 1845 Peel, in a bid to improve relations with Ireland, sought to increase the British government's grant to the Catholic seminary of St Patrick's in Maynooth. This increase from £8,000 to £26,000 annually incurred a prolonged three-day debate during the summer heat.[247] On 5 June Brougham declared that he had had enough. It was intolerable for four-hundred Peers to be packed in a chamber from 10am until 4am every day and be expected to produce efficient legislation. Brougham found support from the Marquess of Normanby who warned his fellow Lords not to blame Reid, but to demand Barry work with increased diligence at Westminster.[248] (Figure 4.10) Brougham believed Barry's assurances were valueless and accused the architect of 'resisting the authority' of the Lords.[249] Both Lord Campbell and the Duke of Wellington supported these sentiments.[250]

Although Brougham was critical of Barry, the press were quick to attack Reid's experiments as the cause of the poor atmosphere within the Lords. *Punch* took up Brougham's concerns and found Reid entirely to blame. Reid's ventilation was not science because it was not practical. In Reid's 'atmospheric catalogue' there was every kind of air 'but one . . . the air of practicability'.[251] Likening Brougham's plight to being 'imprisoned in an exhausted receiver', *Punch* lambasted Reid for treating the Lords as animals and the chamber as a giant air pump.[252] While it was fine to

[246] (Anon.), 'Dr. Reid setting fire to the House of Commons', *The Morning Post*, (London, England), 22 July, 1846; p. 2; Issue 22662.
[247] Port, 'The failure of experiment', p. 225.
[248] House of Lords debate, 5 June, 1845, *Hansard*, 3 Series, Vol. 81, pp. 120–1.
[249] House of Lords debate, 5 June, 1845, *Hansard*, 3 Series, Vol. 81, p. 122.
[250] House of Lords debate, 24 July, 1845, *Hansard*, 3 Series, Vol. 82, p. 1033.
[251] (Anon.), 'Dr Reid's process', *Punch*, Vol. X, (London, 1846), p. 218.
[252] (Anon.), 'Brougham and Reid', *Punch*, Vol. X, (London, 1846), p. 263.

Figure 4.10 Henry Brougham 'imprisoned in an exhausted receiver'

experiment in a temporary House of Commons, to experiment in a permanent House of Lords was considered outrageous. *The Times* echoed these sentiments. Noting analogously that, 'progress with the House of Lords seems to be practically as well as politically difficult', the publication was critical of Reid's infringement on the authority of Barry. Reid's experiments were 'intruding on the province of the architect'. This change in *The Times'* attitude to Reid's work was marked. While once the Conservative leaning journal had praised his experiments, after several years without finding an appropriate system, the paper turned on Reid.

The Times advised its readers to pity Barry when hearing 'that scientific persons have been "assisting" him'.[253] 'Experiments' were the only hindrance to the completion of Parliament, while there was 'not upon the face of the civilized earth a more impracticable set of people than the *savans*'.[254] It was not science as a body of knowledge that was attacked, but science as a method of producing evidence. Referencing knowledge was considered practical, while experiment was portrayed as troublesome. A savant like Reid was to be kept away from the competent profession of architecture and given some unimportant unoccupied building for his experimenting. Reid's orders to Barry were reckoned about as practical as attempting 'to manufacture a moon out of a given quantity of cheese'.[255] In what the *Illustrated London News* asserted to be the greatest architectural work since the fire of London, the only practical mind at work in Parliament was deemed to be Barry's.[256]

After warning that the 'Legislature is evidently not safe in the hands of this aerial Guy Fawkes', *The Times* expressed a distrust of natural philosophers who professed ventilation knowledge. (Figure 4.11) It hoped that if Reid was removed, Goldsworthy Gurney would not be chosen to replace him. Gurney was also a product of 'ventilation mania', favouring continuing experiment, and so was not to be trusted. It was reported how 'Gentlemen with such views as these are not the proper persons to take into consultation on practical matters'. Furthermore, 'until the scientific gentlemen can agree upon . . . a practical atmosphere upon a new plan, the ordinary principle of ventilation . . . shall be applied'.[257] Existing knowledge was to be trusted, rather than dangerous methods for constructing new evidence. According to *The Times* what the public wanted at Westminster was a trusted architect to deliver practical knowledge. Not

[253] (Anon.), 'Poor Lord Brougham . . .', *The Times*, (London, England), 20 March, 1846; p. 4; Issue 19189.
[254] Ibid.
[255] (Anon.), 'The same committee which was appointed . . .', *The Times*, (London, England), 17 August, 1846; p. 4; Issue 19317.
[256] (Anon.), 'National Works', *Illustrated London News*, (London, England), 19 February, 1848; p. 95; Issue 303.
[257] (Anon.), 'Dr Reid has at last found an adherent . . .', *The Times*, (London, England), 27 May, 1846; p. 5; Issue 19247.

Figure 4.11 Reid portrayed as the 'aerial Guy Fawkes', successfully blowing up the House of Commons

the radical suggestions of false natural philosophers who professed to produce new knowledge.

While Brougham's main target was Barry, as delays to the Lords' chamber continued, his enthusiasm for further experiment diminished. All Brougham wanted was a completed chamber. Nevertheless, Reid still found some support in the Upper House. Lord Campbell, who like Brougham had been impressed with Reid's work at the 1834 BAAS meeting in Edinburgh, defended him as 'a man of eminent science'.[258] Lord Sudeley also praised Reid's work in the Commons as the first project of 'systematic ventilation ever carried out' and warned the Peers not to replace the man of science with one 'whose knowledge of the science of ventilation' was unproven.[259] Both Campbell and Sudeley agreed the experiments were 'detestable' but if given time would yield progressive knowledge. However, beyond the discomforts of the Lords, Reid's experiments had publicly lost almost all credibility. Reid's experiments could not be allowed to 'interfere with the practical business of every-day life'. Reid's knowledge of the number of respirations it took to consume a given atmosphere was an amusing 'theory', but hardly aided his boasts of adding twenty per cent to the life-expectancy of legislators.[260] Accompanying this press driven sentiment were concerns raised within scientific circles. At a meeting of the Royal Institution, Faraday, who was assisting Barry in planning his own system of ventilation in the Lords, raised doubts over Reid's experiments and the practicality of his system in the Commons.[261]

While Reid's science struggled for credibility, it is apparent that his own scientific conduct made it difficult for him personally to secure trust. The challenge of transferring Reid's work to Parliament was about much more than questions of authority with Barry. At stake were much wider concerns over scientific method. In Chapter 2, we saw how Barry built Parliament in the context of debates between John Stuart Mill and Whewell over what constituted good science. A great deal of Whewell's criticism of excessive empirical data collecting was aimed at the culture of experimentalism prevalent in 1830s' Edinburgh. In particular, Whewell's writings challenged Brewster and Brougham's commitment to the full-time

[258] House of Lords debate, 24 April, 1846, *Hansard*, 3rd Series, Vol. 85, p. 975.

[259] House of Lords debate, 26 June, 1846, *Hansard*, 3rd Series, Vol. 87, p. 1034.

[260] (Anon.), 'The Lords seem at length determined . . .', *The Times*, (London, England), 6 April, 1846; p. 4; Issue 19203.

[261] (Anon.), 'Dr. Reid's lecture in reply to Dr. Faraday's on ventilation', *The Morning Chronicle*, (London, England), 17 May, 1847; Issue 24200.

pursuit of experimental inquiries.[262] Whewell denounced their obsession with originality and search for truths at the expense of theorizing. Reid displayed a similar enthusiasm for endless experiment and had impressed Brougham and Brewster in Edinburgh. It seems likely then, that for London audiences, Reid's work appeared typical of an Edinburgh scientific framework which Whewell so forcefully condemned. Even Faraday had reservations over the quality of Reid's work. As early as 1831 Faraday received a letter from the chemist Richard Phillips (1778–1851) complaining of Reid's approach to chemistry. Faraday recalled reading this scathing account on a coach to Hastings, finding the portrayal of Reid's experiments so amusing that he left his fellow passengers quite disturbed at his frequent outbursts of hearty laughter. Faraday surmised that Reid produced a few new facts of limited value, but nothing more; no evaluation or analysis.[263] Reid lacked what Whewell would have considered to be the essential quality of imaginative thinking, leading on to scientific theorizing.

Reid's constant experimenting and search for empirical observations was suitable in Edinburgh where men like Brougham and Brewster chased proofs and truth, but in the more diverse setting of Westminster the continual alteration of architectural design subject to experimental findings appeared as a terrifying realization of Whewell's fears.[264] What Barry wanted was not Scottish experiments, but a workable theory on which to practically base his work. He continually showed this throughout the project in relation to ventilation, geology, cast-iron construction, and above all when consulting Faraday. When Barry referenced Faraday he was seeking permanent knowledge, stable and true, for a permanent Parliament. One crucial difference between Reid and Faraday was that while Reid produced chemical knowledge within Parliament, which was itself a place of political performance, Faraday's work went on in the laboratory of the Royal Institution with, as we shall see in Chapter 5, clear spatial divisions between places of knowledge production and display. For Faraday, chemical findings made in his laboratory came into public view once their status as experiments was made stable and their meaning interpreted. Reid on the other hand, in both experimenting and displaying in the House of Commons, appeared unable to clearly distinguish established chemical knowledge from ongoing unstable experiments.

[262] Richard Yeo, *Defining Science: William Whewell, Natural Knowledge, and Public Debate in Early Victorian Britain*, (Cambridge, 1993), p. 91.

[263] 'Letter 515: Michael Faraday to Richard Phillips', (23 September, 1831), in Frank A. J. L. James (ed.), *The Correspondence of Michael Faraday: Vol. 1, 1811–Dec., 1831, Letters 1–524*, (London, 1991), p. 579.

[264] G. N. Cantor, *Optics After Newton: Theories of Light in Britain and Ireland, 1704–1840*, (Manchester, 1983), p. 178.

The types of knowledge Reid produced resembled progressive knowledge which, as Whewell explained, if not handled carefully induced radical instability and dangerous forms of political thought.[265] It is significant then that through depictions in *Punch* and *The Times* Reid was characterized as a Guy-Fawkes-like character that might bring about Parliament's destruction. Radical progressive science threatened the existence of the nation's elected assembly.

Over the next few years the unrelenting campaign against Reid's experiments secured increasing weight. *The Athenaeum* joined in this attack, reporting that in Reid's attempts to ventilate Parliament, 'as in the Niger ships, he has totally and signally failed'.[266] Several anonymous letters appeared in *The Times* praising the paper's efforts in raising public awareness of Reid's work. One supported Barry in noting that both architects and 'Scientific men' considered Reid's system a 'Humbug'.[267] Another letter, signed 'B', declared the press assault to have left 'The Reid ... shaken with the wind' and expected his removal to be imminent. No one trusted his experimenting, 'as he would call it', and Parliament would soon be added to his list of failed ventilation schemes, which already included the Niger expedition ships, the royal yacht, Windsor Castle, and Buckingham Palace.[268] 'A wave or two more' of *The Times*' 'magic pen' and Reid would 'evaporate like a puff of his own foul vapours'. *The Times* was so confident of its power to influence Reid's future that when, after Gwilt's report, three 'scientific gentlemen' were appointed to investigate Reid's system, the publication was sure he would be removed. This appointment was to be a mere formality to tend to the 'vanity' of the chemist. *The Times* believed that Reid would accept being removed if scientific men ordered it.[269] Such judgment would terminate Reid's 'attempt at philosophy'.[270]

The Times was destroying the reputation of not only an Edinburgh experimenter, but of a potentially dangerous political figure. Despite this hostility to Reid, he did manage to maintain his position during the

[265] Laura J. Snyder, *Reforming Philosophy: A Victorian Debate on Science and Society*, (Chicago, 2006), pp. 221, 224.

[266] David Boswell Reid, *Ventilation. A reply to misstatements made by 'The Times' and by 'The Athenaeum' in reference to ships and buildings ventilated by the author*, (London, 1845), p. 21.

[267] 'Antiaeeolus', 'Reid ventilation', *The Times*, (London, England), 25 April, 1846; p. 6; Issue 19220.

[268] 'B', 'Reid ventilation', *The Times*, (London, England), 9 April, 1846; p. 7; Issue 19206.

[269] (Anon.), 'In the course of a discussion ...', *The Times*, (London, England), 27 April, 1846; p. 4; Issue 19221.

[270] (Anon.), 'Dr. Reid has at length ...', *The Times*, (London, England), 13 June, 1846; p. 6; Issue 19262.

late 1840s. Reid provided a sustained defence of his system as well as his reputation. He published an aggressive response to *The Times*, in which he argued that such a journal had little understanding of chemistry or the atmosphere. The popular press lacked 'knowledge', but through 'reckless assertions' had undermined Reid's credibility.[271] Reid believed that only *The Athenaeum's* review was in anyway reputable: the journal, 'in a calm review of a scientific work, of course receives some degree of credit'.[272] However, he endeavoured to show such criticism was misguided. While architecture was a skill of 'ancient days', ventilation was a modern science. Progressive architecture was not to be entrusted to architects like Barry, who in matters chemical displayed 'a deficiency of knowledge', but to men of science like Reid.[273] Architects were to be subservient to those with a chemical understanding of ventilation.[274] Reid's knowledge of the atmosphere and respiration, as well as his observations on the diffusion of gases and links to disease, were all recent discoveries. Reid explained that he really had advanced medical knowledge at Westminster, but that the complaints from members were due to incalculable personal demands, often depending on how much they had eaten or drunk before entering the House. The problem was not the system, but that individual members demanded varying atmospheres at different times. One member might be hot while another was cold.[275] This was a problem without solution; it defied scientific measurement. Reid described how the 'thermometer, so constantly appealed to as a standard of comparison, is of little value as a test of the effect of the atmosphere in communicating the sensation of heat, or that of cold'.[276] Predictably, blaming his inability to secure a comfortable atmosphere on the dietary habits of MPs won Reid few friends within Parliament.

When *The Times* labelled his experiments an 'egregious failure', Reid responded that such trials had provided him with knowledge of how to ensure a healthy atmosphere.[277] Although members and reporters might feel discomfort, Reid adamantly defended the integrity of his method. Reid pointed out that few would trust him or his science if *The Times* kept on printing its critical commentaries. In essence, Reid's argument was that his experiments and system were successful, and that the failings were merely problems of personal preference. Reid subsequently published a weighty catalogue of advocates of his work, which included testimonies from Hawes, Lord Sudelely, Lord Campbell, and architects

[271] Reid, *Ventilation. A reply to misstatements*, pp. 5–6. [272] Ibid., p. 21.
[273] Reid, *Illustrations*, p. xii. [274] Argued in ibid., pp. 70–80.
[275] Reid, *Ventilation. A reply to misstatements*, p. 8. [276] Ibid., p. 10. [277] Ibid., p. 11.

Thomas Brown of Edinburgh and T. Dickson of Manchester.[278] Hawes described Reid's experiments as a 'successful' triumph, while Harvey Lonsdale Elmes, the architect of St George's Hall in Liverpool which Reid ventilated, described Reid's investigations at his 'Chemical Laboratory' in Parliament as progressive.[279] David Stephenson, the Scottish civil engineer, praised the effectiveness of Reid's techniques when employed in lighthouses. What such witness accounts asserted was that Reid could be trusted with the health of the nation's legislature.

Despite this, *The Times* continued to press its claims that Reid's work was inappropriate for Parliament. If indeed Reid did perform scientific experiments, which *The Times* doubted, the new Parliament was just not suitable for such trials. The temporary Houses might be claimed by 'scientific experiment', but to employ the new chambers 'as instruments for advancing the great cause of ventilating philosophy' was egregious. The 'magnificent domicile of a Legislature . . . of the most powerful empire in the world' was not an appropriate place of experiment. It was a place for architectural 'genius' combining artistic inspiration and the successful application of knowledge to building problems, and this Barry provided.[280] The condemnation of Reid's work was enough to lose him control over the ventilation of the Upper Chamber. A separating wall beneath the two chambers was constructed and from 1846 until 1852 two systems of ventilation at Parliament were trialled. In the Lords, Barry drew air in through the Victoria Tower before passing it through water for purification and then forcing it into the chamber via pipes and powered by fans. Air descended into the chamber, but was pre-cooled or warmed in the basement air chamber.[281] Reid's system for the Commons took in air through the Clock Tower before heating or cooling it with water, and then allowing it to enter via holes in the floor of the Commons. This was powered by a current sustained by the central tower and a steam-driven fan.[282]

According to *The Times*, Barry's system of descending air in the Lords brought '"airs from heaven," not "blasts from hell"'.[283] (Figure 4.12) Barry

[278] CRC PS86.8, David Boswell Reid, 'Extracts from official documents, reports, and papers, referring to the progress of Dr. Reid's plans for ventilation', pp. 1, 4, 5, and 7.

[279] Ibid., pp. 4, 7.

[280] (Anon.), 'The new Houses of Parliament', *The Times*, (London, England), 5 April, 1847; p. 3; Issue 19515.

[281] Dale H. Porter, *The Life and Times of Sir Goldsworthy Gurney: Gentleman Scientist and Inventor, 1793–1875*, (Bethlehem: 1998), p. 178.

[282] Ibid., p. 179.

[283] (Anon.), 'The new Houses of Parliament', *The Times*, (London, England), 5 April, 1847; p. 3; Issue 19515; compare this with the architecture of the English electric lighthouse at the 1867 Exposition Universelle which was condemned as unacceptably temporal, in Patricia Mainardi, *Art and Politics of the Second Empire: The Universal Expositions of 1855 and 1867*, (New Haven, 1987), pp. 146–47; thanks to Jane Garnett for this.

Figure 4.12 Honourable Members of Parliament enjoying refreshment from Reid's aerial brewery

was constructed as a trustworthy producer of well-ventilated architecture, as he employed existing knowledge in his designs, referencing respectable authorities, like Faraday. Simultaneously Reid's credibility was deconstructed because his science was not a consistent body of knowledge, but a method, considered untrustworthy and dangerous. While Barry's science appeared solid and consistent, Reid's was apparently ever changing and subject to varying results produced through experiment. Barry's knowledge promised a coherent plan and ordered approach to complete Parliament. Reid's held no promise of completion and risked perpetual disorder at Westminster. For five years architect and chemist ran competing systems as a trial under the scrutiny of a committee featuring authorities in matters of practical science, including Faraday, Wheatstone, and the naval architect, John Scott Russell.[284] When the Commons moved into their new chamber they complained so much of Reid's system that John Manners, then Commissioner of the Board of Works, was compelled to sack the chemist and replace him with Barry's engineer, Meeson.[285] An 1852 select committee reported on Barry's relative success compared to Reid's endeavours. Despite this, concerns remained about Barry's lighting which produced so much heat as to undermine his

[284] *First report from the select committee on new House of Commons*, PP. 1850 (650, 650-II), p. iii.

[285] Porter, *The Life and Times of Sir Goldsworthy Gurney*, p. 179; *Ventilation and lighting of the House committee. Report of the standing committee on the ventilating and lighting the House of Commons*, PP. 1852–53 (570), p. 1.

ventilation scheme.[286] Neither architect nor chemist produced a system which won unanimity. At the suggestion of Barry's work in the Lords being a success, Reid stipulated that Barry had merely imitated his own practices. Reid recalled how in 1838, Barry had visited his experimental lecture room in Edinburgh and made detailed observations of his ventilation practices.[287] Nevertheless such allegations of plagiarism provided scant consolation for the Edinburgh chemist.

Conclusion: A Tale of Two Cities

The challenges of moving science from Edinburgh to Westminster and then from the temporary to permanent Parliament buildings were technical and social, but they were also methodological. Reid's epistemological framework was as controversial as the integrity of his ventilation apparatus. Experimenting in Westminster was very different to experimenting in Edinburgh. While in Scotland Reid found an audience sympathetic to his continual data collecting and commitment to finding new truths, at Parliament his constant experimenting aroused fear and condemnation. To conduct ceaseless experiments in an Edinburgh classroom in the hunt for truth was one thing, but to do the same within the walls of the Palace was quite another. At stake were very different models of what constituted good knowledge. While Barry worked in accord with what he considered to be permanent, stable knowledge, Reid appeared a conveyor of temporal evidence, which might change at any given moment subject to new experimental findings. Barry's permanent Parliament required permanent knowledge. Of course, this is not to say experiment was unacceptable altogether. Such practices contributed to Barry's report on stone and also to the evidence which Faraday provided. However, in each case experiment provided material which supported a fixed body of knowledge which could be referenced once established. Barry helped first to produce the report on stone, and it was then consulted once agreed on. Faraday provided evidence that had been tested at the Royal Institution and then interpreted into stable ideas. Reid provided ideas, but these regularly changed and rarely attracted any kind of wider consensus. As will be shown in Chapter 5 experimental work at Westminster could succeed, but only if managed in a way that preserved enough confidence in Parliament.

[286] *Second Report from the Select Committee on Ventilation and Lighting of the House; together with the proceedings of the committee minutes, of evidence, appendix and index*, PP. 1852 (402), p. vi.

[287] Reid, *Narrative*, p. 25.

It should come as little surprise that *The Times*, Conservative in its sympathies, was the main critic of Reid. Reid was committed to an Edinburgh programme of experiment which, particularly for Whewell, carried dangerous political implications. Knowledge made through continual experiment for practical applications was typical of an Edinburgh epistemological model which Reid epitomized. As already shown, both these attributes were hallmarks of the radical science of John Stuart Mill. If we place Reid within this experimental Edinburgh programme then it becomes apparent that his work was not only scientifically controversial but politically radical. It is not too much to suggest that when Barry undermined Reid's reputation, or Reid appeared in *Punch* or *The Times* as a nineteenth-century Guy Fawkes threatening to destroy Parliament through his experiments, he was being characterized as a politically dangerous figure. While radical politicians wanted reform and an end to the established political order, Reid, through his ventilation schemes, was literally threatening to do this. In his hands, science became a means to radical ends. Utilitarians might talk of reforming Parliament and ending aristocratic privilege but Reid actually risked the physical integrity of the Palace of Westminster and the health of MPs. At Parliament science appeared a radical force, not just for its content, but politically. The Edinburgh enthusiasm for experiment seemed capable of doing what no amount of popular unrest could.

5 Enlightening Parliament: The Bude Light in the House of Commons and the Illumination of Politics

Light, seeking light, doth light of light beguile.

Shakespeare, *Love's Labour's Lost*

In May 1850 *The Lady's Newspaper* reported on a lecture at the Royal Institution entitled 'the philosophy of "a lamp"'. In the presence of Prince Albert, along with a distinctly non-specialist audience, Michael Faraday (1791–1867) delivered an account of the latest developments in artificial lighting. Particular attention was paid to the recent practice of using oxygen gas to create increasingly bright lamps for lighthouses and streets. Faraday drew the audience's attention to the application of this new practice to the interior of the House of Commons. *The Lady's Newspaper* informed readers that,

> The combustion of the carbon is more perfectly accomplished by the aid of pure oxygen gas, the effect of which in increasing the brilliancy of flame was exhibited. It was by this plan that Mr. Goldsworthy Gurney lighted the House of Commons with what was termed the Bude light.[1]

Gurney's illumination at Parliament had been the culmination of over a decade of experimenting in the Commons, during which he had endeavoured to create a system of healthy lighting for members. It had been a demonstration of how science could bring optical order to the place of British government. Yet it was also a claim for what sort of science was suitable to guide the construction of the Houses of Parliament. Faraday's explanation of Gurney's light at the Royal Institution, and its subsequent description in a publication focused at a female readership, demonstrates that not only were new powerful means of creating light of interest to broad Victorian audiences, but that the controlled lighting of the Commons was itself worthy of attention. This chapter shows how the application of Gurney's light at Westminster, and the government's use of men of science as authorities to guide this project, unfolded within the context of controversial new understandings of how light operated.

[1] (Anon.), 'Royal institution', *The Lady's Newspaper*, (London, England), 18 May, 1850; p. 2, Issue 177.

Chris Otter has shown that technologies of artificial lighting were often, especially for liberals, intrinsically associated with notions of progress: to turn night into day was the pinnacle of reordering nature.[2] His detailed study of Victorian optics argues that illumination engendered scientific and political-cultural conceptions of 'modernity' and connects a history of vision with a history of freedom and liberalism. He provides an analysis of lighting which is simultaneously a history of politics. Illumination was symbolic of political, as well as scientific, enlightenment. This understanding of the importance of light in the nineteenth century heightens the validity of studying the interior illumination of the Commons as a place where science and politics were entangled. With light associated with modernity, finding a reputable system of lighting was constructed as essential to creating a modern and efficient Parliament. With the pre-1834 Parliament often criticized for lacking social vision and resisting political change, and concomitantly ridiculed as a place of physical darkness, Britain's new legislature was to have brilliant light to accompany the nation's newly reformed political system.

This chapter is about how the space of Parliament was conscripted into political and scientific schemes but it is also about how the rebuilding presented a unique moment for scientific individuals to enhance their own authority. So far we have seen how different forms of science were imported to Westminster, how building Parliament could stimulate new bodies of knowledge, and how Parliament presented challenges for those seeking to employ science in its construction. In this section I look at the use of science in Parliament's construction as part of wider political and scientific controversies. Parliament's building was a chance for a variety of protagonists to demonstrate their principles. The introduction of gas lighting involved a range of differing agendas. Joseph Hume believed efficient powerful illumination was important to securing a modern efficient legislature. However, the scheme also represented a chance to demonstrate the utility of experiment in a manner consistent with the writings of fellow Utilitarian John Stuart Mill. Gurney and his key supporter, Faraday, saw Parliament as a place to fulfil their own experimental programmes, while David Brewster (1781–1868) recruited the work into his own highly contested understandings of how light operated. Brewster's influence in the scientific world was waning by the late 1830s, however, providing evidence before a Parliamentary select committee was for him a way of re-establishing his credibility and salvaging his career.

[2] Chris Otter, *The Victorian Eye: A Political History of Light and Vision in Britain, 1800–1910*, (Chicago, 2008), p. 2.

Just as Reid found with his ventilation, issues of trust and context were important in the selection of Gurney, but this chapter is different because unlike Reid, who constantly struggled to build trust at Parliament, Gurney was trusted before he introduced the Bude Light at Westminster. While Reid laboured to transfer his work from Edinburgh to Westminster and struggled to have his experiments accepted in the permanent buildings, Gurney owed his appointment to Hume with whom he had already secured trust through his work on lighthouses during the preceding decade. In part one, this relationship is explored. Why did the government choose Gurney, Faraday, and the Bude Light at Westminster, rather than traditional candle lighting or alternate gas illumination apparatus? I then examine the proceedings of the 1839 select committee to investigate the lighting of Parliament and show how Hume, as chairman of this committee, interpreted Gurney's work as a trustworthy approach to internal illumination. Much of Gurney's credibility with Hume was due to his cooperation with Faraday. In contrast to Reid's ventilation which faced a constant battle to secure trust, an influential network promoted the Bude Light.

Parliament was therefore used in a different way. Not as a place to manufacture trust and knowledge, but rather as a place to emphasize a political agenda. Hume relied on the authority of Faraday because this secured Gurney credit, but he sought to place the work in a wider context. Lighting Parliament was about showing how experimental practices were useful and that Parliament could be politically enlightened. However, this raised different problems to those which Reid faced. Parliament as a chance to promote a certain approach to science or politics had the potential to backfire. Using the space in this way opened it up to a range of interpretations. Brewster enrolled Gurney's work within his own optical research, therefore implicating it in larger debates over how light should be investigated. Yet to use science to build the assembly credibility risked other radical interpretations, in which the Bude Light appeared as a dangerous analogy for demonstrating Parliament's lack of integrity.

Lighting Sea and State

In 1839 Barry wrote to Lord Duncannon requesting that he appoint an assistant who was knowledgeable on the subject of lighting.[3] The urgent concern for Barry was providing lighting in the House of Commons, which had traditionally been lit by wax candles. Yet these produced heat, irregular light, often dripped on members, and were liable to be

[3] Dale H. Porter, *The Life and Times of Sir Goldsworthy Gurney: Gentleman Scientist and Inventor, 1793–1875*, (Bethlehem: 1998), p. 169.

disrupted by attempts to ventilate the chamber. Furthermore, Barry's ornate ceiling created increased shadow. To solve these problems Barry and Duncannon wanted a trustworthy authority to direct the establishment of a system of gas lighting for the Palace.[4] They envisaged the chamber illuminated with cutting-edge apparatus which would offer regular powerful lighting, as an alternative to the practice of candle illumination.

Despite Reid's attention to lighting during his work on ventilation, the Board of the Office of Woods and Forests pursued the advice of Joseph Hume, who recommended the lighthouse illumination authority, Goldsworthy Gurney (1793–1875). Hume had chaired a lighthouse committee which involved examining Gurney's work for Trinity House and, in particular, his new light. Gurney had developed a lamp which combined an oil flame with an injected stream of oxygen. The 'oxygen' or 'Bude' light, on account of where Gurney lived and experimented, caused the flame to burn exceedingly bright.[5] Gurney explained how the introduction of oxygen to the interior of a flame transformed a usual light into an exceptional light. A common oil flame burnt only in its exterior, but by introducing oxygen to the interior, it caused burning there too.[6] Instead of 'common air', Gurney's lamp burnt oxygen created by heating manganese and channelling this through a tube to beneath the flame. Gurney asserted that manganese could be obtained in sufficient quantities from the mines of Devon and Cornwall.[7] Working at his home in Bude he consulted local Cornish mine proprietors to secure a cheap supply of the element. Gurney claimed his system was easy to manage, economical, and safe.

Born near Padstow and educated at Truro Grammar School, Gurney had as a boy witnessed the mining engineer Richard Trevithick (1771–1833) performing experiments, which made a lasting impression.[8] Interested in chemistry, Gurney trained as a doctor before practising as a surgeon in London from 1820. In 1822 Gurney secured appointment as lecturer in chemistry and natural philosophy at the Surrey Institution, where he had time to perform experiments on heat, electricity, and various gases including the production of oxygen from black oxide of

[4] Denis Smith, 'The building services', in M. H. Port (ed.), *The Houses of Parliament*, (New Haven, 1976), pp. 218–31, 229.

[5] Joseph Hume, *Report from the select committee on lighting the House; together with the minutes of evidence, appendix and index*, PP. 1839 (501), p. 6.

[6] Ibid., p. 1. [7] Ibid., p. 2.

[8] G. B. Smith, 'Gurney, Sir Goldsworthy (1793–1875)', rev. Anita McConnell, *Oxford Dictionary of National Biography*, Oxford University Press, 2004 [http://ezproxy.ouls.ox.ac.uk:2117/view/article/11764, accessed 21 September 2013].

manganese.[9] After his 1823 publication, *The Elements of Science*, Gurney experimented with steam power for road transport and after enduring great financial loss returned to Bude.[10] There Gurney devoted himself to experimenting on lighting, eventually presiding over the North Cornwall Experimental Club, which paid particular attention to questions of agriculture and mining.[11] Along with experiments on reflecting light from one lamp through his house using a series of mirrors, Gurney worked on applying chemical compounds to burning flames to generate increased illumination. Gurney's lighting apparatus was very much a product of this industrial Cornish context, where there was a local scientific culture which emphasized rudimentary experiment and observation as practices for agricultural and mining improving. Yet in transferring his work to Westminster, Gurney moved his local provincial knowledge to a national scene.

Despite growing up among Tory gentlemen and Anglican clergy, Gurney became increasingly associated with liberals, radicals, and Whig reformers. On hearing of the 1834 lighthouse commission, Gurney wrote to Hume, drawing attention to his development of a light incorporating an oxyhydrogen blowpipe.[12] The description of the Bude Light impressed Hume's committee and it recommended Trinity House employ Gurney to improve lighthouse illumination nationally. Since 1514 the responsibility for many of Britain's lighthouses fell to the Corporation of the Elder Brethren of Trinity House. It was for this Corporation that Gurney conducted most of his experiments which later were to underpin his promises to illuminate Parliament. From 1836 until the 1839, he worked alongside Faraday, the 'Scientific advisor in experiments on lights to the Corporation', to develop a light powerful enough for general lighthouse use.[13]

Hume received copies of Faraday's reports for Trinity House regarding the Bude Light which reported that Gurney's lighting could produce 140 times the power of the standard Argand lamp presently used in lighthouses.[14] This meant that at the same time that Hume was raising

[9] T. R. Harris, *Sir Goldsworthy Gurney, 1793–1875*, (Penzance, 1975), p. 19.

[10] Ibid., p. 22; on Gurney's steam carriage, see Crosbie Smith and Ben Marsden, *Engineering Empires: A Cultural History of Technology in Nineteenth-Century Britain*, (Basingstoke, 2005), pp. 132–33.

[11] (Anon.), 'North Cornwall Experimental Club', *The Cornwall Royal Gazette, Falmouth Packet and Plymouth Journal*, (Truro, England), 31 October, 1845; Issue 4176.

[12] Porter, *The Life and Times of Sir Goldsworthy Gurney*, p. 138.

[13] 'Letter 884: John Henry Pelly to Faraday', (4 February, 1836), in Frank A. J. L. James (ed.), *The Correspondence of Michael Faraday: Volume 2, 1832–December 1840, Letters 525–1333*, (London, 1993), p. 332.

[14] Porter, *The Life and Times of Sir Goldsworthy Gurney*, p. 139.

concerns over the form of the new Parliament, he was also reviewing the results from these Trinity House experiments. While he had been increasingly alarmed at the poor light experienced in the Commons, Hume simultaneously witnessed a demonstration of how science could provide powerful new means for illumination. Crucially this research for Trinity House was conducted in a way which met with Hume's approbation. Hume appreciated Faraday's emphasis on experiment over theory; he preferred to obtain the

> results of a practical nature under the form of experiments, which might supply direct evidence to the observer, than to trust altogether to deductions from acknowledged principles . . . [Faraday] was thus led insensibly into very numerous experiments.[15]

Comparing the Bude Light with candles and other lamps, Faraday tested practices of lighting in his laboratory using an artificially manufactured mist. He measured each light by employing tracing paper and a small cylindrical chamber. Faraday observed the layers of tracing paper required to keep out the light of a candle, producing 'a certain constant, and standard degree of obscurity' equal to one candle.[16] This paper created the effect of cloud or fog. He made lenses out of the paper for the chamber of between one and fifty sheets. Faraday suggested that this provided a means of accurate measurement of light power. Applying paper lenses until light was completely obscured allowed Faraday to calculate the relative 'penetrating power' of the Bude Light and its competitors. Faraday adapted this apparatus several times, fearing that 'the eye could not be trusted alone as the judge of equal degrees'.[17] He attempted to remove human error from his examination of the Bude Light to enhance its claims of superiority over alternate lighting apparatus.

In the Royal Institution's theatre, Faraday performed experiments using this self-made light meter, and presented the results to Trinity House, before Hume inspected them. (Figure 5.1) By recording light readings at a distance of twenty-eight feet, Faraday reported a standard Trinity House Argand oil lamp, patented in 1780, required twenty-seven degrees of paper to be completely obscured. The Bude Light registered a more impressive thirty-two degrees.[18] Although Faraday recognized this was a very minimal increase in penetrative power for lighthouses at sea, it did suggest lighting superiority. Faraday trialled an Argand, equivalent to

[15] 'Letter 4884 (2:943a): Faraday Report to Trinity House', (Oct., 1836), in Frank A. J. L. James (ed.), *The Correspondence of Michael Faraday: Vol. 6, Nov., 1860–Aug., 1867, undated letters, and additional letters for Volumes 1–5: Letters 3874–5053*, (London, 2012), pp. 645–59, 646.

[16] Ibid., p. 647. [17] Ibid., p. 649. [18] Ibid., p. 654.

Figure 5.1 Michael Faraday delivers one of his celebrated Christmas lectures in the lecture theatre of the Royal Institution. Permission from the Royal Institution, via the Science Photo Library, 2017

roughly six candles, and the Bude Light over a period of forty-nine days and found Gurney's to produce the light of 20.86 standard Argands.[19] This was at a cost increase of just two pence more than a standard Argand. In October yet another report followed, citing David Brewster as an authority in the establishment of Gurney's light for Britain's lighthouses.[20] Interestingly by the time Gurney was trialling his light in the House of Commons, Faraday was replicating his experiments performed at the Royal Institution off the coast of Orford Ness where there were two lighthouses. Faraday placed the Bude Light in one lighthouse and an Argand in the other. He then rowed out to sea at night and compared the intensity of each lamp in a scaled-up experiment of those performed in London.[21] Hume appreciated such experiment: it was empirical observation which promised to provide applications of utility. The produce of

[19] 'Letter 4894 (2:1057a): Faraday Report to Trinity House', (15 January, 1838), in James (ed.), *The Correspondence of Michael Faraday: Vol. 6*, pp. 673–78, 673.

[20] 'Letter 4898 (2:1114): Faraday Report to Trinity House', (29 October, 1838), in James (ed.), *The Correspondence of Michael Faraday: Vol. 6*, pp. 683–94, 694.

[21] 'Letter 4908 (2:1190a): Faraday Report to Trinity House', (12 August, 1840), in James (ed.), *The Correspondence of Michael Faraday: Vol. 6*, pp. 698–708, 701–02.

Faraday's experiments suited Hume's Utilitarian demands of improvement.

When describing the Bude Light to the 1839 committee, Gurney cited this work for Trinity House along with Faraday's 1838 lighting report as empirical evidence for the practicality of the device. Hume thus promoted the Bude Light for Parliament, confident that the method of illumination was weightily supported by experimental trials. By recommending Gurney implement his light apparatus in the Commons and perform experiments to adapt and perfect its application, Hume was claiming that the work of Faraday and Gurney, which Trinity House commissioned, could be replicated and advanced at Westminster. After proposing Gurney's light to the members of the Commons in 1839, including the Speaker of the House and the Chancellor of the Exchequer, Hume arranged for Gurney to trial the Bude Light at Parliament, securing funding from the Treasury for experiments.[22] The same practices lighting up Britain's oceans of darkness would enlighten the home of British political representation.

For Hume, the importance of science in securing the progressive integrity of Parliament was paramount. Hume's political views, explored in Chapter 1, which stressed 'progress' and utility, were the sort likely to make him sympathetic to Gurney. Gurney embraced the Whig cause of gradual political reform. Yet importantly, Hume had an understanding of natural science consistent with Gurney's experimental approach. Hume studied natural science as a boy before reading medicine at Edinburgh University in the 1790s. As shown, Edinburgh was a place where experimental chemistry was projected as a practical discipline, in a manner not dissimilar to that in which Gurney worked in Cornwall.[23] By the 1830s Edinburgh was at the heart of an 'empire of science', including Whigs like Brougham and experimentalists like Brewster.[24] The cultural values of this Edinburgh network shaped Hume's promoting of Gurney.

Hume's Utilitarian counterparts included James Mill, and he had been one of the tutors of James' son, John Stuart Mill. In this network, the organization of science was an urgent concern. Hume's student, in his *System of Logic*, explained the reasoning principles he deemed essential for

[22] Charles Lemon, *Report from the select committee on lighting the House; together with the minutes of evidence, and appendix*, PP. 1842 (251), p. 9.

[23] On eighteenth-century Edinburgh chemistry, see Jan Golinski, *Science as Public Culture: Chemistry and Enlightenment in Britain, 1760–1820*, (Cambridge, 1992), pp. 11–49.

[24] Steven Shapin, '"Nibbling at the teats of science": Edinburgh and the diffusion of science in the 1830s', in Ian Inkster and Jack Morrell (eds.), *Metropolis and Province: Science in British Culture, 1780–1850*, (London, 1983), pp. 151–78, 152.

the progress of science.[25] In examining the importance of experiment, Mill cited Hume's, and his father's, old Edinburgh University tutor, Dugald Stewart, noting that experiment was vital to investigate nature, revealing phenomena through observation and ascertaining the order of the Universe. The basis of all knowledge, Mill contended, was induction, which he defined as the operation of discovering and proving general propositions.[26] Merely observing was inference, but to observe and then propose a law, was induction: for Mill the predictive function of induction was its great utility.[27] Mathematics was deduction, which could still be valuable if produced from known facts, but experiment was exceptionally useful for discovering natural phenomena.[28] Mill distinguished observation from experiment by suggesting the latter was 'an immense extension of the former'. Experiment allowed the observation of laws concealed in nature, such as the isolation of a gas like oxygen.[29] Therefore to experiment was to isolate a part of nature. Mill was not averse to hypothesis and deduction, believing them to be important forms of human imagination.[30] Hume's understanding of experiment was not identical to Mill's, but both men were at the heart of British Utilitarianism in the 1830s. John Stuart Mill's work is helpful because it provides a Utilitarian interpretation of experiment in science. Brougham and Brewster were decidedly not Utilitarians, but it is intriguing how over the question of illumination in Parliament, they shared Hume's support of experimental work.

As a Utilitarian, Hume valued the material improvement which Faraday and Gurney promised. As a graduate of Edinburgh, and Fellow of the Royal Society alongside Brougham and Brewster, he appreciated their experimental methodology. Hume's constant recommendation and approval of the Bude Light for both lighthouses and Parliament was not a reaction to a self-evidently progressive technology, but a culturally shaped response to work invested with values which Hume trusted. Faraday and Gurney were not just authorities because they did experiments, but because Hume considered their work credible. However, it would seem that Gurney was an acceptable compromise character. His emphasis on experiment evidently pleased Edinburgh alumni like Hume, but it was not as alienating to Barry as Reid's work. As already shown Barry was far happier trusting Gurney's opinion and considered his knowledge more

[25] John Stuart Mill, *System of Logic Ratiocinative and Inductive: Being a Connected View of the Principles of Evidence and the Methods of Scientific Investigation*, (London, 1886), p. 6; on Hume and John Stuart Mill, see Don A. Habibi, *John Stuart Mill and the Ethic of Human Growth*, (London, 2001), p. 8; for context, see Laura J. Snyder, *Reforming Philosophy: A Victorian Debate on Science and Society*, (Chicago, 2006), p. 206.

[26] Mill, *System of Logic*, pp. 185–86. [27] Ibid., p. 191. [28] Ibid., pp. 187, 250.

[29] Ibid., pp. 249–50. [30] Ibid., p. 322.

reliable. Unquestionably this was because of Gurney's association with Faraday. Barry was sure that Gurney was much preferable to Reid.[31]

Hume's choice of Gurney was part of his wider ambitions to ensure Parliament's buildings embodied scientific learning and became an enlightened place of governance. He believed Gurney and Faraday's inductive experimental methodology was the correct way to secure this reputation. To some degree this projection was interpreted in this way. Soon after Gurney's appointment at Parliament, the *Mechanics' Magazine* noted this analogous transferal of practices from lighthouses to government. After witnessing an evening exhibition in which Gurney illuminated the Commons by Bude Light for the first time, the journal explained how, 'The light is brought into the House by catoptric [reflective], and diffused by dioptric [refractive] principles, so that while it is exceedingly brilliant and effective, it is soft and pleasant'.[32] After Gurney performed experiments of diffusing the light through various media, including a particularly spectacular use of crystal octahedron facets, combined with prisms, the commentator acknowledged the transferal of science from the nation's lighthouses, to its Parliament: both were important concerns for the reading public of the world's foremost mercantile nation in the politically reformed climate of the late 1830s. It was understood that the Bude Light, 'which may be regarded as a new era in the science of illumination, is due to the public spirit of the Elder Brethren of Trinity House, at whose expense the Bude light has been perfected for the greater security of our merchantmen and navy'.[33] Gurney's science was protecting the tools of Britain's safety and commerce, and now promised to secure the illumination of the nation's new bastion of reformed representative government.

The Politics of Light

Hume invited Gurney to trial his light in the temporary Commons. He was to experiment in private when the House was empty, but also to use his experiments as performance in illuminating evening debates. This immediately provoked concern. It was not just the quantity of light which was at question, but its quality and safety. In April 1839 the Tory MP for Scarborough, Frederick Trench, demanded 'a report be obtained from

[31] 'Letter 3473: Charles Barry to Michael Faraday', (6 July, 1858), in Frank A. J. L. James (ed.), *The Correspondence of Michael Faraday: Vol. 5: Nov., 1855–Oct., 1860, Letters 3033–3873*, (London, 2008), p. 405.

[32] (Anon.), 'The new light in the House of Commons', *Mechanics' Magazine*, (London, England), 11 May, 1839, Vol. XXXI, p. 96.

[33] Ibid., p. 96.

competent scientific authorities, whether the adoption of the oleo-oxygen Bude light may not expose this House and its Members to the dangers of explosions'.[34] Trench recalled how the Donovan lighthouse lamp had exploded, blowing off the roof of the building.[35] Trench favoured maintaining the status quo of candles, fearing that gas lighting might emit poisons. The Bude Light was 'dangerous, disagreeable, and inconvenient', while Gurney's claims of economy were inappropriate: 'where the health and comfort of the Representatives of the people of the British Nation is at stake, I think then those petty questions of comparative economy paltry and contemptible'.[36] Lighting involved more than illumination, but the well-being of the nation's elected legislature. Trench's concerns matter because what they show is that the adoption of Gurney's Bude Light in 1839 was a radical, rather than obvious, choice. If it was accepted that Parliament should be built as a scientific edifice, then precisely what that science entailed was unclear. For Hume, Gurney and Faraday's work signified improvement, but for Trench it presented a massive risk.

Within days of Trench's complaints, Hume sought Faraday's advice. Hume was keen that Gurney should have a 'fair opportunity of making his experiment', despite the fears raised, and wanted Faraday's opinion as to the safety of the Bude Light for illuminating Parliament.[37] Faraday replied that he felt the experiment safe and attached a report he had made on Gurney's lamp in 1838.[38] Faraday's opinion carried weight. When it came to questions of experiment, he was a most trusted authority for the government throughout the 1830s and Gurney's lamp was typical of how Faraday could be referenced in experimental concerns. Within central government's expansive collecting of information, Faraday was frequently requested to offer guidance.[39]

Faraday was born in Surrey, the son of a blacksmith belonging to a small Christian sect known in England as the Sandemanians.[40] As an apprentice bookbinder in London, Faraday attended four of Humphry Davy's lectures in 1812, before Davy, as a lecturer of the Royal Institution, employed him as his amanuensis. From 1813 Faraday served as a laboratory assistant at

[34] (Anon.), 'Parliamentary business', *The Morning Post*, (London, England), 22 April, 1839; p. 3, Issue 21310.

[35] PP. 1839 (501), p. 69. [36] Ibid., p. 72.

[37] 'Letter 1167: Joseph Hume to Faraday', (29 April, 1839), in James (ed.), *The Correspondence of Michael Faraday: Volume 2*, p. 581.

[38] 'Letter 1179: Faraday to Joseph Hume', (4 June, 1839), in James (ed.), *The Correspondence of Michael Faraday: Volume 2*, p. 587.

[39] Geoffrey Cantor, *Michael Faraday: Sandemanian and Scientist. A Study of Science and Religion in the Nineteenth-Century*, (Basingstoke, 1991), pp. 154–55.

[40] Frank A. J. L. James, 'Faraday, Michael (1791–1867)', *Oxford Dictionary of National Biography*, Oxford University Press, 2004; online edn, January 2011 [http://ezproxy.ouls.ox.ac.uk:2117/view/article/9153, accessed 21 September 2013].

the Institution, working closely with Davy, who in 1820 became President of the Royal Society. Davy provided Faraday with an apprenticeship in chemistry. In May 1821, Faraday was appointed acting superintendent of the house of the Royal Institution, before earning fame for his work on electromagnetic rotation. By the end of the decade he had a reputation as a meticulous experimentalist, before establishing celebrity in 1831 for his *Experimental Researches in Electricity*.[41] During the 1830s Faraday's research output was industrious, being a frequent reference in civic science concerns. He became increasingly known for his ability to communicate experimental principles to his audiences, which apart from the government included the wider public, along with Queen Victoria and Prince Albert.[42] Among the subjects in which Faraday demonstrated his authority was light, conducting lectures accessible to audiences including children.[43]

Geoffrey Cantor has demonstrated how central, as a Sandemanian, Faraday's religious values were to his life, shaping his understanding of science. Faraday made his confirmation of faith in 1821 before marrying into a leading Sandemanian family and was, by the 1840s, an elder of the church.[44] Sandemanianism envisaged a literal interpretation of the Scripture and taught that life should be lived through Christ's example.[45] It emphasized neutrality from worldly politics but encouraged loyalty to the crown and rule of law.[46] Ignoring politics conversely entailed civic responsibility to the state; public duty was consistent with his Sandemanian convictions.[47] This commitment to the state and rejection of party politics made Faraday an ideal employee for the Royal Institution. Faraday endeavoured to make the Institution's lecture theatre into a place where scientific authority could be witnessed. Living above the theatre and performing research in the laboratory beneath it, Faraday used the Institution to hold public displays in which he replicated experiments previously enacted. He established the Institution as a public space where scientific advice could be obtained easily.[48]

[41] Geoffrey Cantor, David Gooding, and Frank A. J. L. James, *Michael Faraday*, (New York, 1991), pp. 12–13.
[42] Ibid., pp. 16, 5.
[43] See Michael Faraday, *The Chemical History of a Candle: A Course of Lectures Delivered by Michael Faraday*, (London, 1904).
[44] Geoffrey N. Cantor, 'Reading the book of nature: the relation between Faraday's religion and his science', in David Gooding and Frank A. J. L. James (eds.), *Faraday Rediscovered: Essays on the Life and Work of Michael Faraday, 1791–1867*, (Basingstoke, 1985), pp. 69–81, 69.
[45] Cantor, *Michael Faraday*, p. 6. [46] Ibid., p. 155.
[47] Ibid., p. 160; compare with, L. Pearce Williams, *Michael Faraday: A Biography*, (London, 1965).
[48] David Gooding, '"In nature's school": Faraday as an experimentalist', in Gooding and James (eds.), *Faraday Rediscovered*, pp. 105–135, 108.

In Faraday's lecture theatre, natural phenomena could be observed seemingly free from human action. Faraday appeared to make 'experiments transparent': audiences witnessed nature working, rather than experimental practice, as Faraday replicated original research from the laboratory to the theatre.[49] By showing experiment through public displays, he appeared to allow nature, or at least an audience's observing of nature, speak for itself. Faraday laboured to show his work to be honest and free of human manipulations. Experiment was research, but could also be performance. Such demonstrations were claims that natural phenomena, such as electromagnetic rotation, were self-evident rather than humanly constructed. As Forgan has noted, in the 1820s and 1830s, Faraday's experimental spectacles provided a rare source of scientific reference in London and made the Institution a customary place where government could consult on scientific concerns.[50] Faraday provided credible experimental knowledge to a public audience. Combined with his moral obligation of loyalty to the state and avoidance of financial gain, this made Faraday the government's favoured chemical authority.

Faraday's practice of replicating experiments through displays has consequences for our understanding of his recommendation that Gurney's lighting was safe. In writing to Faraday, Hume sought the advice of someone experienced in experimenting. In backing the Bude Light, Faraday asserted that Gurney's experiments were conducted in an appropriate manner. He claimed that the light's ability to illuminate the House of Commons safely was not mere hypothesis, but proven through trial. Faraday was trusted because his approach to experiment, along with regular public displays, was considered to be honest reporting of natural phenomena. Faraday had first-hand experience of the oxygen light through his reports for Trinity House. In light of the concerns surrounding Gurney's lamp, a select committee superintended Gurney's work at Westminster. Hume was appointed chair and publicly cited Faraday's endorsement. In a sense, choosing Gurney to experiment, on Faraday's recommendation, before agreeing to install the Bude Light permanently, echoed Faraday's own approach at the Royal Institution. Experiments that were performed in private could secure trust if repeated publicly. This process was similar to Reid's approach of performing experiments in his Edinburgh classroom and then replicating this work in front of MPs at Westminster. Gurney's addition of oxygen to a gas lamp was also to be replicated and displayed before MPs, before permanent adoption. It was

[49] Ibid., pp. 107, 106.

[50] Sophie Forgan, 'From servant to savant: the institutional context', in Gooding and James (eds.), *Faraday Rediscovered*, pp. 51–67, 57.

important that the natural phenomenon of increased light from burning pure oxygen was witnessed.

In response to Trench's criticisms, Hume's 1839 committee to superintend Gurney's illumination of the Commons invoked experimental knowledge. Both Gurney and Faraday were examined over their experiments on the Bude Light, before an assortment of 'eminent scientific and practical men' were questioned; Hume's committee thus emphasized both induction and utility.[51] Faraday reported to the committee how Trinity House favoured Gurney's lamp above all alternatives as it provided 'remarkable steadiness for eight hours', could be managed by 'a carpenter of ordinary ability', and was from Faraday's 'own careful experiments', proven to be cost effective.[52] It was a quality light, both of suitable quantity and safety. He explained that as the Corporation's 'servant' he was preparing to employ the light at Orford Ness. Faraday and Gurney asserted that their work for the Corporation was 'the results of a great many experiments made with the object of getting out the real truth; one or two experiments may vary, so as not to arrive at the truth'.[53] Importantly, during the proceedings of this government committee, it was deemed prudent to accurately define a link between truth and experiment. Gurney and Faraday thus defined truth as an observation which could be consistently made through systematically repeated experiment. Through such truth, knowledge of how to achieve controlled illumination could, they contended, be obtained: 'The adjustment of the supply of fuel, and then of the oxygen, enables you to have a great command over the lamp'.[54]

Along with Faraday and Gurney, Hume's committee assembled a selection of witnesses to support the work within the Commons. Neil Arnott, prominent in Benthamite circles, felt the Bude Light would 'serve perfectly the purpose of lighting The House', despite warning to ensure the bright light was not within the eye-line of members.[55] He suggested a white roof for the chamber should be trialled, as well as an orange tint to the lamp. He was confident that the light would be a 'great improvement'.[56] Dr Andrew Ure, the Scottish chemist and author of an article on light in the *Dictionary of Chemistry* (1820), made experiments on the Bude Light in his own house the night before, reporting it to be a triumph of illumination. Using Wheatstone's recently designed photometer, which measured the intensity of a light, he compared shadows

[51] PP. 1839 (501), p. iii; on science as a utility, see Robert Bud and Gerrylynn K. Roberts, *Science versus practice: chemistry in Victorian Britain*, (Manchester, 1984), in particular pp. 11, 165.

[52] PP. 1839 (501), pp. 7–8. [53] Ibid., p. 9. [54] Ibid., p. 8. [55] Ibid., p. 9.

[56] Ibid., p. 11.

produced from Bude and Argand lamps, as well as candles.[57] He concluded that the Bude Light was thirty times more powerful than candles and three times more so than an Argand. Ure's sole reservation was that the experimental light was poorly positioned in the Commons. Set in the centre of the roof, it produced the effect of a 'tropical sun'.[58] By trialling side lighting, Ure believed Gurney might replicate 'our own sun' in the chamber; something less dazzling for members. After Ure the Committee questioned Wheatstone, who had witnessed Ure's experiments with his photometer and confirmed the superiority of the Bude Light.[59] University College London's ex-professor of natural philosophy, Dionysius Lardner (1793–1859), supported these claims of the Bude Light's superiority, but echoed calls for further experimenting on the light's positioning. (Figure 5.2) He emphasized the importance of using the Commons for such tests, rather than a small room, noting that 'the nearer you can bring the circumstances of the experiment to the circumstances to which it is to be practically applied, the better'.[60] Together these authorities presented a consensus of understanding that experiment was what was suitable to solve the problems of illumination in Parliament. What is most interesting, however, is the comparison between Gurney's support and Reid's. While the place of Reid's experiments was contentious, with individuals such as Robert Smirke raising doubts over the appropriateness of using the Commons for experiments, there seems to have been accord that the only place where Gurney could work to establish a lighting system worthy of Parliament, was inside the building itself.

Trench remained unconvinced and attempted to discredit the evidence of the committee. Although he felt Faraday's opinion to be greatly respected, Trench believed the evidence of Wheatstone and Ure to be absurd and poorly substantiated, relying on a new and unproven photometer.[61] Even if the Bude was safe it would only take 'a drowsy Servant', responsible for the lighting, to cause disaster. Claims contrary to this were, Trench believed, the assertions of the 'friends of the Bude', who he was sure were all Whigs and radicals.[62] Trench portrayed the light, and the experiment substantiating the claims of its efficaciousness to be political, not in terms of party, although Trench was a Tory, but in terms of personality. Trench complained that the committee had been 'attended only by those Gentlemen who were its [the Bude Light's] avowed Favourers' and that Hume and Henry Warburton had undermined Trench during his questioning of the witnesses.[63] Indeed of the committee, Thomas Acland, Charles Lemon, and William Molesworth

[57] Ibid., p. 15. [58] Ibid., p. 17. [59] Ibid., p. 20. [60] Ibid., p. 29.
[61] Ibid., pp. 69–70. [62] Ibid., p. 71. [63] Ibid., pp. 72–73.

Figure 5.2 Gurney's Bude Light as installed to illuminate Trafalgar Square

were all MPs of constituencies local to Bude. In short, Trench endeavoured to show that the committee had been driven not by science, but by politics, and in doing so he questioned the authority of the examined witnesses and the experiments they cited. Of Gurney's work, Trench felt it unlikely that 'his experiment is likely to make converts of the men of Science'.[64] By men of science, he referenced David Boswell Reid, George Birkbeck, and William Brande, whom Trench felt to be unbiased regarding the Bude Light. He demanded Brande be called and 'subjected to Cross-examination of the Friends of the Bude Light' and cited Brande's publicly stated concerns surrounding the light. In doing so, Trench attempted to discredit the committee by showing it as selectively examining only those sympathetic to Gurney's work and pursuing a political, rather than scientific, agenda.

Experiment was not simply reasoning, or revealing nature's working, but a form of dialogue and display in securing approval for Gurney's illumination of the Commons. It held cultural and political significance. To address Trench's concerns, the committee concluded that Gurney should continue to experiment in the Commons by repositioning and trialling new lenses and glass covers for the light. The concerted message of the Whig dominated committee was that Gurney's light was credible because of experimental evidence, and that to secure its suitability for Parliament required yet more experiment. Political values and specific notions of what it was to be scientific shaped this choice. Being experimental and seeking scientific authority was politically and socially contingent.

Interpreting the Bude Light

As seen with William Smith's promotion of geology, Barry's conceptions of professional architecture, and Reid's plans for practical chemistry, the use of science at Parliament could fit within varying agenda. For Hume, Parliament's lighting was part of a wider project of improvement which included Reid's ventilation and, less successfully, a change of site. For Gurney, Parliament presented an opportunity to showcase the practical nature of his work. However, it was not just those involved in the internal selection of the Bude Light who utilized Parliament's illumination to their own ends. What is most surprising about the 1839 investigation into Parliament's lighting was the selection of David Brewster as a witness. For many, Brewster was a troublesome figure. He was the epitome of the Edinburgh experimental approach to science which aroused such concern from William Whewell and seemed so dangerous in the hands of Reid. Brewster considered himself a marginalized figure and so his selection for the 1839 committee presented a rare chance to reclaim some scientific authority in wider society. This section examines how beyond

[64] Ibid., p. 74.

Parliament, protagonists recruited the Bude Light to fit within their own discourse and agenda. Gurney's apparatus could be placed within wider intellectual debates surrounding light, but could also be manipulated to produce politically scathing accounts of British politics.

In the late 1830s Brewster was an isolated figure who felt his ideas unvalued and his work maligned by a powerful group of Cambridge men of science. On no subject were the differences between Brewster and Whewell greater than light. They contested what light was, how it travelled, and how it could be controlled. Since the eighteenth century, Newton's 'corpuscular' theory, which held that light was made up of particles (corpuscles) travelling in straight lines, dominated optics.[65] However, during the early nineteenth century this view, also known as 'projectile' theory, came under increasing criticism for its apparent failure to explain the phenomena of light diffraction, polarization, and interference. Critics of corpuscular theory contended that light was better explained through the mathematical work of Augustin Fresnel (1788–1827).[66] A French military engineer, Fresnel's optical papers between 1815 and 1827 posited that light was a wave. In Britain a heated debate over the acceptance of wave, or undulatory, theory raged through the 1820s and 1830s.[67] Light as a wave was mathematical, it was theoretical, and it was problematic. Such a wave appeared to obey no laws of dispersion, while its movement suggested the existence of a medium to travel through which could not be clearly defined.[68] At stake was not just the question of light, but also wider notions of scientific methodology. Newtonian natural philosophy asserted induction as the means of true scientific discovery; deductions were to be taken from observations absent of pre-experiment hypothesizing.[69] Wave theory however, appeared deeply hypothetical and apparently untestable.

In Britain the Cambridge mathematicians George Biddell Airy, Whewell, and John Herschel (1792–1871) made the loudest appeals for wave theory to be accepted.[70] Their theory had 'the same claims' as the theory of gravitation, 'namely, that it is certainly true, and that, by mathematical operations of general elegance, it leads to results of great interest'.[71] Leading the case for the projectile theory of light against the Cambridge mathematicians, were

[65] G. N. Cantor, *Optics After Newton: Theories of Light in Britain and Ireland, 1704–1840*, (Manchester, 1983), p. 3.

[66] Ibid., pp. 147–58. [67] See ibid., pp. 159–72.

[68] Ibid., p. 167; also see Geoffrey Cantor, 'The changing role of Young's ether', *The British Journal for the History of Science*, Vol. 5, No. 1 (June, 1970), pp. 44–62.

[69] Geoffrey Cantor, 'The reception of the wave theory of light in Britain: a case study illustrating the role of methodology in scientific debate', *Historical Studies in the Physical Sciences*, Vol. 6 (1975), pp. 109–32, 109.

[70] Cantor, *Optics After Newton*, p. 174.

[71] George Biddell Airy, *On the Undulatory Theory of Optics*, (London, 1866), p. vii.

the Edinburgh experimentalists David Brewster and Henry Brougham.[72] Brewster argued that wave theory relied on a luminiferous ether which was completely speculative. He instead provided explanations for the phenomenon of light drawn from induction, maintaining that light was a projection of particles. Brewster experimented on light absorption and concluded that if two rays, identical accept in wavelength, entered a medium, only for one to be transmitted and one absorbed, this disproved wave theory.[73] Light would have a material difference which Airy or Fresnel's mathematics could not explain. At the 1842 and 1845 BAAS meetings, Brewster presented his results which challenged Airy's mathematical theory, and from 1849 to 1853 he assisted Brougham in delivering papers at the Royal Society, BAAS, and Paris Académie, refuting wave theory.[74] Unlike Brougham however, at the heart of Brewster's science was a Calvinistic conviction that experiment was the sole means of exposing the truth of God's world. To observe God's creation was honest discovery, while to hypothesize risked the sin of pride by speculating on what God might have designed.[75]

Of course, mathematical models of light were not free of experimental investigations, but Brewster was eager to portray mathematical approaches as ignorant of experimental observations.[76] After a Church of Scotland education, Brewster had attended University of Edinburgh from 1794 and began optical experimenting in the early 1800s. A reform Whig, Brewster was a committed advocate of Newtonian induction. He had become a close friend of Brougham while at University and it had been Brougham who first advised Brewster to investigate optics in 1798. Both Brougham and Brewster upheld Newton's maxim of *hypotheses non fingo*, rejecting conjectural theories.[77] Importantly both philosophers were the last prominent fellows of the Royal Society to remain committed to corpuscular theory against what they regarded as absurd hypothetical wave theory.[78] Furthermore, like Hume, both had attended University of Edinburgh in the 1790s and endorsed the experimentalist manifesto, the *Edinburgh Review*. Through the 1830s Brewster's observations on the structure of the eye accompanied his work in establishing the BAAS. Yet by the 1833 BAAS meeting in Cambridge, many condemned Brewster for his refusal to accept optical wave theory.[79]

[72] Cantor, 'The Reception of the Wave Theory of light in Britain', p. 111.
[73] Cantor, *Optics After Newton*, p. 168. [74] Ibid., pp. 176–77. [75] Ibid., pp. 178, 181.
[76] Ibid., pp. 130–31, 180.
[77] G. N. Cantor, 'Henry Brougham and the Scottish methodological tradition', *Studies in History and Philosophy of Science*, 2, no. 1 (1971), pp. 69–89, 73.
[78] Ibid., p. 82.
[79] A. D. Morrison-Low, 'Brewster, Sir David (1781–1868)', *Oxford Dictionary of National Biography*, Oxford University Press, 2004; online edn, October 2005 [http://ezproxy.ouls.ox.ac.uk:2117/view/article/3371, accessed 21 September 2013].

Brewster was convinced that the attack on his experimental research was a question of nationality.[80] It seemed to him that the mathematical Cambridge faction targeted his work because he was Scottish, believing Whewell omitted Scottish writers in his *History of the inductive sciences*.[81] Brewster was indeed a target of Whewell's writings. Whewell wanted universities to emphasize moral and mental training, while Brewster's commitment to originality was a direct challenge to this image of science. Brewster's investigations, like Reid's, were the sort that if introduced to untrained students could have radical consequences. To teach students to be original and question academic authority might lead to them challenging the broader authority of the nation's political institutions. At the same time Whewell disliked Brewster's adoration of technical applications of scientific knowledge.[82] Both Whewell and Airy regarded Brewster with open contempt, and after he failed to secure a teaching post at the University of Edinburgh in 1833, Brewster's influence seemed diminished. In the same year, he saw his role within the BAAS disintegrate, despite being crucial in establishing it in 1831. Then in 1841 he had a paper rejected from the Royal Society, which of course he attributed to the anti-Scottish Cambridge faction, and it was not long before he excluded from the society's publications altogether.[83]

Brewster found his situation incredibly painful, feeling that he had dedicated his life to honest laboratory labours. It is also clear that Whewell offended Brewster's Calvinism. Whewell's speculating on what might be, namely the existence of an ether for light to travel through, presented an affront to God by arrogantly ignoring what could be observed. Whewell seemed like a dogmatic speculator, committing the sin of pride, while being part of a group who excluded Brewster's honest investigations.[84] With this background, the selection of Brewster as a witness for the lighting of Parliament was dramatic. It is remarkable that he was selected given these intellectual tensions. However, it seems that while he was increasingly isolated in scientific networks, for political audiences he was still a valued opinion. After all he was the inventor of the kaleidoscope. Brewster's public image was still positive enough that he could be called before Parliament as a respectable witness. Yet given his diminished scientific influence, it is also apparent that he viewed the committee as a unique opportunity to try to claw back some authority. A

[80] See David Brewster, *The Martyrs of Science, or the Lives of Galileo, Tycho Brahe, and Kepler*, (London, 1841).

[81] Cantor, *Optics after Newton*, p. 175.

[82] Richard Yeo, *Defining Science: William Whewell, Natural Knowledge, and Public Debate in Early Victorian Britain*, (Cambridge, 1993), pp. 91–92, 225.

[83] Cantor, *Optics after Newton*, pp. 175–77. [84] Ibid., p. 181.

strong performance could help re-establish his credentials and perhaps win him some much-needed credibility. The chance that illuminating Parliament presented was one that Brewster was desperate to take advantage of, and he did his best to exert his views over the work which Gurney conducted.

Before Hume's 1839 committee, Brewster claimed to have discovered the principle of Gurney's method of lighting several years previous. He recalled recommending the application of the 'oxygen Gas light to Oeconomical purposes' at a committee in 1832 investigating northern lighthouses. In private correspondence to Faraday, Brewster lamented that his role in creating the Bude Light, as with so much of his work, had been sidelined.[85] He had performed experiments 'in general successful' on the use of oxygen and endorsed the practice. That the light produced by Gurney was perhaps too bright was not a problem; via 'experiment' the correct number and position of burners would be found.[86] From his own experiments he explained how one single light would be best, as induction showed that two sources would irritate the retinas of members. Having found that a retina rendered itself insensible when placed under a source of light, Brewster asserted that a second source of light would create a potentially damaging 'new centre of insensitivity'.[87] Two sources of light in the House would produce damage to MPs' retinas; 'no eye' would stand such strain for 'any length of time'. Brewster warned that the 'structure of The House is unfit for the distribution of light', and advised that the shape of the roof should be changed, the walls and ceiling painted white, the floor kept light, and white seats be installed for members.[88] This was an extremely radical proposal, but even if not made permanent, Brewster was keen that such a white chamber should at least be trialled. Brewster stipulated that he understood optics; he knew the effect of light on the retina and how it varied for the left and right eyes. Brewster surmised that,

> the present mode of lighting The House is absurd, and such as no person at all acquainted with the physiological action of light on the retina, and the principles of its distribution, could have adopted.[89]

Brewster's advice was that Gurney continue experimenting, trialling alternate lamp positions, coloured filters, wall and roof materials and angles, white calico on the walls, and furniture colours. The Edinburgh Whig experimentalist believed that these were the correct means of

[85] 'Letter 1387: David Brewster to Faraday', (15 March, 1842), in Frank A. J. L. James (ed.), *The Correspondence of Michael Faraday: Vol. 3, 1841–Dec., 1848: Letters 1334–2145*, (London, 1996), p. 55.

[86] PP. 1839 (501), p. 12. [87] Ibid., p. 13. [88] Ibid., p. 14. [89] Ibid., p. 14.

securing suitable lighting for Parliament. Instead of merely presenting evidence to the committee, Brewster seized the chance lighting Parliament presented by proposing a bold scheme for a very different looking chamber.

The committee's other witnesses agreed with Brewster's sentiments. Ure endorsed the philosopher's evidence and agreed that Brewster's claims of retina damage had experimental credibility. Both candles and multiple sources of Bude lighting would be injurious to vision; as the 'luminiferous focus of irradiation ought not to be seen by the eye', Ure recommended light be reflected into the House.[90] Crucially, Faraday consented with Brewster's approach to optics. Although Faraday's views on wave theory were undecided in the 1830s, he did suppose light was probably a substance rather than a wave at the time of the committee.[91] What really mattered though was that Faraday believed experiment to be the most suitable means of acquiring optical understanding. He felt any explanation for light's movement should be based not purely on mathematical calculations, but on observations of light's properties. He agreed with Brewster's claims that reflected light did not behave in accordance with Fresnel's wave theory and disliked the reliance of the theory on hypothesis and mathematical theorems.[92] Regardless of whether light was wave or matter, optical theory should, he believed, derive from experimental observations: induction was 'supreme'.[93]

Only Charles Barry raised doubts as to the appropriateness of Brewster's prescribed experiments and these were made on architectural, rather than scientific, grounds. Barry was concerned that Gurney's light produced too much gloom due to the decorative ribbed roof of the Commons. When Hume proposed Brewster's suggestion of trialling a plain smooth ceiling, Barry was ardent that it would 'not be desirable in point of taste that the ceiling should be plain and colourless'.[94] As with Reid's ventilation, when Barry was threatened with the prospect of Edinburgh style experiments at Westminster, he reacted defensively. For Barry, the nightmare scenario of having not just Reid experimenting in the Commons, but Brewster as well, was too much. Barry advised that the light be lowered and warned Hume to curtail such fancies of experiment. Despite this, between 1839 and 1842 Gurney's work at Parliament, performed with Faraday and Brewster's endorsements, typified this experimental approach to optical theory. The illumination of Parliament was a demonstration of the practicality of

[90] Ibid., p. 18.

[91] Frank A. J. L. James, '"The optical mode of investigation": light and matter in Faraday's natural philosophy', in Gooding and James (eds.), *Faraday Rediscovered*, pp. 137–61, 139–40.

[92] Ibid., pp. 144, 149. [93] Ibid., p. 154. [94] PP. 1839 (501), p. 11.

experiment. The credibility of experiment was in part attributable to its two-dimensional character. Experiment was private trial and observation, but could also be public performance. Adding pure oxygen to a flame was a test, but its replication in front of an audience in the Commons was also a display. Such an exhibition could be problematic, but could also secure credibility. Gurney's displays fit within the context of early nineteenth-century discourse on the roles of hypothesis and induction. Brewster's involvement with the lighting, and Hume and Brougham's concern with making Parliament scientific both through lighting and ventilating, reveals a propensity to associate experiment with scientific progress at Westminster within this network of Edinburgh alumni.

Gurney's subsequent work adopted some of Brewster's recommendations. In May 1842, a select committee reviewed Gurney's endeavours and presented a description of the project. Evaluating the 'scientific value and originality of his Inventions', it reported on three years of continual trial and observation.[95] Charles Lemon led the committee, which included Benjamin Hawes, John Russell, and Trench. It stated that at first

> Mr. Gurney tried various experiments to enable him to light The House by illuminating the whole of the sloping ground-glass roofs, from the outside; and passing the Light generally through the whole surface. The effect was pleasing, but the Light lost by passing through so large a field, rendered the expense too great to be persevered in.[96]

After this Gurney tried to establish seven centres of light on each side of the roof, 'but this experiment was abruptly terminated by a Vote of The House; and the use of candles was again had recourse to'.[97] Despite members immediately rejecting any further trials considered harmful or inconvenient, a motion was carried on a division that 'Mr. Gurney's experiments should be resumed'. Gurney responded by trialling the recommendation of the 'scientific men' examined by the 1839 committee, 'that pendants below the roof should be lighted and the Light broken down and softened by ground glass'. Hume felt Gurney's established illumination to be a successful vindication of internal gas burners and moved in the House 'That the mode of Lighting then in use be continued, and that the arrangements which were at that time experimental should be perfected'.[98] This motion was withdrawn, yet the 1842 committee felt the lighting met with a 'very general expression of approbation'. As a result, Gurney was permitted to carry on his experimental work and illumination of the Commons, in the government's pay, until March 1843.

[95] PP. 1842 (251), p. 3. [96] Ibid., p. 2. [97] Ibid., p. 4. [98] Ibid., p. 5.

One problem Gurney did encounter was securing a supply of pure oxygen at Westminster. Manganese proved difficult to obtain cheaply beyond the mineral riches of Cornwall, and so Gurney trialled new practices and apparatus to create illumination 'brilliancy'. He applied

> the knowledge which he had derived from experiments, commenced in 1822, and repeated in 1839 ... aimed at the conclusion, that an equally powerful Light ... could be obtained by the combustion of coal gas, purified by a process of his own, in combustion with definite proportions of atmospheric air.[99]

Gurney was confident in this practice, but felt it disadvantageous to remove the use of pure oxygen altogether. Nevertheless, he believed this 'result of a long series of experiments' was suitably powerful for the Commons. Gurney passed the gas through a vessel 'containing certain chemical salts' which removed much of its ammonia and sulphur. Tests showed that this purification caused the coal gas to burn with greater intensity at a lower temperature. Gurney trialled various coal gases to attain a brighter light, comparing common London coal gas, with Glasgow and Edinburgh gases. He found the latter two to contain less sulphur and produce a more intense flame.[100] A source of oxygen was kept on standby to increase the light intensity when desired.

After inviting a group of MPs, including the Speaker of the House, to witness experiments with this method, Gurney cited Faraday's approval.[101] The 1842 committee lauded the success of this scheme and endorsed its extension into the libraries and lobbies of Parliament. The committee asserted that Gurney's work was successful, and that because of his experimental evidence, 'All the scientific gentlemen examined concurred ... that a great increase of Light may be obtained by burning a given quantity of gas, according to Mr. Gurney's method'.[102] Gurney explained how the gas light system, with oxygen as a reserve, worked consistently well through the 1840, 1841, and 1842 sessions. He believed that within the Commons, he had observed the true principles of powerful lighting. Brightness was inherently dependent on the amount of atmospheric air consumed, and where exactly this was applied to the flame. Gurney informed the committee that this knowledge allowed him completely to understand how to regulate intense illumination: 'the importance of this accuracy was not suspected until it came out in experiments'.[103] Gurney requested Wheatstone perform independent 'photometric experiments' in the crypt of the Commons, comparing the light intensities of a wax candle, a common gas-light, and a modified atmospheric Bude Light. Wheatstone did so and defended Gurney's techniques, noting that whereas a gas light

[99] Ibid., p. 5. [100] Ibid., p. 13. [101] Ibid., p. 12. [102] Ibid., p. 6. [103] Ibid., p. 15.

emitted the equivalent of 10.36 wax candles, the new Bude Light projected the light of 68.39.[104] Wheatstone's photometer suggested that for four times the fuel consumption, the Bude Light produced over six times the light.

On concluding his evidence, Gurney aptly placed his experimentally attained observations in the context of Brewster's own research on using oil lighting and polyzonal lenses in lighthouses, as well as Fresnel's work in French lighthouses; both had trialled lenses made of a series of sections to produce intense light with a reduced lens thickness.[105] After Gurney's evidence, the committee interrogated a medical pupil of Gurney's, who had studied under him in 1822, William Keene. Keene recalled how it was Gurney's practical demonstrations of inflamed coal gas combined with oxygen made to students which had initiated Gurney's conception of the Bude Light; experiments made to his students two decades previous had shaped the practices of illuminating Parliament. Yet when Charles Lemon asked if the experiments in the Commons had been 'necessary' to advance the development of Europe's premier light, Keene felt that there could be 'No doubt of it'.[106] Gurney's experimental programme appeared to have been vindicated.

Employing the Bude Light in the Commons in the context of this intellectual controversy not only shows how individuals in government, like Hume, sought scientific knowledge for Parliament, but provides an important example of how the space of Parliament could itself be enrolled as part of wider controversies. The choice of Gurney was a statement that regardless of new theories of how light operated, what Hume wanted was experiment. Poor lighting was so closely associated with the old pre-reformed Parliament that to have gas lighting appeared to signal a radical break with the past. The Bude Light was, for Gurney and Hume, a symbol that the new Parliament was different from that of 1834. Where the old Parliament governed in darkness, the future was a legislature in glorious light.

Nevertheless, using Parliament in this way was dangerous. Morrell and Thackray have shown that in the political uncertainty of the 1830s and 1840s, nature was used to secure political authority. Political commentators sought to maintain stability by showing how economic and political order was consistent with order in nature, such as geological stratification. As Morrell and Thackray put it, 'To the politician and theologian science became a means of bolstering those of their claims which could be understood in terms of the natural or ordained place of man'.[107] I have shown in

[104] Ibid., p. 25. [105] Ibid., p. 17. [106] Ibid., p. 27.

[107] Jack Morrell and Arnold Thackray, *Gentlemen of Science: Early Years of the British Association for the Advancement of Science*, (Oxford, 1981), pp. 30–33.

Chapters 1, 2, and 3 that this was true at Parliament. Using science through geology, architecture, and chemistry were ways of building government an image as enlightened. The Bude Light, however, demonstrates that this could be a risky strategy. Parliament was a very public space and although access to the Commons was restricted, what happened at Parliament was in the public eye. So, to do anything at Westminster which involved science opened it up to external interpretation. The result of this was, in the case of the Bude Light, rather ironic. Instead of securing universal praise as a symbol of enlightened governance, it provided an unfavourable comparison with the government's daily conduct. It brought the apparent failings of MPs into sharp focus.

While Hume envisaged the lighting as a beacon of reformed government, publically the Bude Light was developed into a powerful medium for highlighting government's, and indeed society's, failings. In the summer of 1841 Melbourne's Whig government fell from power and Peel formed a new Conservative ministry. During the same summer, an anonymous observer launched a satirical journal entitled *The Bude Light.* (Figure 5.3) This publication, although only running from June to August, portrayed itself as an analogy of Parliament's new internal lighting and claimed to be inspired by its eponym. It bombastically proclaimed that 'In the Bude Light we present to society the great new lustre of the time – the freshest and most enduring flame that has of late risen from the crucible of genius – the mighty illuminator . . . the burning interpreter, and expositor, and expounder of the world's secrets'.[108] It offered to be a literary light conveying 'rays of truth, censure, and satire', and intended to illuminate 'the whole surface of society'.[109] It boasted that it had a truly 'enlightened publisher'.[110]

Readers were informed of the analogy of introducing light into Parliament, with the Bude Light becoming a symbol of ending the privileged position of the Lords and Commons. While it had long been known that few could 'storm the breach' of privilege and enter Parliament, it was also understood that 'the "Bude Light" should be introduced into Parliament; and now it is there, and has latterly been blazing away in the very heart, centre, and focus of M.P.ism'.[111] Illumination in Parliament was not just about optics, but actually revealing the conduct of MPs in their bastion of privilege. The satirical journal appeared to have no political affiliation, pouring scorn equally on Whigs, Conservatives, and indeed all society, while having little sympathy for popular movements like the Chartists who sought political reform.[112] At the end of Melbourne's

[108] (Anon.), *The Bude Light: A Social, Satirical, Farcical, Fashionable, Personal, Political, Musical, Poetical, Attical, Dramatical, Tart, Smart, Courting, Sporting, Literary, Skiterary, Monthly Illuminator,* (London, 1841), p. 7.

[109] Ibid., pp. 10–11. [110] Ibid., p. 201. [111] Ibid., pp. 16–17. [112] Ibid., p. 223.

Figure 5.3 An Enlightened Publisher: The *Bude Light* journal took full advantage of the satirical potential of Gurney's scheme. Reproduced by kind permission of the Bodleian Libraries, the University of Oxford. (Reference Per.2706 g.8, page 201)

ministry, *The Bude Light* ridiculed the government's increased expenditure and inability to manage duties on sugar, corn, and timber. As for Queen Victoria, it lampooned her inability to conduct basic arithmetic, claiming she could not comprehend the concept of fixed duties or understand constitutional questions.[113] Recalling the scandal of the Queen's refusal to remove any of her ladies of the bedchamber sympathetic to Melbourne's government in 1839, which had meant Peel refusing to take office, *The Bude Light* joked that as Melbourne's administration dissolved in the summer of 1841, so too had the Queen.[114] It likened Melbourne to a vender of gin, believing him to have lived 'by the "crown"' for too long. As the 1841 election raged, *The Bude Light* declared the Whigs to be 'In a minority, In a mess, In doubt, In debt . . . and In-significant', while Peel's Conservatives were 'Out of luck, Out of temper . . . and, at present, are Out a canvassing'.[115]

[113] Ibid., pp. 42–45. [114] Ibid., pp. 220–21. [115] Ibid., pp. 135–6.

Nor was the journal constrained to politics, considering the army, navy, church, and legal system to all be well within its purview. All established society was open to its gaze, including 'trade, science, art, music, and the theatres'.[116] Sometimes its observations were erratic. For example, it accused Whewell of little more than possessing an unpronounceable name, noting that it was 'more easily *whistled* than pronounced'.[117] This peculiar sense of humour found some sympathetic audiences, with its inspiration from Gurney's work not lost on readers. While one advertisement celebrated its assessment of the Whigs in Parliament, another felt its name to be most appropriate, observing that 'the superior brilliancy of the Bude Light is produced by introducing oxygen *into* the flame, so the Editor of this little Miscellany seeks to add to our enjoyment by turning all things to humorous account – wit being the oxygen of its intellectual atmosphere and the follies of the day the carbon for combustion'.[118] In typical fashion, *The Bude Light* included its own readers as subjects for amusement. In August 1841, it reckoned that its account of British society was so scathing that women would be shocked, claiming that Mrs Backbite, who desired a book of scandal, had stayed up reading all night, 'Till she burnt her own nose with the flaming "Bude Light"'.[119] Another female reader, described only as a milliner or mantua-maker, and judging from her handwriting 'to be addicted to excessive tea-drinking', was lampooned for her interest in 'socialism'. Of this political creed, the journal claimed that no one knew its meaning, but maintained its own purpose was to mock and make humorous the conduct of all polite society.[120]

Ultimately the journal's claims of 'audacity' were short lived. However, it is important that it took an apparatus heralded as a sign of reformed government and turned it into an analogy for political incompetence. It matters because it shows how dangerous the use of science at Parliament could be. Politicians might control who and what they chose to build Parliament, but they could not control how the works were interpreted. Embracing science in government so as to appear enlightened was a process which carried no little risk. Displaying science in Parliament's construction could present society with an allegory of government, but what the lesson of this would be was open to interpretation. For Hume modern lighting was clearly a sign of a reformed legislature, but for *The*

[116] Ibid., p. 5. [117] Ibid., p. 211.

[118] (Anon.), 'Notices of the press', *The Musical World: A Journal and Record of Science, Criticism, Literature, & Intelligence, Connected with the Art*, Vol. XVI – New Series, Vol. IX, (1 July, 1841), p. 16; "J. H. F.", 'New books', *The Mirror of Literature, Amusement, and Instruction*, Vol. 37, No. 1062, (5 June, 1841), p. 367.

[119] (Anon.), *The Bude Light*, p. 253. [120] Ibid., pp. 287–88.

Bude Light it served to bring the government's perceived lack of ability into comparison with the integrity of the science which illuminated its deliberations. *The Bude Light* claimed to dispel misconceptions and gossip from politics with a 'legitimate focus of all brilliancy'.[121] It provided an imagined political voice to Gurney's invention, turning the Bude Light into an eye within the walls of government.

Conclusion: A Parliament of Vision

It is tempting to see the illumination of the Commons as analogous to the lighting of lighthouses. Hume's choice of Gurney was a deeply metaphorical conception: if the product of Gurney's experiments could illuminate expanses of dark ocean and guide ships through treacherous paths and stormy seas, then those same experiments could light up the Commons and navigate the nation's legislators through political turmoil and progressive reform. For the radical Edinburgh educated doctor, such induction based lighting was essential to guide the body in its endeavours through the storms of social change. As shown in Chapter 1, a chief concern with the pre-1834 Parliament, apart from its resistance to political reform, had been its poor lighting. Through Faraday's guidance, Brewster's reference, and Gurney's lamp, the House of Commons was constructed as a realization of the principles of induction and experimental optics. Among the controversies of early nineteenth-century optical theory, Parliament's lighting was a confident assertion of the practical, even Utilitarian, nature of experiment. The Bude Light epitomized this, but through further trials promised still greater illumination. At stake was not only an experimental programme, but a distinct political agenda in which the physical structure of the nation's legislature employed the latest apparatus for improved efficiency and health.

The Bude Light's adoption in Parliament was part of political and experimental agenda. Its use at Westminster attracted attention and this meant interpretation. Unlike Reid, who saw the credibility of his experiments rapidly evaporate like the smoky air of one of his public demonstrations, Gurney received greater support. Although he employed a very similar experimental approach to Reid, combining private trial and public performance, he was at the centre of a powerful network of supporters. As a political sponsor, Hume remained committed to supporting Gurney in the face of Trench's challenge, while Faraday provided a constant reference of approval. Politically and experimentally Gurney's position was maintained. While Reid himself was trialled in Parliament, Gurney had

[121] Ibid., p. 4.

cut his teeth on lighthouses and this stood him in good stead at Westminster, while Faraday's support was enough to protect him from Barry's contempt. Nevertheless, this chapter has demonstrated that even with extensive credibility and support, science at Parliament was subject to interpretations beyond the control of politicians and experimentalists. To employ science at Westminster was to open it up to national judgement. Sometimes this involved recruitment to larger problems of natural philosophy, such as Brewster's performance at the 1839 committee. The Bude Light presented Brewster with an opportunity to say that his approach to optics was practical and to try to salvage his reputation. While Brewster had been sidelined by a powerful group of Cambridge mathematicians, the rebuilding of Parliament presented an opportunity to secure renewed credibility. Yet Parliament was a venue in which non-specialist audiences could evaluate and employ the Bude Light in less consistent ways to Hume's agendum. It could become a dangerous political analogy to highlight Parliament's lack of competence or legitimacy in society.

Not all interpretations were so controversial. Thirteen days after the publication of the 1839 committee minutes, Faraday received a letter praising his experimental endeavours in the science of lighting. Isaac D'Israeli (1766–1848), a Jewish author and father to a future Prime Minister, informed Faraday that he was 'illuminating the world in all ways, of Enlightment, that of the Bude light and the other light'.[122] His own sight failing, D'Israeli lamented that 'philosophy' had not 'yet invented an artificial eye for the blind'. Nevertheless, experimental philosophy promised that the nation's body of representatives, which would one day include a certain Benjamin Disraeli (1804–1881), and to whom so much responsibility fell, would work towards social progress in perfect illumination.

[122] James Ogden, 'D'Israeli, Isaac (1766–1848)', *Oxford Dictionary of National Biography*, Oxford University Press, 2004; online edn, May 2008 [http://ezproxy.ouls.ox.ac.uk:2117/view/article/7690, accessed 21 September 2013]; and 'Letter 1197: Isaac D'Israeli to Faraday', (20 August, 1839), in James (ed.), *The Correspondence of Michael Faraday: Volume 2*, p. 603.

6 Order in Parliament: George Biddell Airy and the Construction of Time at Westminster

> There! Out it boomed. First a warning, musical; then the hour, irrevocable.
>
> Virginia Woolf, *Mrs Dalloway*

> It is very desirable that all the clocks in the New Palace . . . should keep strictly the same time . . . I know no method by which the movements of one clock or system of clocks can be made strictly to accompany those of a leading clock except by galvano-magnetic agency . . . and there is no place in the world where it could be introduced efficiently for the first time with so great a prophet of success and with so much benefit from it as in the New Palace.[1]

George Biddell Airy's comments, made in 1851 regarding the Westminster clock in St Stephen's Tower, reveal three salient concerns regarding the keeping of time at Parliament. Primarily it appears that of all the places in the world, Airy perceived government at Westminster to be the most demanding of an ordered system of time. He also felt that this time should be synchronized throughout the building, regulating the daily routine of government. Finally it is evident that to be accurate, he believed that this governing time system should be maintained by galvanic power. His ideal time system was one which would bring order to the place of British government. Between 1846 and 1853 Airy, as Astronomer Royal, worked to implement this system at Westminster. The Westminster clock was to be of incomparable mathematical accuracy and a work of horological science. Airy envisaged it generating an electric time signal to control every clock in Parliament; through committee rooms, private accommodation, and both the upper and lower Houses. The Westminster clock itself was to be regulated by a telegraphic signal from the Royal Observatory at Greenwich and so was to be under Airy's immediate supervision.

The importance of Airy's proposal is evident from both its political and scientific contexts. Technologically, during the 1840s telegraphy was a

[1] Royal Observatory Greenwich Papers, Cambridge University Library (RGO) 6/608, 'Letter from George Airy to T. W. Phillips', (21 February, 1851), pp. 18–19.

troublesome apparatus. The future of telegraphic communication was uncertain and its potential remained unfulfilled well into the 1850s. So, for Airy to propose its use for Parliament was not just radical, but presented the chance to salvage a technological scheme which was coming increasingly to appear as a failure. However, Airy's work at the Palace was also important politically. Ryan Vieira has cogently shown how questions over 'efficiency' and time dominated Parliament in the mid-nineteenth century. In context of railway and industrial expansion Parliament came to appear as a slow moving, de-synchronized, conservative organ.[2] The popular press characterized the Prime Minister John Russell as 'The Sluggard', while constantly scrutinizing the industry of MPs. With the rapid increase of Parliamentary business in the 1840s and rising public pressure for efficiency, a network for time management was an important consideration for the new Parliament. However, this context meant that efficient time regulation was not only technologically radical, but potentially politically revolutionary.

E. P. Thompson has shown how public displays of time embody cultural authority. During the seventeenth and eighteenth centuries, public clocks were set by sundials, often projecting time from buildings reflecting the cultural values of society, including churches and town halls.[3] As Britain moved into an age of 'industrialization', so the practices of disciplining labour changed too; in the nineteenth century, increasing emphasis was placed on mechanically synchronized labour.[4] Iwan Morus has explained how electric telegraphy in the mid-nineteenth century became 'the body politic's nervous system'.[5] The electric telegraph provided instantaneous information over vast expanses of space. The apparatus offered a powerful means of constructing a national system of time discipline, but I argue that though this system was regulated from Greenwich, Parliament became the most prominent symbol of controlled national time.[6] The Westminster clock was a display of authority, symbolic of Parliament's control and, as such, it was important that it have an

[2] Ryan A. Vieira, *Time and Politics: Parliament and the Culture of Modernity in Britain and the British World*, (Oxford, 2015), pp. 72–74.

[3] E. P. Thompson, 'Time, work-discipline, and industrial capitalism', *Past and Present*, No. 38 (December, 1967), pp. 56–97, 63.

[4] Ibid., p. 69; on the problems of the term 'industrialization', see p. 80; for a revision of Thompson, which challenges the Marxist notion that clock time was a single entity, inherently linked to the advance of science and regulating economic activity, see Paul Glennie and Nigel Thrift, *Shaping the Day: A History of Timekeeping in England and Wales, 1300–1800*, (Oxford, 2009), pp. 47–49.

[5] Iwan Rhys Morus, '"The nervous system of Britain": space, time and the electric telegraph in the Victorian age', *British Journal for the History of Science*, Vol. 33, No. 4, On Time: history, science and commemoration (December, 2000), pp. 455–75, 458.

[6] Ibid., pp. 465–69.

accuracy which embodied the increasingly industrialized society which it was intended to regulate.[7]

Airy's time system in the Royal Observatory was often likened to a 'factory' and this metaphor is useful when considering Parliament as the nation's timekeeper. Wise and Smith have noted that in the second half of the nineteenth century, British mathematical theorizing was often considered characteristic of a national 'factory mentality' where measurement and industry were guiding values in concepts of work.[8] Airy's attempts to eradicate error from time at Parliament allow us to place the Westminster clock in a context of increasing concern about the analytical regulation of society. The essence of a 'factory system' was the mechanical removal of human intelligence, and therefore error, from manufacturing.[9] Machines were subservient, unremitting devices which embodied controversial notions of control and order. During the 1820s and 1830s London was a central stage for these debates, and it is within this context that the construction of the Parliamentary time system took place. Airy's conception of a clock, mathematically grounded, projecting time maintained through seemingly non-human techniques, sought to secure Parliament authority in regulating a modern society.

Traditional narratives of the clock's construction exaggerate the role of the lawyer, horologist, and amateur architect, Edmund Beckett Denison (1816–1905). McKay's excellent account of the project tends to emphasize Denison's role at the expense of Airy's. He explains that Denison designed the clock 'single-handedly', while the clock's escapement 'could only be the result of logic and scientific thought', of which 'Denison is the only person who could be credited with such faculties'.[10] Although eventually Denison did take control over the clock's construction, this was not without extensive debates with Airy over several of the clock's mechanisms. Despite collaborating on the project, these disagreements were the result of two differing programmes for securing authority over

[7] The links between Foucault's concept of 'governmentality' (the way in which a state exercises control over its people) and automatic machinery is shown in Simon Schaffer, 'Enlightened automata', in William Clark, Jan Golinski, and Simon Schaffer (eds.), *The Sciences in Enlightened Europe*, (Chicago, 1999), pp. 126–65, 129.

[8] M. Norton Wise and Crosbie Smith, 'Measurement, work and industry in Lord Kelvin's Britain', *Historical Studies in the Physical and Biological Sciences*, Vol. 17, No. 1 (1986), pp. 147–73, 147–48.

[9] Simon Schaffer, 'Babbage's intelligence: calculating engines and the factory system', *Critical Inquiry*, Vol. 21, No. 1 (Autumn, 1994), pp. 203–27, 209; on clockwork in London, also see Simon Schaffer, 'Babbage's dancer and the impresarios of mechanism', in Francis Spufford and Jenny Uglow (eds.), *Cultural Babbage: Technology, Time and Invention*, (London, 1996), pp. 53–80.

[10] Charles McKay, *Big Ben: The Great Clock and the Bells at the Palace of Westminster*, (Oxford, 2010), pp. 75, 147.

Parliament's time. Each had their own conception of how the clock should be made accurate.[11] Simon Schaffer has shown how the rise of quantification in science is not an inevitable process.[12] Rather, systems of knowledge organization are constructed as values of measurement are chosen. Regimes of measurement do not imitate universal values, but are part of the social-cultural context in which they are produced.[13] Accuracy then, is a human construction, rather than self-evident measurement of nature, and so too is precise time. During the implementation of Parliament's time system, the leading horological authorities involved wrestled with this problem. To Airy's system of galvanic time regulation and mathematical horology, there were alternatives, such as the scheme which Benjamin Lewis Vulliamy (1780–1854) put forward.

Airy's system was a model of his wider plans for Greenwich time on a national scale. He used Westminster time as a claim for his proposed telegraphically-projected national time system. During this period, the Observatory increasingly became the heart of British imperialism, with Airy confident that Britain's commercial superiority was partly attributable to the disciplined work conducted at Greenwich.[14] Under Airy, Parliament would be constructed as a model for Greenwich time on a national and international scale. Yet in showing how controversial his Parliamentary system was, I also suggest how problematic national uniformed time would prove. Derek Howse explained that as rail and steamship transport expanded in the 1840s, it created the demand for uniform time. As the 'need for a standard time became pressing, the means of satisfying that need became available'; in effect, the railways 'forced a uniform time on a not-unwilling population'.[15] My account of the Westminster time system controversy shows that the rise of national Greenwich time was anything but a simple question of supply and demand. Time was not a universally accepted measurement, reliably entrusted to the Astronomer Royal, but a culturally shaped construct, contested and controversial. Accurate time was debated through mechanical, mathematical, galvanic, and telegraphic practices and

[11] On accuracy, see Simon Schaffer, 'Metrology, metrication, and Victorian values', in Bernard Lightman (ed.), *Victorian Science in Context*, (Chicago, 1997), pp. 438–74; See M. Norton Wise, 'Introduction', in M. Norton Wise (ed.), *The Values of Precision*, (Princeton, 1995), pp. 3–13; Graeme J. N. Gooday, *The Morals of Measurement: Accuracy, Irony, and Trust in Late Victorian Electrical Practice*, (Cambridge, 2004), pp. 1–2.

[12] Simon Schaffer, 'Astronomers mark time: discipline and the personal equation', *Science in Context*, Vol. 2, No. 1 (March, 1988), pp. 115–45, 118.

[13] Ibid., pp. 138–39.

[14] William J. Ashworth, 'John Herschel, George Airy, and the roaming eye of the state', *History of Science*, 36:2, (1 June, 1998), pp. 151–78, 161.

[15] Derek Howse, *Greenwich Time and the Longitude*, (London, 1997), pp. 89, 92.

apparatus. Conceiving of the rise of absolute mathematical time as becoming systematically homogeneous between 1880 and 1914 is misleading. Time is inherently problematic, even keeping the accuracy of a 'true universal day'.[16] Time systems are not self-evident products to meet needs determined by 'progressive' transport developments. As Marsden and Smith observed, the 'Royal Observatory's authority in the later nineteenth century as the great regulator of the nation's time – and thus of its trade, transport, commerce and industry at home and across the world – had been neither inevitable nor self-evident'.[17]

Part one explores Airy's proposed time system for Parliament and shows how he emphasized the role of mathematical theory in horology, as well as galvanic and telegraphic apparatus. Part two details how problematic Airy's system was by outlining Vulliamy's alternative mechanical time system, while also demonstrating the problems of telegraphic and galvanic instruments. I conclude by showing how Airy failed to implement his own system. It was Denison who effectively took over from Airy in 1853 and he not only reduced Greenwich's influence over the clock, but seemingly abandoned Airy's galvanic projection of time through the Palace. This diminishing of Airy's planned authority for the Observatory fits within a wider context of resistance to Greenwich standard time. Denison's rejection of direct galvanic control was prophetic of the objections to a single national time in favour of existing traditional local times, encountered in the years after Parliament's construction.[18]

It has been shown how Parliament embodied, employed, and stimulated different forms of science, but Airy's instruments and know-how provided ways of disciplining Parliament and bringing it within Greenwich's time regime. In the previous chapters science appears as something which politicians sought to control and impose on Parliament. Science could be enlisted to show Parliament was enlightened, or to materially improve the architecture and conditions of the building. In this chapter, I show how Airy's notion of science went much further. Science could be imported as a body of knowledge or constructed as a cultural resource but, at Parliament, it could also take control. Airy's work was about placing Parliament within a wider regime of disciplined time. While politicians sought to govern science and use it to their own ends, Airy's science was a governing power over the assembly. Science was constructed and displayed in Parliament, but it could also order and

[16] Stephen Kern, *The Culture of Time and Space, 1880–1918*, (London, 1983), pp. 11–2.
[17] Ben Marsden and Crosbie Smith, *Engineering Empires: A Cultural History of Technology in Nineteenth-Century Britain*, (Basingstoke, 2005), p. 17.
[18] Ibid., pp. 20–1; on concerns surrounding telegraphic time, see Morus, '"The nervous system of Britain"', pp. 463–69.

regulate business, both in the Lords and Commons, and beyond the Palace's walls.

Airy's Time System for Parliament

Absent from Charles Barry's original design for Parliament was the Clock Tower. A late inclusion to the Palace, its construction began in 1843. Within a year Barry was in search of a clock to mount in the tower. Believing that the project's scale and prestige called for authority in horology, Barry applied to Benjamin Vulliamy of Pall Mall for plans and calculations.[19] Vulliamy agreed to design and build the clock for two-hundred guineas in 1844, but in November 1845 the chronometer maker Edward John Dent (1790–1853), armed with a testimonial from Airy, applied to the Office of Woods and Forests to tender designs for the clock.[20] When in January 1846 Charles Canning (1812–1862) became First Commissioner of Woods and Forests, he immediately faced the challenge of building the new clock.[21] Serving in the final few months before the fall of Robert Peel's Conservative administration, Canning was the son of George Canning, who had briefly held the office of Prime Minister in 1827 after several years as Foreign Secretary. Rather than leave the arrangements in the hands of Barry, Charles Canning requested Airy provide advice on the best way to build a clock suitably accurate for the nation's legislature.

Canning informed Airy that 'the Clock which is to be placed in the clock-tower of the new Houses of Parliament should be the very best which British Science and Skill can supply'. He stressed the public service Airy would provide in guiding the design of a clock for the national assembly. The clock's accuracy was a question which, Canning believed, 'the public are much interested'. Airy's advice would 'be the safest and most satisfying guide which the Commissioners of this Board can follow'.[22] Canning asked for a recommendation as to who should design and build the clock under Airy's supervision. This request illustrates that the Westminster clock was a matter of importance beyond the private dealings of Barry; its accuracy was a question of Parliament's reputation. Vulliamy had promised to design 'the most powerful eight-day clock ever

[19] Alfred Barry, *The Life and Works of Sir Charles Barry*, (London, 1867), p. 171.

[20] M. H. Port, 'Barry's last years: the new palace in the 1850s', in M. H. Port, *The Houses of Parliament*, (New Haven, 1976), pp. 142–72, 169.

[21] Thomas R. Metcalf, 'Canning, Charles John, Earl Canning (1812–1862)', *Oxford Dictionary of National Biography*, Oxford University Press, 2004; online edn, January 2008 [http://ezproxy.ouls.ox.ac.uk:2117/view/article/4554, accessed 24 August 2013].

[22] RGO 6/607, 'Letter from Charles Canning to George Airy', (20 June, 1846), p. 6.

made in this country', yet Canning wanted the government's most trusted mathematical authority to oversee the project.[23]

Born in Alnwick, Northumberland, George Biddell Airy's (1801–1892) early career had been prolific. Attending Trinity College Cambridge from 1819, he graduated as Senior Wrangler achieving the top marks in the University's Mathematical Tripos examinations of 1823 and was elected a college fellow in 1824.[24] At Cambridge, Airy had been through the system of mathematical training developed during the 1810s and 1820s which emphasized personal tuition.[25] After 1815, Cambridge became a centre of mathematical education, and within this culture Airy had excelled under the private coaching of the English mathematician, George Peacock (1791–1858).[26] Airy had been Peacock's most celebrated student and the mathematical star of the 1820s; by 1826 he had published his influential *Mathematical Tracts*, which was to serve as the textbook for the Cambridge Mathematical Tripos exams.[27] In the same year he was appointed the Lucasian Professor of Mathematics, before taking over the newly established Cambridge University Observatory in 1828. Through these roles Airy was part of a defining generation of Cambridge mathematicians who sustained a revival in English mathematics between 1815 and 1840. Along with Charles Babbage (1791–1871), William Whewell, and John Herschel, Airy took what had been a French academic preserve and transformed Cambridge into a bastion of mathematical learning.[28]

In Victorian Britain, mathematics was not supported by a professional community until the close of the century. Mathematical research was usually part of broader works in science, often connected to mechanical concerns.[29] Finding a common nineteenth-century definition of

[23] Vulliamy to Barry, in *Clocks (New Houses of Parliament)*, PP. 1852 (500-I), p. 5.

[24] Allan Chapman, 'Airy, Sir George Biddell (1801–1892)', *Oxford Dictionary of National Biography*, Oxford University Press, 2004; online edn, January 2011 [http://ezproxy.ouls.ox.ac.uk:2117/view/article/251, accessed 24 August 2013].

[25] Andrew Warwick, *Masters of Theory: Cambridge and the Rise of Mathematical Physics*, (Chicago, 2003), p. 50.

[26] Ibid., pp. 72–74.

[27] George Biddell Airy, *Mathematical Tracts on Physical Astronomy, the Figure of the Earth, Precession and Nutation, and the Calculus of Variations*, (Cambridge, 1826); also see David B. Wilson, 'The educational matrix: physics education at early-Victorian Cambridge, Edinburgh and Glasgow Universities', in P. M. Harman (ed.), *Wranglers and Physicists: Studies on Cambridge Mathematical Physics in the Nineteenth Century*, (Manchester, 1985), pp. 12–48, 15.

[28] I. Grattan-Guinness, 'Mathematics and mathematical physics from Cambridge, 1815–40: a survey of the achievements and of the French influences', in Harman (ed.), *Wranglers and Physicists*, pp. 84–111, 84.

[29] For an overview, see Raymond Flood, Adrian Rice, and Robin Wilson (eds.), *Mathematics in Victorian Britain*, (Oxford, 2011); however, this work brands British mathematicians inferior to their Continental colleagues, and fails to appreciate that the subject was not usually pursued for its own sake, but for practical mechanical

mathematics is difficult, with different practitioners claiming varying definitions of the discipline.[30] In the case of Airy, the subject frequently involved the use of calculations in relation to machinery. 'Applied mathematics' entailed the employment of mathematics as an investigative tool of the physical world in natural philosophy and mechanics.[31] British mathematics in particular was often part of wider concerns regarding machinery and mechanical problems, but the question of whether the discipline could be applied to practical matters was one which raised controversy and, during his career, Airy earned a reputation for employing mathematical theory for practical solutions. To solve the problem of unreliable clock pendulums caused by the Earth's irregular gravitational pull around the world, Airy attempted to calculate the Earth's mean density. In the 1820s and 1850s, Airy conducted experiments to compare the speed of pendulum swings at different levels in mine shafts to ascertain the varying attraction of gravity at different depths in the Earth's crust.[32] As Astronomer Royal he attempted to solve the problem of compass deviation caused by an iron ship's magnetism using magnetic 'correctors'. Here, again, Airy deployed mathematical theory to calculate the angle and position of magnets in relation to a ship's compass.[33] As Winter shows, Airy's mathematical solutions raised profound questions over authority in Victorian science.[34] Yet Airy's determination to employ mathematics practically, shaped his work at Greenwich and Westminster.

Appointed Astronomer Royal of the Royal Observatory at Greenwich in 1835 Airy became, in effect, the government's 'general scientific advisor' offering guidance over weights, measurements, and railway gauges.[35] (Figure 6.1) In the years following his appointment at Greenwich, the Observatory secured an eminent place in Victorian society, as Joseph Conrad highlighted when he made the institution the target of an

applications, as shown in Joan L. Richards, 'Mathematics in Victorian Britain by Raymond Flood; Adrian Rice; Robin Wilson. Review', *Isis*, Vol. 104, No. 4 (December, 2013), pp. 853–55.

[30] As astutely argued in, Richards, 'Mathematics in Victorian Britain', p. 854.

[31] A. D. D. Craik, 'Victorian "applied mathematics"', in Flood, Rice, and Wilson (eds.), *Mathematics in Victorian Britain*, pp. 177–98, 178.

[32] Allan Chapman, 'The pit and the pendulum: G. B. Airy and the determination of gravity', *Antiquarian Horology* (Autumn, 1993), pp. 70–78, 71.

[33] Charles H. Cotter, 'George Biddell Airy and his mechanical correction of the magnetic compass', *Annals of Science*, Vol. 33, No. 3 (1976), pp. 263–74, 266; also see A. E. Fanning, *Steady as She Goes: A History of the Compass Department of the Admiralty*, (London, 1986), pp. xxxvi–xxxvii.

[34] See Alison Winter, '"Compasses all awry": the iron ship and the ambiguities of cultural authority in Victorian Britain', *Victorian Studies*, Vol. 38, No. 1 (Autumn, 1994), pp. 69–98; although dated, see Charles H. Cotter, 'The early history of ship magnetism: the Airy-Scoresby controversy', *Annals of Science*, Vol. 34, No. 6 (1977), pp. 589–99.

[35] Wilfrid Airy (ed.), *Autobiography of Sir George Biddell Airy*, (Cambridge, 1896), p. vi.

Figure 6.1 *Punch's* gently mocking portrayal of George Biddell Airy, here aloft the Greenwich time-ball

anarchistic plot to provoke middle-class Britain to demand destabilizing retribution in his 1907 novel, *The Secret Agent*.[36] Airy's role at Greenwich involved not only mathematical skill, but huge administrative efforts in

[36] Marsden and Smith, *Engineering Empires*, p. 17.

data collection; he was an industrial handler of technical information and organiser of astronomical knowledge.[37] As Airy put it, the study of astronomy was 'pre-eminently the science of order'.[38] In his role at Greenwich, his guiding principle was that the work of the Observatory should always be of public service. As the institution was state-funded, he believed it should produce knowledge which would be of value to the government. By examining the controversial search for the planet Neptune, Allan Chapman has demonstrated how the Royal Observatory under Airy was not a research institution, but a site of government science.[39] Airy's mandate was not to advance scientific knowledge, but to be of service to the government and nation. While discovering new planets promised little practical benefit, Airy envisaged the organization of a centre of science, offering a time service and mathematical guidance in naval and industrial matters.[40] It is within this context that Canning appealed to Airy regarding the Westminster clock.

Airy responded to Canning's letter with a detailed proposal for the clock. He provided fifteen 'specific conditions applying to the accurate going of the clock', including wheels of hard bell-metal, jewelled pendulum pallets, a compensated pendulum, a going fuse (gear mechanism), and vitally, a dead-beat escapement driven by a remontoire action wheel train.[41] As the clock was to be a symbol of national pride he asserted that its workings should be accessible for foreign visitors to view.[42] To design and build the clock along these specifications, Airy recommended Vulliamy, Dent, and John Whitehurst (1788–1855) all compete to tender.[43] Importantly each of these competitors was to work to Airy's instruction. Airy cited the example of the Royal Exchange clock, situated in the financial heart of London opposite the Bank of England, as

[37] Allan Chapman, 'George Biddell Airy, F. R. S. (1801–1892): a centenary commemoration', *Notes and Records of the Royal Society of London*, Vol. 46, No. 1 (Jan., 1992), pp. 103–10, 105–06.

[38] Ashworth, 'John Herschel, George Airy, and the roaming eye of the state', p. 160.

[39] Allan Chapman, 'Private research and public duty: George Biddell Airy and the search for Neptune', *Journal for the History of Astronomy*, 19, (Cambridge, 1988), pp. 121–39, 122.

[40] Ibid., p. 123.

[41] Airy, in Vaudrey Mercer, *The Life and Letters of Edward John Dent: Chronometer Maker and Some Accounts of His Successors*, (London, 1977), pp. 348–49; there were four problems involved in building a clock. First, the clock's drive, which was the source of power, in the form of either a spring, or a weight. Second, the clock required a regulator, or escapement, to slow and control the release of power from the drive with regularity. Then there was the going-train, which was the means of keeping count of the movement which this release of power caused. Finally, there was signalling, which involved the indicating of the time a clock kept.

[42] Ibid., p. 349.

[43] Wilfrid Airy (ed.), *Autobiography of Sir George Biddell Airy*, p. 180.

evidence of the practicality of such an arrangement. Under Airy's guidance, Dent had designed and built the clock for the Royal Exchange, which opened in 1844. Airy boasted that the result was 'that a clock has been mounted which . . . is superior even to most astronomical clocks in the steadiness of its rate'.[44] 'As to time', it was accurate to within a second of every hour. He noted that 'a person standing on the pavement can take time from the face without an error of one second'. If Airy was allowed to have control over the Westminster project, he promised that 'the clock which shall be mounted shall be creditable to the nation'.[45] Airy's supervision offered Parliament a clock so accurate, that the authority of the time it projected would be unanimously acknowledged. The Palace of Westminster would become a world-renowned site of time regulation.

Airy's promises of accuracy carried weight. Since taking over the Observatory he had worked hard to build accuracy into the daily routine of reading astronomical time, as well as testing chronometers. The Observatory had long checked chronometers for the Admiralty; accurate timekeepers were vital tools of navigation, and problems of temperature change and pendulum distortion hindered reliable astronomical calculations for longitude.[46] Airy maintained this service at the Observatory, but took it much further. Using mathematical theory, he sought to produce increasingly accurate clocks. The problem of accuracy in clocks involved the maintaining of a constant regular force from the pendulum, or spring, to drive the clockworks. Any irregularity to the movement of the pendulum, often caused by changes in temperature which disturbed the composite metal, could reduce the regulation of measured seconds and eventually render the clock unreliable. To solve this problem, clocks were equipped with various escapements. These were devices for securing the consistency of a pendulum's force and measuring its movement into regulated seconds.

As early as 1825, Airy submitted a paper to the Cambridge Philosophical Society on the teeth of clock wheels, followed in 1826 by a paper on theoretical horology detailing his mathematical investigation of pendulum disturbances caused by air resistance.[47] He analysed various escapements to develop an ideal 'theoretical' escapement. The main challenge for horologists was circular error, which was the difference in movement between a theoretical pendulum and a pendulum in operation. Mathematical formulae for clock-building assumed a pendulum moved

[44] Airy to Charles Canning, in PP. 1852 (500-I), p. 10. [45] Ibid., p. 10.

[46] See David W. Waters, 'Nautical Astronomy and the Problem of Longitude', in John G. Burke (ed.), *The Uses of Science in the Age of Newton*, (Berkeley, 1983), pp. 143–69.

[47] J. A. Bennett, 'George Biddell Airy and horology', *Annals of Science*, Vol. 37, No. 3 (1980), pp. 269–85, 270–71.

in a small arc through the air, however, in practice a pendulum swung in an arc varying by several degrees. This error could be caused by temperature changes which altered the length of the pendulum, or external impulses affecting the mechanism.[48] This error was what defined the accuracy of a clock. The solution was to make the pendulum out of either wood, or varying metals which reacted differently to temperature changes and compensated each other. To keep constant pendulum beats, an escapement mechanism could be employed and this was what Airy focused on.

At Greenwich, Airy transferred this theoretical work to practical concerns. During 1836 he directed Dent to produce a model of his own escapement, later implementing it in a full clock design in 1840. In 1841, he developed a clock escapement for Dent's two regulators at the Pulkovo Observatory near St Petersburg. This 'remontoire dead beat escapement' consisted of two identical escape wheels, one driven by the clock's train, and the other by a remontoire spring. This remontoire spring was regularly wound up by the clock's main source of power, a falling weight, and maintained a constant driving force for the clock's pendulum. The mechanism served to even out the irregularities in force transmitted from the pendulum driving the clock's workings.[49] Airy's design was calculated to provide Dent's clock with a consistent impulse from the swinging pendulum which ensured accurate timekeeping. It was conceived to bring control to a clock subjected to the freezing temperatures of Pulkovo. Airy worked with Dent to improve this design through the 1840s. Between 1842 and 1844 they constructed the Royal Exchange clock, before embarking on the Westminster clock project.

While Airy recommended three clockmakers to Canning, he favoured Dent. Not only had Dent experience of converting Airy's mathematical designs into mechanical devices, but Airy appreciated his ability to produce accuracy. Born in Westminster, Dent had secured a reputation as a fine chronometer maker. In 1829, his Chronometer No. 114 won the Royal Observatory's premium award, while his experiments to reduce temperature distortion with mercury-filled clock pendulums had secured him fame at the 1838 BAAS meeting.[50] Dent specialized in pendulums of varying metals which evened out irregularities caused by temperature change. His compensating pendulums used two different metals which had alternate rates of expansion, so while one expanded down, the other moved up, keeping the pendulum's bob in a constant position.[51] In the

[48] McKay, *Big Ben*, p. 55. [49] Bennett, 'George Biddell Airy and horology', p. 272.

[50] G. C. Boase, 'Dent, Edward John (1790–1853)', rev. Anita McConnell, *Oxford Dictionary of National Biography*, Oxford University Press, 2004; online edn, May 2007 [http://ezproxy.ouls.ox.ac.uk:2117/view/article/7512, accessed 24 August 2013].

[51] McKay, *Big Ben*, pp. 155–56.

1840s, Dent favoured a zinc tube with a central iron rod. When Airy had worked to ascertain the longitude of Europe's premier observatories using chronometers set to Greenwich time, he had employed Dent's instruments. The two men worked well together, with Dent ably bringing Airy's mathematical escapements to mechanical life.[52] So when Airy offered himself as the Westminster clock's supervisor, part of the promise was that Dent could realize Airy's theoretical horology. Before Dent applied to tender designs for the clock, he received Airy's approval; he believed Dent's Royal Exchange clock 'to be the best in the World as regards accuracy of going'.[53]

When Dent was asked to submit designs in late 1845, before Airy's involvement, he refused to work within any specifications other than from the Astronomer Royal. He would take no advice from an 'authority less eminent'.[54] Airy reciprocated this acclaim consistently throughout the project. After a comparative inspection of Dent and Whitehurst's factories in 1847, Airy reported on Dent's 'excellence of fine workmanship' and 'great accuracy'. His 'tools are somewhat superior to Mr. Whitehurst's', while his chronometers had been 'subjected to the severest examination in the rating at the Royal Observatory'.[55] Whitehurst, himself from a family of eminent clockmakers, was a competent 'mechanic' and a renowned turret clockmaker, but lacked Dent's accuracy. While Dent could provide the accuracy of an astronomical clock, Whitehurst's 'works have never been subjected to the severe examination as to regularity of rate to which chronometers of a maker competing for the Government purchase are subjected'.[56] Airy felt that the practice of chronometer making ensured the highest possible horological accuracy. It was an accuracy scientifically measured at Airy's Observatory, and of a sort, he asserted, ordinary clockmakers could not replicate. Airy's definition of accuracy included the precise mechanical practices of chronometer production. Dent encapsulated this when he claimed that if he was selected to work under Airy, 'science, stability of fixing, and engineering will be united in producing a national monument of finished mechanism and accurate performance'.[57]

[52] Mercer, *The Life and Letters of Edward John Dent*, p. 123. [53] Airy, in ibid., p. 346.
[54] Dent, in ibid., p. 347. [55] Airy to A. Milne, in PP. 1852 (500-I), pp. 47–48.
[56] Ibid., p. 48; Whitehurst inherited the business, via his father, from his grandfather John Whitehurst. On his grandfather, see Denys Vaughan, 'Whitehurst, John (1713–1788)', *Oxford Dictionary of National Biography*, Oxford University Press, 2004; online edn, September 2013 [http://ezproxy-prd.bodleian.ox.ac.uk:2167/view/article/29295, accessed 2 September 2014].
[57] Edward Dent to the Commissioners of Her Majesty's Woods and Forests, in PP. 1852 (500-I), p. 45.

Although Airy's proposals focused on the construction of a clock of incomparable accuracy, his claims were for much more than simply a mounted time-piece. From his earliest involvement Airy was actually promoting an entire time 'system' for Parliament, with the Westminster clock as its central feature. In his initial response to Canning, Airy recommended that the clock's attendant should be provided with a pocket chronometer, regularly compared to the Royal Observatory's time-signal ball. This process would, Airy believed, ensure the clock was always correct to Greenwich time. Airy suggested this chronometer as a temporary measure before the establishment of a direct telegraphic link between Greenwich and the clock. Crucially, Airy specified that when the connection was established, the signal would be sent from Greenwich to Westminster, effectively governing the clock in accordance with Greenwich time. He wanted the clock under his control: 'The severity of the check upon the going of the clock thus placed in the power of the Astronomer Royal would exceed anything that has ever been proposed'.[58] In coming under the 'power' of Airy it is possible to conceive of his plans for the clock as both an extension of his time system at the Observatory, and as a microcosmic model of how Greenwich time might become a national time system. It was not enough that the clock keep a regular action, but that it should accurately maintain time taken to Airy's standards.

In this way, Airy envisaged the Westminster clock as an extension of his Greenwich time regime. What then did this system entail? Apart from competent data handling, one of the defining features of Airy's tenure as Astronomer Royal was his penchant for 'physical objects which did things'.[59] At the Observatory he endeavoured to improve the level of accuracy of recording time by placing an increased reliance on instruments rather than humans. Both the practice of time-signalling and astronomical transit observations were to be subject to a 'factory mentality'.[60] At the Observatory, time was recorded by observing the transit of stars across a wire grid through a telescope (a micrometre); this was the observation of the Earth's rotation.[61] To secure the accuracy of time observed in this way, Airy designed and installed a new transit circle at Greenwich in 1851, capable of recording with a critical accuracy

[58] Airy to Charles Canning, in PP. 1852 (500-I), p. 10.

[59] Chapman, 'George Biddell Airy', p. 108.

[60] Allan Chapman, 'Sir George Airy (1801–1892) and the concept of international standards in science, timekeeping and navigation', *Vistas in Astronomy*, 28, (London, 1985), pp. 321–28, 322–23.

[61] A. J. Meadows, *Greenwich Observatory, Volume 2: Recent History, (1836–1975)*, (London, 1975), p. 30.

of up to 0.06 seconds.[62] Yet accurate time recording remained problematic. Until 1854, transit recording through the 'eye-and-ear method' consisted of coordinating a transit observed through a micrometre along with the sound of a clock. The actual time recorded thus depended on the reaction time, or 'personal equation', of the recording astronomer. To compensate for this problem Airy developed a test for his observers, of modelling the transit of an 'artificial star' and recording the time taken to initiate a galvanic signal measuring the transit. This quantified an observer's 'personal equation'.[63]

Further to reduce error, Airy implemented a new observation method where-by an observer recorded the transit by pressing a button and registering a galvanic signal.[64] Together with the transit simulations and new transit circle, Airy constructed a new regime of precise time recording. He claimed his system was accurate because the observations of nature recorded would, theoretically, be identical for all observers.[65] Through galvanic signalling and calculations for personal equations, Airy believed that his Greenwich system was accurate because it relied on mechanical and electrical instruments, rather than human efforts; the Observatory was structured like a factory for producing precise measurements of time.[66] Electrical devices promised to render nature 'self-recording', and Airy was convinced that through electromagnetism and telegraphy, both time recording and transmitting would become instantaneous, independent of the human nervous system, and this secured accuracy.

With this time regime established in the private space of Greenwich during the early 1850s, Parliament was conceived of as a projector of these reliable transit observations in the more public space of Westminster. The Westminster clock was to be an extension of the Greenwich system. Airy explained that 'the great clock, which is to be the standard of time for London, and is to have a galvanic communication with the Royal Observatory, to report its own performance to the Astronomer Royal daily'.[67] As Airy had believed that telegraphic and

[62] Allan Chapman, *Dividing the Circle: The Development of Critical Angular Measurement in Astronomy, 1500–1850*, 2nd ed., (Chichester, 1995), pp. 121–22; J. A. Bennett, *The Divided Circle: A History of Instruments for Astronomy, Navigation and Surveying*, (Oxford, 1987), pp. 176–77.

[63] Schaffer, 'Astronomers mark time', p. 117. [64] Ibid., p. 130.

[65] Ibid., p. 138; on professionalism and amateurism in astronomy, see Allan Chapman, *The Victorian Amateur Astronomer: Independent Astronomical Research in Britain, 1820–1920*, (Chichester, 1998), pp. 3–5.

[66] Chapman, 'Sir George Airy (1801–1892) and the concept of international standards', p. 323.

[67] Frederick Dent to William Molesworth, in *Westminster New Palace. Copies of all papers and correspondence relating to the great clock and bells for the new Palace at Westminster (in continuation of Parliamentary Paper, no. 500, of session 1852)*, PP. 1854–5 (436), p. 13.

electromagnetic instruments could reduce the reliance of transit observations on the human nervous system, so too did he consider such apparatus for connecting Westminster to Greenwich, even as early as his 1846 specifications. Having such a connection would ensure the clock was trustworthy in displaying Greenwich time: it would secure the clock a reputation as a reliable timekeeper. As Airy put it, it was 'desirable (in order to secure to the clock the credit which it ought to have) that there should be a galvanic connexion between the clock at Westminster and the Royal Observatory'.[68] If Dent could produce a clock of consistent time regulation, Airy would ensure that the time displayed was precisely that recorded through the Greenwich transit circle. For the clock to project absolute time as governed by the Earth's rotation, it was proper 'that the correctness of the clocks performance should be subjected to the check of daily official examination at Greenwich'.[69]

Airy planned to use the South Eastern Railway's new telegraphy network to disseminate time through the nation, and it was this same system which he hoped could be employed to link Greenwich to Westminster. From 1849 Airy was in communication with the Telegraph Superintendent of the South Eastern Railway.[70] He explained how with the South-Eastern Railway, he would make

> arrangements for galvanic communication, by which signals automatically given at every hour by the great clock would be seen at Greenwich, and signals given from Greenwich, at 30 minutes after every hour, would be seen in the great clock room, and that a person must watch these signals at least once a day for regulation of the clock.[71]

While Airy's establishment of a national time signal began in 1851, it is significant that his original conception of telegraphic communication with Parliament was as early as 1846. Parliament was an early claim for his uniformed Greenwich time projection. Telegraphic communication promised to disseminate instant time across London, without depending on a human to carry it.[72] Airy hoped that eventually the clock itself would be completely regulated from Greenwich, thus reducing human error further.

Yet if Parliament was to be an extension of Airy's Greenwich time system, it was also itself to be a model of Airy's wider ambition to establish

[68] RGO 6/608, 'Letter from George Airy to Lord Seymour', (27 December, 1851), p. 48.
[69] RGO 6/608, 'Letter from George Airy to Benjamin Hall', (9 October, 1855), p. 101.
[70] Derek Howse, *Greenwich Time and the Discovery of the Longitude*, (Oxford, 1980), p. 89.
[71] George Airy to Lord Seymour, in *Clocks (New Houses of Parliament)*, PP. 1852 (500), p. 10.
[72] On manual time projecting, see David Rooney, *Ruth Belville: The Greenwich Time Lady*, (London, 2008), especially pp. 35–51.

a national time system. In November 1846, Airy raised the question of an electrically powered system of time signalling for Parliament with Charles Wheatstone, who had presented his findings on a 'telegraph-clock' to the Royal Society in 1840. Airy proposed 'that a magnet shall be made to enable us, either now or at some future time, to communicate galvanically to Greenwich the first striking-blow of each hour'. If the sticking hammer of the clock's bell fell with a force of 150-200lbs, Airy supposed it might be used to break a circuit on contact. Some of the hammer's 'power might be used to push a magnet within a coil of wire', sending a signal to Greenwich. Yet this impulse would also be used to control every clock in the Palace of Westminster. Airy asked Wheatstone if through the hammer's strike, 'the great clock could be made to move, galvanically, all the other clocks in the buildings'.[73]

Wheatstone sent Airy his Royal Society electromagnetic clock paper. The application of such apparatus to Parliament was precisely the sort of role Wheatstone had originally envisaged for his clock. He believed systematic timekeeping would be beneficial 'in public offices and large establishments' where one good clock could 'serve the purpose of indicating the precise time in every part of the building where it may be required, and accuracy ensured which it would be difficult to obtain by independent clocks'.[74] Wheatstone explained how his system involved one master clock and as many secondary as desired. After reading this Royal Society paper, Airy met up with Wheatstone at King's College London in December and discussed the systems application in Parliament.[75]

After these discussions, Airy decided that, on the 'principle recommended by Mr. Wheatstone', the hour striking hand would break contact with a powerful magnet, to form a magneto-electric current. The clock's hour wheel was to carry a ratchet-shaped succession of cams, breaking contact with the magnet up to once a minute to produce a current. This galvanic signal would regulate all the clocks in the new Palace.[76] The spread of time through government would be instant and automatic. For Airy, it would be accurate, relying on apparatus for projection, rather than being carried manually around the Palace from clock to clock. In a sense, Airy sought to remove the 'personal equation' from transmitting time in government.

[73] RGO 6/607, 'Letter from George Airy to Charles Wheatstone', (13 November, 1846), p. 248.
[74] Charles Wheatstone, 'Description of the electro-magnetic clock', *Abstracts of the Papers Printed in the Philosophical Transactions of the Royal Society of London*, Vol. 4, (1837–1843), pp. 249–50.
[75] RGO 6/607, 'Letter from George Airy to Charles Wheatstone', (27 November, 1846), p. 251.
[76] Airy to A. Milne, in PP. 1852 (500-I), p. 38.

Airy wondered if batteries could be used to sustain the circuit broken on striking. Furthermore, he suggested that if such a system could drive clocks in Parliament, then it could be replicated at the Observatory and used to power 'the Greenwich signal' throughout the nation. Airy believed that Wheatstone's electric clock offered the potential to replicate the accuracy of an astronomical 'perfect clock' in many places. In employing a galvanic pulse in a system of clocks, Wheatstone's object was 'that of enabling a single clock to indicate exactly the same time in as many different places, distant from each other, as may be required'.[77] With reference to Wheatstone, Airy's attempt to drive every clock in Parliament with a galvanic signal from the Westminster clock, in turn telegraphically controlled from Greenwich, was an attempt to project standard Greenwich time through the rooms of government. Airy reported that

> the result of my communication with Mr. Wheatstone is, that I am convinced that it is most desirable that advantage should be taken of the construction of this powerful clock for regulating all the clocks in the Houses of Parliament, committee-rooms, lobbies, &c.[78]

Using galvanic power to bring about ordered astronomical time in Parliament was a model of national time.

Airy is perhaps best understood as a modeller, in that he liked to try out practical ideas on a small scale before performing them in full after analysing their success or failure.[79] At Parliament he worked on a system of time which could be perfected, enlarged, and applied nationally, and then internationally. This was consistent with much of his other work. His pendulum experiments on the Earth's density assumed that if the phenomena of gravitational variance could be modelled under absolute control, they might be applied to chronometer deviations the world over.[80] His experiments to calculate the personal equation consisted of setting up a rotating drum with a light in, and recording an observer's reaction time. Once again Airy 'modelled' the transit of a star, in a controlled environment, and replicated this in full scale celestial observations.[81] Under Airy, the Royal Observatory was transformed into what he supposed to be 'the archetypal scientific corporation'.[82] It was a controlled environment, which Airy used to 'extend Greenwich science, like Greenwich time,

[77] RGO 6/607, 'Letter from George Airy to Charles Wheatstone', (13 November, 1846), p. 249.
[78] Airy to A. Milne, in PP. 1852 (500-I), p. 37.
[79] Thanks to Simon Schaffer for suggesting this notion of Airy as a 'modeller', and relating it to his work at Parliament.
[80] Chapman, 'The pit and the pendulum', p. 71.
[81] Schaffer, 'Astronomers mark time', p. 117. [82] Winter, 'Compasses all awry', p. 74.

over the globe, thereby exerting on a global scale the control which experiments exerted over nature in Greenwich'.[83] If we understand Airy in this way, as a designer of controlled systems to be replicated on a greater scale, then we can see how Parliament's time system was a model for national Greenwich standard time. 'Diffusing correct and uniform time throughout public buildings' had implications for diffusing time across wider spaces.[84]

Apparatus and Alternatives: The Problems of Airy's System

Airy's proposals came at a crucial moment for British politics. Since the 1830s, Parliament had witnessed a dramatic boom in its workload, while the pressures on its time had never seemed greater. (Figure 6.2) The efficient management of Parliament's time was already a concern in 1846, but during the 1847–48 session this problem was exasperated. From 18 November 1847 to 5 September 1848 Parliament dealt with 200 public bills and 255 divisions, while Hansard recorded 3,402,004 words spoken. Compared with 1846, this represented a 22 per cent rise in public bills, 50 per cent rise in divisions, and 20 per cent increase in words recorded. This massive expansion in work, which *Punch* labelled 'the monster session' and tested even Queen Victoria's patience, meant Parliament had to sit for longer than ever before, sparking fears that all this work was wasting the energy and health of MPs.[85] These concerns were accompanied by growing public scrutiny of MPs. The political culture of the mid-nineteenth-century Britain was coming rapidly to emphasize that MPs should be industrious to justify their position.[86] Airy's scheme for an efficient system of Parliamentary time regulation came at an important political moment.

However, Airy's time system was deeply problematic. His claims for galvanic regulation, telegraphic signalling, and mathematically grounded horology were not self-evidently advantageous. What he proposed was not an inevitable system based on universally accepted principles of technological advancement and quantifiable scientific measurement, but a time regime embodying his own cultural values. It was a product in keeping with the factory-like practices of the Royal Observatory. The emphasis on chronometer precision and reduced human interference were cultural choices practiced at Greenwich. There were alternatives

[83] Ibid., p. 72.
[84] RGO 6/607, 'Letter from Samuel McClellan to George Airy', (2 December, 1847), p. 119.
[85] Vieira, *Time and Politics*, pp. 66–67. [86] Ibid., p. 11.

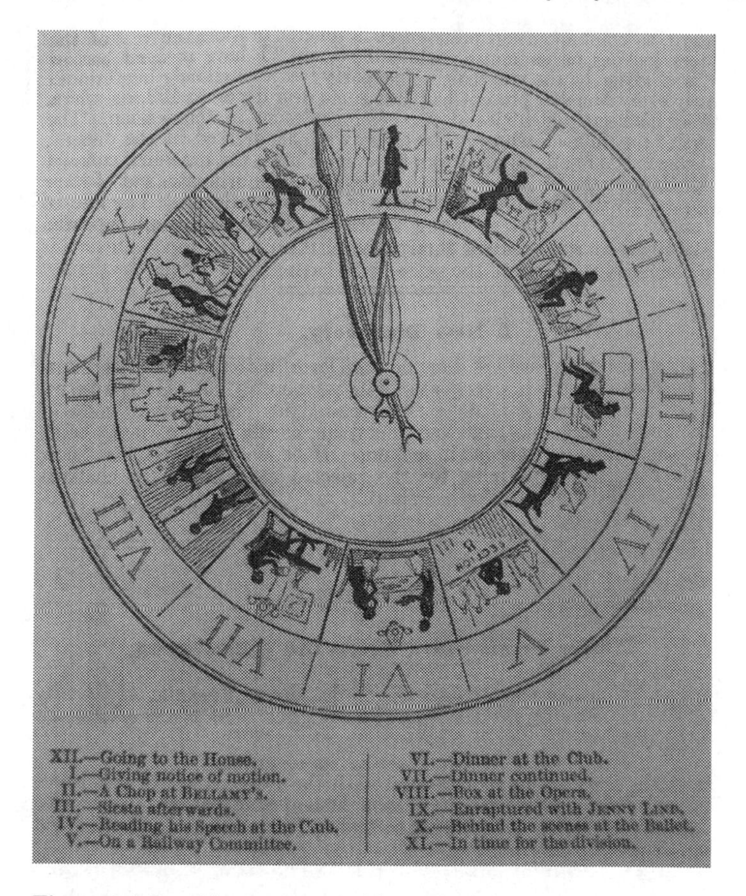

Figure 6.2 *Punch's* own proposal for a clock to regulate Parliamentary business, with time allotted for dining, the opera, and enjoying pleasant company

to, and difficulties with, Airy's proposals. These divided into three broad categories. The first was the mechanical problem of building the Westminster clock. Airy's specifications, including those for his own mathematically-grounded escapement and his wish that Dent build it, came into conflict with Vulliamy's designs based on practical clock-making experience. Second was the apparatus of telegraphy, which although promising instant time communication, was new and unreliable. Finally, the problem of galvanically controlled 'sympathetic-clocks' threatened to ruin Airy's schemes. Maintaining both a current from batteries and a constant time signal with electricity was difficult. Telegraphy and galvanism were experimental, rather than experimentally grounded. This distinction meant that Airy had to labour to have his system accepted.

Several years of bitter dispute followed Canning's decision to overrule Barry's invitation to Vulliamy and his selection of Airy as referee for the project. On the one side was Airy, backing Dent's craftsmanship, and on the other was Vulliamy promoting his own approach to timekeeping at Parliament. Between 1846 and his death in 1854, Vulliamy mounted a sustained challenge to Airy's authority and Dent's ability. In 1857 Edmund Beckett Denison claimed on behalf of Dent and Airy that the whole controversial affair had been indicative of 'the non-progressiveness of the art of clockmaking, during the time that all other branches of mechanical science have been making greater progress than in any former period of the world'.[87] He asserted that horology was changing and that to ignore mathematical learning meant falling behind. Denison alleged that progress in horology meant greater attention to mathematical theory concerning mechanics, but this was not simply a case of Airy's 'science' against Vulliamy's ignorance of mathematics. It was a contest over what progress in horology was; both alternatives claimed to provide accuracy and neither was as self-evident as Denison asserted. Vulliamy proclaimed his accuracy to be reliable and experience proven, and endeavoured to show that he could produce a device which was modern, 'progressive', and accurate. Despite Denison judging him to be 'unscientific', Vulliamy was a reputable craftsman, not only from a hundred-year family tradition of clockmaking, but the clockmaker to the Queen.[88] He boasted the Duke of Wellington's 'unqualified approbation' for a recently completed clock at Windsor Castle.[89]

For the Westminster clock, Vulliamy proposed a clockwork mechanism based on a smaller clock he was building at Somerleyton Hall in Suffolk for Samuel Morton Peto (1809–1889), the managing builder of the Houses of Parliament.[90] It would have no clock driving spring, this being the 'principle source of error' in other clocks; without one, the effects of temperature fluctuation, which distorted a spring's regularity, were removed. His clock would instead be driven by a 'two-seconds pendulum', making 'one vibration in two seconds of time, with a bob weighing from 450 to 500lbs., suspended without a spring, and requiring consequently no compensation'.[91] He would use a dead escapement,

[87] Edmund Beckett Denison, *Clocks and Locks. From the "Encyclopaedia Britannica". Second Edition; with a Full Account of the Great Clock at Westminster*, (Edinburgh, 1857), p. x.

[88] McKay, *Big Ben*, pp. 58–59. [89] Vulliamy to Barry, in PP. 1852 (500-I), p. 7.

[90] McKay, *Big Ben*, p. 70.

[91] Vulliamy's proposals attached to letter from Barry to A. Milne, in PP. 1852 (500-I), p. 14.

which would drive the wheel trains of the clock by a falling weight, providing a regular single revolution per six hours over a period of eight days.[92] This use of gravity would regulate 'the maintaining power' of the clock.[93] Altogether the device would strike accurate, mechanical time, 'much louder and differently from any clock in the metropolis'.[94]

Vulliamy had visited the recent Exposition de L'Industrie Français to obtain details of the latest mechanical developments in Europe, and acquired the support of Robert Willis and William Henry Smyth, the President of the Astronomical Society.[95] Both approved of Vulliamy's work and provided support to his bid to build the Westminster clock. Vulliamy reported how they had made 'valuable suggestions' and offered to replace Airy as referee for the clock. Vulliamy, who had devoted thirty years of his life 'to the improvement in House clocks', completely rejected Airy's specifications and his authority as referee, arguing that Willis was a superior judge in mechanical matters.[96] Vulliamy's choice to cite Willis' support was tactful. It was an attempt to show that Denison's claims that Vulliamy's work was not acceptable to those of mathematical training were unfounded. Willis was, after all, Cambridge University's Jacksonian Professor of Natural and Experimental Philosophy. He was continually concerned with machinery and had implemented a syllabus on mechanics at the University during the 1830s and 1840s, the focus of which had been the practical application of mathematical theory to engineering problems.[97] In referencing Willis' authority, Vulliamy was claiming that his work could satisfy the demands of an authority equal to Airy.[98] Indeed Vulliamy felt Willis to be the most qualified person in Europe to oversee the Westminster clock's construction, unlike Airy, who he considered to be too theoretical.[99]

Vulliamy also cited the support of the Company of Clockmakers of the City of London. After the government awarded Dent and Airy the contract for the clock, the Company published a complaint against Vulliamy's exclusion. They claimed that the Board of Works' rejection of Vulliamy had appeased Airy's underhand scheme to take complete control of the project.[100] Airy could not deliver a clock 'worthy of the

[92] Ibid., p. 15. [93] Ibid., p. 24. [94] Ibid., p. 13. [95] Ibid., p. 12. [96] Ibid., p. 13.

[97] Ben Marsden, 'Willis, Robert (1800–1875)', *Oxford Dictionary of National Biography*, Oxford University Press, 2004; online edn, October 2009 [www.oxforddnb.com/view/article/29584, accessed 7 March 2014].

[98] Ben Marsden, '"The progeny of these two "fellows"": Robert Willis, William Whewell and the sciences of mechanism, mechanics and machinery in early Victorian Britain', *British Journal for the History of Science*, Vol. 37, No. 4 (December, 2004), pp. 401–34, 403.

[99] Alexandrina Buchanan, *Robert Willis (1800–1875) and the Foundation of Architectural History*, (Woodbridge, 2013), p. 258.

[100] S. Elliott Atkins, *Copy of the Memorial presented to Her Majesty's Commissioners of Works and Public Buildings by the Clockmakers' Company of London, respecting the great clock to be*

building and of the art', because his only experience of a turret clock was that of the Exchange. They believed that Airy lacked knowledge of European clocks, including what the Company regarded as 'the first clock in Europe', that of the Hotel de Ville in Paris.[101] Instead of Airy, the Company recommended new referees be appointed, including Barry and a 'limited number of the profession' of clockmakers. If the Westminster clock was to replace the Hotel de Ville as the world's premier timekeeper, which they felt it should do, the guidance of the project should be removed from Airy and placed in the hands of those who understood the art of mechanical clockwork.[102]

Denison patronizingly labelled the Company the 'old brass wheel and wooden pendulum' interest and felt Vulliamy to be 'their master', but their alternative proposal for the Westminster clock had some support in Parliament.[103] Lord Brougham, the constant advocate of experimentalism, found Airy's theoretically devised clock unconvincing. Brougham instead felt Vulliamy's designs to be plausible; they were based on practical experience of clock-building. It was not so much a question of induction versus deduction that struck Brougham, as the sense of injustice he felt at Vulliamy having been offered the contract only to have it taken away. Brougham brought Vulliamy's grievances to the attention of Parliament in 1848 and 1852, and he became Vulliamy's greatest supporter in the Upper Chamber.[104] In 1848 Brougham, 'who has taken great interest in all that relates to the Houses of Parliament', had copies of correspondence surrounding the clock circulated throughout Westminster.[105] Brougham hoped this would reveal Airy's hankering for power. Brougham argued that Airy's failure to recognize Vulliamy's brilliance, combined with his desire for 'unrestricted control' over the project, revealed the Astronomer Royal to be poorly 'qualified for the high position to which he aspires'.[106] While Vulliamy orchestrated the attack on Dent and Airy's proposals, in Brougham he found a supporter who shared in his scepticism of Airy's theoretical approach to horology.

The obstacle to this claim that Airy's work was mere hypothesis was that Airy had already worked with Dent to produce the Royal Exchange clock. It therefore became a vital component of Vulliamy's attack to undermine the integrity of this apparatus. In doing so he not only tried

erected at the new Palace at Westminster; together with the answer thereto, PP. 1852 (415), p. 1.

[101] Ibid., pp. 3, 2. [102] Ibid., p. 3. [103] Denison, *Clocks and Locks*, p. 109.

[104] McKay, *Big Ben*, p. 75.

[105] (Anon.), *A portion of the papers relating to the Great Clock for the New Palace at Westminster, printed by order of the House of Lords*, (London, 1848), p. 6.

[106] Ibid., pp. 7, 11.

to sabotage Airy's horological authority, but also Dent's reputation as a constructor of accurate clocks. Dent was, he agreed, merely 'an eminent maker of marine chronometers', with only three years of experience in building public clocks, while as for Airy 'there are other individuals certainly as well, if not better, qualified to offer an opinion on the subject'.[107] According to Vulliamy, Airy and Dent's Royal Exchange clock was inaccurate; Airy's mathematical escapement had not created perfect regulation. Airy's claim that it was the 'best in the world' was 'a mere figure of speech, an assertion without proof'.[108] Vulliamy wanted the clock examined through a rate performance test to ascertain just how reliable the apparatus was. The entire work was an 'experiment', yet an experiment 'unauthenticated by any publication of the clock's rate and performance'.[109]

While Vulliamy attacked Airy and Dent, Airy endeavoured to damage Vulliamy's claims of mechanical accuracy. According to Airy, Vulliamy's designs were an 'impossible' choice, as they lacked a suitable degree of precision. The clock's pendulum suspension appeared suspect, while he believed that Vulliamy did not understand the precision required in chronometer production. His proposed clock 'would be a village clock of very superior character, but would not have the accuracy of an astronomical'.[110] Between 1846 and 1852 Denison persistently lobbied the government in favour of Airy's 'scientific' approach, contrary to Brougham's objections. He claimed that Barry and Vulliamy had 'been acting in concert ... to get the Astronomer Royal's judgement set aside', and warned Viscount Morpeth, First Commissioner of Woods and Forests from 1846 to 1850, to beware of Barry's horological dealings. As he put it, 'you are not very familiar with the art of clockmaking, and Mr. Barry may succeed in persuading you that he is'.[111] Denison suspected Barry's support of Vulliamy had more to do with Vulliamy's son than Barry's horological knowledge; from 1836 until 1841 Barry provided employment for George John Vulliamy.[112]

By 1852, Dent was firmly in place as the clock maker. Vulliamy's initial refusal to work under Airy's specifications and Airy's subsequent 1847 report on his plans left Vulliamy discredited. As Vulliamy refused to work to Airy's specifications, he was disqualified from competing against Dent and Whitehurst. Vulliamy's promise of a clock of mechanical excellence fared poorly against the expectation of Airy's astronomical accuracy. As Denison summarized, Vulliamy and the Clockmakers

[107] Vulliamy to Barry, in PP. 1852 (500-I), p. 11. [108] PP. 1852 (415), p. 2.
[109] Ibid., pp. 2–3. [110] Airy to A. Milne, in PP. 1852 (500-I), p. 49.
[111] Denison to Morpeth, in PP. 1852 (500), p. 4.
[112] Mercer, *The Life and Letters of Edward John Dent*, p. 344.

Company's 'eulogy . . . of the thirty-hour clock of the Hotel de Ville . . . as the best in Europe, merely proves that they are not only innocent, but ignorant of all the improvements in the construction of large clocks, which have been introduced since 1781, when that clock was made'.[113] Vulliamy's notion of a modern clock did not present the Astronomer Royal with an overly troublesome challenge.

The importance of time at Westminster, and who implemented it, was demonstrated in a private letter to Airy criticizing his choice of Dent. The passions the controversy aroused revealed the significance of regulating time in the national assembly. Appearing just weeks after Airy's damning report on Vulliamy's designs, the anonymous critic had little time for gentlemanly pleasantries in condemning Dent's reputation:

> You damned old scoundrel speaking against a respected man as a clock maker much more so than Dent . . . A more perfect humbug than Dent don't live – that you should disgrace yourself by supporting him I am as every one else surprized. The years the Exchange has been done & you & Dent figured away praising it before it was erected & now it is the laughing stock of all who come to change. You deserve to have an action brought against you for libelling a respectable man. I don't know much about clocks but I know this humbug you praise has been long enough about setting the chimes going at the change.[114]

Aside from this warning of the dangers of praising a clock while unconstructed, the author informed Airy that 'to suppose you are bribed upon the affair . . . is the general feeling now'.

Managing who was actually going to build the clock was the most public of two problems Airy faced. The second was that of telegraphic communication. During the early 1850s, Airy employed galvanically powered telegraphy for determining longitudes beyond Greenwich and, from 1853, projecting Greenwich time.[115] The Observatory was connected via telegraph wires through Britain's railway network, as well as to the Electric Telegraph Company in the Strand. In 1855 the Post Office applied for a direct line to Greenwich, and by the 1870s was projecting time beyond the confines of the railways to wider audiences.[116] With a cross-Channel cable to France established in 1851, and a trans-Atlantic line in place by 1866, the telegraph, which some historians have anachronistically labelled the 'Victorian Internet', promised to annihilate distance and communicate Airy's time signals instantaneously anywhere in the world.[117] (Figure 6.3)

[113] Denison to John Manners, in PP. 1852 (500), p. 26.
[114] RGO 6/607, 'Anonymous letter to George Airy', (12 July, 1847), p. 280.
[115] Meadows, *Greenwich Observatory*, pp. 61–2. [116] Ibid., p. 70.
[117] Tom Standage, *The Victorian Internet: The Remarkable Story of the Telegraph and the Nineteenth Century's Online Pioneers*, (London: 1998), p. 2; on this anachronism, see Morus, 'The nervous system of Britain', p. 456.

Figure 6.3 Britain and France connected by submarine telegraphy for the first time

However, in the 1840s and 1850s telegraphy appeared as a '*failed* technology'.[118] Early problems with insulating cables and frequent ruptures dogged telegraphic enterprises. Even the eventually triumphant Channel cable endured calamity after its initial laying in 1850. It did not take long before several overly keen French fishermen hauled up the cable, breaking it. Although engineers re-established the connection in 1851, raising fresh capital proved troublesome.[119]

While submarine telegraphy was at best challenging, overland telegraphic communication was far from easy to establish. The little understood phenomenon of delayed electric signal, or 'retardation', encountered in underground cables, ensured the apparatus remained unreliable throughout the 1850s.[120] With early batteries it was difficult

[118] Bruce Hunt, 'Scientists, engineers and Wildman Whitehouse: measurement and credibility in early cable telegraphy', *British Journal for the History of Science*, Vol. 29, No. 2 (June, 1996), pp. 155–69, 156.

[119] Marsden and Smith, *Engineering Empires*, p. 203.

[120] Hunt, 'Scientists, engineers and wildman whitehouse', p. 157; on submarine telegraphy, see Bruce J. Hunt, 'Doing Science in a global Empire: cable telegraphy and electrical physics in Victorian Britain', in Bernard Lightman (ed.), *Victorian Science in Context*,

to sustain a magneto-electric current strong enough to support a substantial telegraphic network, while the measurement of galvanic signalling was inaccurate. Around the same time that Airy proposed telegraphic communication between Westminster and Greenwich, wire faults and insulation gaps plagued London's newly laid underground telegraphy network. Without devices to accurately measure magneto-electric current, and problems of locating underground faults, telegraphy in the 1840s was expensive and experimental.[121] Moving telegraphy beyond the laboratory and showroom was difficult. With Isambard Kingdom Brunel's backing, the Great Western Railway (GWR) established a telegraphic connection from Paddington to West Drayton in 1839, extending this line to Slough in 1843. While the system of wires suspended from iron poles won great publicity, by 1849 the GWR abandoned the exercise. Telegraphy had been promoted as a way of regulating the railways and reducing accidents, but this application did not arouse much enthusiasm from railway companies.[122]

Of Airy's specifications, which included telegraphic signalling, Vulliamy was quick to draw attention to their experimental character. Under Vulliamy's direction, the Clockmakers' Company warned that 'a clock of the importance of that originally projected for the New Palace at Westminster, ought not to be made a subject for experiment'.[123] This line of argument echoed that deployed against Reid's attempts to make the permanent building a subject of experiment. As the clock would be a permanent feature of the building, making it a work of experiment seemed wholly inappropriate. Despite such concerns, Airy consistently reiterated that if the clock was to be 'the standard time for London', a galvanic communication with the Observatory was essential.[124] Although there were apparent failings with telegraphic apparatus, in 1851 Airy continued to demand that provisions be made for a time signal to Westminster. The 'idea of galvanic connexion with the Royal Observatory ... [was] still entertained'.[125] However, Airy noted that this was not so much an accepted requirement, as an enthusiastic hope. Far

(Chicago, 1997), pp. 312–33; Crosbie Smith, *The Science of Energy: A Cultural History of Energy Physics in Victorian Britain*, (London, 1998), pp. 275–77.

[121] Bruce J. Hunt, 'The ohm is where the art is: British telegraph engineers and the development of electrical standards', *Osiris*, 2nd Series, Vol. 9, Instruments (1994), pp. 48–63, 50; on measuring electrical resistance, see Simon Schaffer, 'Late Victorian metrology and its instrumentation: a manufactory of ohms', in Robert Bud and Susan E. Cozzens (eds.), *Invisible Connections: Instruments, Institutions, and Science*, (Washington, 1992), pp. 23–56.

[122] Marsden and Smith, *Engineering Empires*, pp. 194–96. [123] PP. 1852 (415), p. 3.

[124] Frederick Dent to William Molesworth, in PP. 1854–55 (436), p. 13.

[125] RGO 6/608, 'Letter from George Airy to T. W. Phillips', (21 February, 1851), p. 18.

from ensure a telegraphic signal at Westminster, in 1852, Airy's critics noted that he had not even established a sustained link from Greenwich to the already completed Royal Exchange clock.[126]

Intrinsically linked to the trouble of telegraphy was the problem of galvanically controlled electric clocks. To sustain a current to project a time-signal involved apparatus only recently tried; to employ such instruments would also be experimental. Airy raised concerns over its use, noting that in suggesting

> the practicability of employing the spare power of the Large Clock to excite a magneto-electric current, by means of which many or all the smaller clocks in the Palace could be kept in motion simultaneously with the Large Clock ... I expressed a doubt (founded on my own experiment).[127]

Airy was worried that he did not have a sufficient amount of 'accurate information' on the subject of galvanically powered clocks to be able to introduce them on an extended scale in Parliament.[128]

To bring about uniform time in the Palace, Dent and Airy agreed that they should produce every clock within it. This was such a fundamental criterion for their time system, that when it was discovered that Barry had asked Vulliamy to provide several internal room clocks, Airy and Dent threatened to abandon the project.[129] It was, in their view, vital Dent compete to build all the clocks in Parliament; if not he would withdraw his designs for the large clock.[130] After receiving Dent's threats, Alexander Milne, co-commissioner of the Office of Woods and Forests, instructed Barry to order no more clocks from Vulliamy, despite him 'having great faith' in the clockmaker.[131] Vulliamy's clocks, manually regulated in accordance with chronometers and the chimes of the Westminster clock, were insufficient to provide the accurate time system which Airy envisaged. His clocks threatened to 'prevent the application of this principle' of instantaneous time.[132]

However, if Vulliamy's trusted system of human regulation provided unspectacular accuracy, Airy's galvanic time system required much experimental inquiry. Opposing Vulliamy's designs was easier than providing a credible alternative. The major difficulty with a galvanic telegraphic current, capable of transmitting a reliable time signal, was the

[126] PP. 1852 (415), p. 2.
[127] RGO 6/607, 'Letter from George Airy to Charles Gore', (22 September, 1847), p. 117.
[128] RGO 6/608, 'Letter from George Airy to T. W. Phillips', (21 February, 1851), p. 19.
[129] RGO 6/607, 'Letter from George Airy to Morpeth', (27 March, 1847), p. 51.
[130] RGO 6/607, 'Letter from Edward Dent to the Commissioners of H. M.'s Woods and Forests', (3 July, 1847), p. 162.
[131] RGO 6/607, 'Letter from A. Milne to Edward Dent', (14 July, 1847), p. 164.
[132] RGO 6/607, 'Letter from George Airy to Charles Gore', (22 September, 1847), p. 117.

problem of galvanic batteries. To generate a current which could drive several clocks simultaneously involved the use of a series of galvanic cells connected together to form batteries, but the metallic surface of these were liable to oxidize.[133] In attempting to construct 'sympathetic' clocks for the Royal Observatory, Dent and Airy had experimented on various batteries during the 1840s. Both the 'Smee' battery (a fluid type consisting of one part sulphuric acid to seven parts water, with a positive platinum plate and a negative zinc plate) and the 'Sand' battery (containing siliceous sand moistened with sulphuric acid) were available to Airy in the 1840s, but neither offered the reliability he sought.[134]

After consulting Wheatstone, Airy asked Dent to experiment on employing a clock's striking hammer to generate a powerful time signal.[135] Working alongside Wheatstone, Dent found the power of a hammer to be insufficient; a finding which Airy initially found discouraging.[136] However, through the summer of 1847 he enacted his own experiments which allayed these fears. In February 1847 Airy applied to Watkins and Hill of Charing Cross, the 'Mathematical, Philosophical & Chemical Instrument Makers, to Her Majesty', for a 'Magneto Electric Inductive Machine' with a lever handle operating a horizontally mounted contact breaker.[137] This modelled the action of a clock hammer breaking an electric circuit to generate a signal. Using this device, Airy attempted to show 'the effect of breaking magnetic contact ... in producing a magneto-electric current'.[138] This apparatus generated a current sustained with battery power. Practice with this device convinced Airy that it was possible to use galvanically controlled clocks at Parliament, and after a visit to St Petersburg at the end of the summer, during which he witnessed some limited demonstrations of galvanic clock control, he was sure of its practicality.[139] When in 1851 the commissioners of the Office of Woods and Forests, still unconvinced of Airy's galvanic system, attempted to open up the manufacture of internal clocks to competition,

[133] John A. Chaldecott, 'Platinum and the Greenwich system of time-signals in Britain: the work of George Biddell Airy and Charles Vincent Walker from 1849 to 1870', *Platinum Metals Review*, 30 (1), pp. 29–37, 29.

[134] Ibid., pp. 31–32.

[135] RGO 6/607, 'Letter from George Airy to Charles Wheatstone', (27 November, 1846), p. 251; RGO 6/607, 'Letter from George Airy to Edward Dent', (26 February, 1847), p. 149; on Wheatstone, see S. P. Thompson, 'Wheatstone, Sir Charles (1802–1875)', rev. Brian Bowers, *Oxford Dictionary of National Biography*, Oxford University Press, 2004; online edn, January 2011 [http://ezproxy.ouls.ox.ac.uk:2204/view/article/29184, accessed 1 September 2014].

[136] RGO 6/607, 'Letter from Edward Dent to George Airy', (27 February, 1847), p. 150.

[137] RGO 6/607, 'Letter from George Airy to Messers Watkins and Hill', (17 February, 1847), pp. 244–5.

[138] Ibid., p. 240.

[139] RGO 6/607, 'Letter from George Airy to Charles Gore', (22 September, 1847), p. 117.

Airy implored the Board to delay this course until the subject of electric clocks was better understood.[140] Even in 1851 the apparatus was problematic, but Airy still felt 'the adoption of this agency through the clocks of the New Palace' could be made a reality. If only one clock could be 'kept going by another' for a year without error, 'the problem may be considered as practically solved'.[141]

The solution to the problem was eventually found in Charles Shepherd's (1830–1905) system of sympathetic clocks. Airy had refused the Greenwich clockmaker Samuel McClellan's proposed system in 1847 after McClellan boasted that he could provide 'time in every room governed by a common sized 8 day clock'.[142] After inspecting the system Airy felt the plan unsuitable for the Palace and ended communications with McClellan.[143] In 1851 however, Airy witnessed a system he believed to be perfect for Parliament, which he later implemented at Greenwich. Shepherd's principle of clocks galvanically controlled by one single 'master' clock was displayed at the Great Exhibition. Airy visited Shepherd to discuss this system in June 1851. At Shepherd's shop at 53 Leadenhall Street, Airy was impressed by the displayed system of simultaneous clock movement. He then saw the system's practicality demonstrated in a draper's warehouse in St Paul's Churchyard. There Airy observed the principle applied 'in constant use', driving eight different clocks in different parts of the factory. This application of galvanically regulated time in a factory environment was precisely what Airy wanted to implement at Westminster. All eight clocks were powered 'by one battery and regulated by one pendulum'.[144] Airy wrote to Lord Seymour, Commissioner of Works, recommending Shepherd be brought in to establish a similar system of factory time in the Palace; Shepherd could make the Westminster clock the 'master' of a time regime.[145]

While at Parliament Airy waited on Seymour's instruction, at Greenwich he worked with Shepherd to introduce this system. Shepherd chose twenty-eight Smee batteries to employ in a system of sympathetic clocks constant with Greenwich astronomical time, in the Observatory.[146] From 1852 Shepherd's galvanic apparatus was used to

[140] RGO 6/608, 'Letter from George Airy to T. W. Phillips', (21 February, 1851), p. 15.
[141] Ibid., pp. 19–20.
[142] RGO 6/607, 'Letter from Samuel McClellan to George Airy', (2 December, 1847), p. 119; McClellan had a bid to build the Westminster clock rejected in 1848.
[143] RGO 6/607, 'Letter from Samuel McClellan to George Airy', (21 December, 1847), p. 120; RGO 6/607, 'Letter from George Airy to Samuel McClellan', (23 December, 1847), p. 121.
[144] RGO 6/608, 'Letter from George Airy to Lord Seymour', (14 June, 1851), p. 27
[145] Ibid., p. 28.
[146] Chaldecott, 'Platinum and the Greenwich system', p. 33; on later battery developments see p. 35.

connect the Observatory's transit clock with the Greenwich time-ball and the Observatory gate clock to project accurate time to a public audience, as well as regulate time within the Observatory.[147] Although Airy appeared to have found a viable means of projecting Greenwich time at Westminster galvanically, beyond the controlled confines of the Observatory, the apparatus lacked the political support to be accepted within Parliament. While Airy might administer regulation at Greenwich under his own direction, attempting to introduce Shepherd's system in Parliament remained laborious.

Restraining the Astronomer Royal and Building the Westminster Clock

Airy's endeavours to establish a time system at Westminster reveal the challenges in implementing a regime of regulated order. Perhaps the greatest demonstration of these difficulties was that in Parliament, Airy's proposals were only followed in a very limited form. Following a disagreement with Denison in 1853, Airy had little involvement at Westminster, beyond a final review of the clock and bell on completion. Instead, the time system built into Parliament was Denison's conception and what he produced was very different to Airy's scheme. Denison had little time for galvanic time regulation and was loath to place the Westminster clock under the direct 'power' of the Astronomer Royal. His clock would itself keep accurate time with only limited reference to Greenwich's authority. Yet while Denison's proposals differed to Airy's, he still claimed to be scientific. His clock was intended to emphasize mathematical accuracy, combined with experience, and it would project time through the Palace: not by galvanic power, but by the chimes of the clock's bells.

Since 1848, Denison had supported Dent's application to build the clock. Educated at Eton and Trinity College Cambridge, Denison was a self-proclaimed authority on horology, architecture, and ecclesiastical matters. In 1841, he began a career as a lawyer, collaborating with his father, MP for Doncaster and vice-chairman of the Great Northern Railway Company, to frame private bills for establishing new railways. An evangelical with strong Anglican connections, in 1845 he married Fanny Catherine Lonsdale, the daughter of the Bishop of Lichfield.[148]

[147] Bennett, 'George Biddell Airy and horology', p. 281.

[148] L. C. Sanders, 'Beckett, Edmund, first Baron Grimthorpe (1816–1905)', rev. Catherine Pease-Watkin, *Oxford Dictionary of National Biography*, Oxford University Press, 2004; online edn, May 2007 [http://ezproxy.ouls.ox.ac.uk:2117/view/article/30665, accessed 24 August 2013].

As well as expressing his dissatisfaction with Gothic architecture at the Royal Institution, Denison was a consistent commentator on horological matters, publishing *A rudimentary treatise on clock and watchmaking* in 1850, as well as articles on clocks, watches, and bells for the *Encyclopaedia Britannica*.[149] Pursuing mathematics at Cambridge, Denison graduated as thirtieth wrangler in the 1838 Tripos and through the 1840s earned a reputation sufficient to be appointed chairman of the Great Exhibition's horological jury.[150]

From 1848 until its completion, Denison was the permanent guiding figure in establishing a Westminster time system.[151] (Figure 6.4) As early as 1846 Denison was in communication with Airy, proposing his own escapement for the clock, which Airy received amicably.[152] Together, Denison and Airy consulted each other and developed an escapement with a remontoire action and innovative rollers on the clock wheel to reduce friction from the pulse of the pendulum.[153] Denison was well aware that Airy's approval as referee was vital to be included in the project and worked hard to win his approbation.[154] He fully supported both Airy and Dent in the controversy with Vulliamy, applauding their scientific zeal as the correct course for establishing time at Parliament. Reflecting on Airy's rejection of Vulliamy's proposals, Denison considered the case as an emblematic struggle between modern accurate astronomical time, and inferior antiquated mechanical accuracy. The way Denison saw it, while Airy had demanded an accuracy worthy of Parliament, England's clockmakers had stood idly by, ignoring the Astronomer Royal's 'progressive' innovations. For the Exchange clock, Airy had demanded 'scientific improvement' and Dent had delivered it.[155] Dent was 'the first horologist in the world', succeeding in the face of 'the genuine English hatred of improvements'.[156] Denison felt Airy's demand for the first striking blow of each hour to be accurate to within a second of time to be a entirely appropriate for the clock. While Dent provided this, Denison believed Vulliamy's plans were 'founded on mistaken and unscientific notions'.[157] When in March 1853 Dent died and Vulliamy launched one final attempt to secure the clock contract, Denison insisted that all of

[149] See Edmund Beckett Denison, *On some of the grounds of dissatisfaction with modern gothic architecture. A lecture delivered at the Royal Institution of Great Britain*, (London, 1859).
[150] Peter Ferriday, *Lord Grimthorpe, 1816–1905*, (London, 1957), pp. 7, 20.
[151] Port, 'Barry's Last Years', p. 169.
[152] RGO 6/607, 'Letter from George Airy to Edmund Denison', (15 December, 1846), p. 173.
[153] RGO 6/607, 'Letter from Edmund Denison to George Airy', (23 December, 1846), p. 175.
[154] Ibid., p. 176. [155] Denison, *Clocks and Locks*, p. 75. [156] Ibid., pp. 76–77.
[157] Ibid., p. 103.

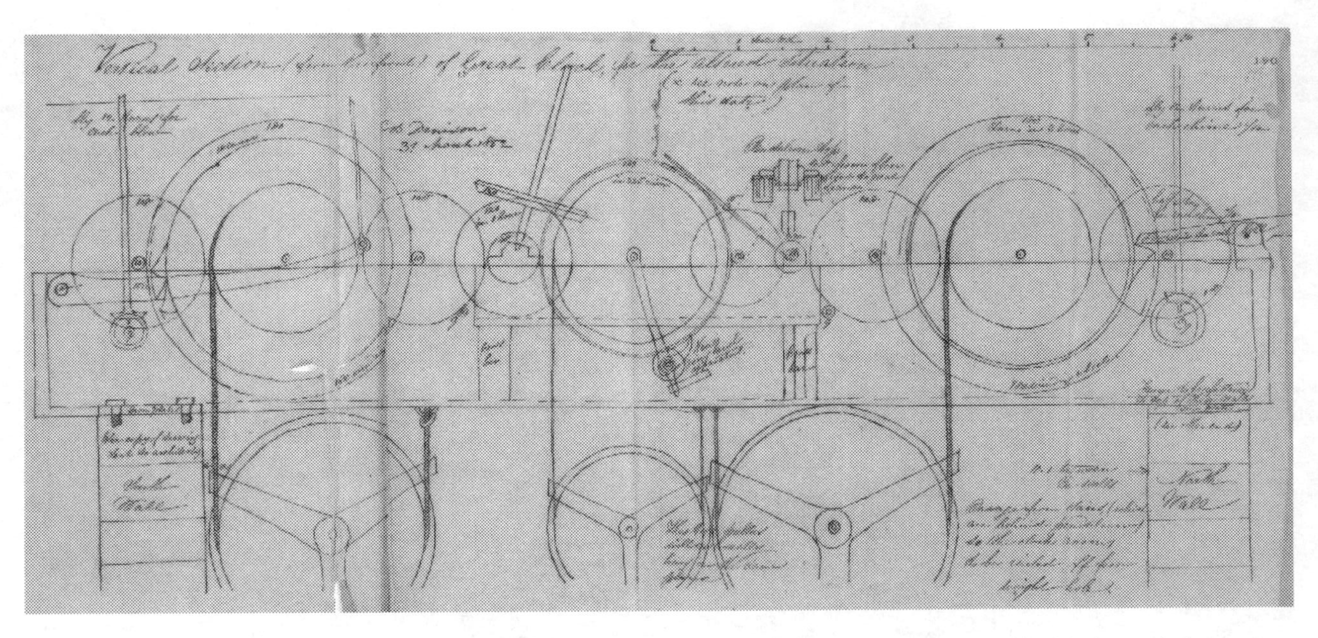

Figure 6.4 Denison's original plan of the vertical section of the Westminster Clock. Reproduced by kind permission of the Syndics of Cambridge University Library

Dent's 'scientific business' be passed on to his stepson Frederick Rippon Dent.[158]

Denison poured praise not only on Dent's skill, but Airy's horological authority. On smooth running teeth for clock wheels, the 'most comprehensive view of the whole theory' was Airy's.[159] The aim of this theory was to 'uniform' the relative velocity of the contact of teeth with wheels driving the clockwork. In November 1851, he consented to join Airy in the construction of the clock.[160] At first it appeared a mutually beneficial arrangement. Airy expressed his respect for Denison when he explained how 'Every opinion of yours shall have *at least* as much weight with me as my own'.[161] However, although Denison was eager that the Westminster clock should embody the 'progressive' science of the age, and lauded both Dent's precision and Airy's demands of accuracy, he differed from the Astronomer Royal's conception of science and, with it, his understanding of horology. Denison had publicly criticized Airy's 1826 work on clock escapements which held that gravity escapements were inaccurate. A gravity escapement consisted of two arms which were alternatively raised to a certain height and then allowed to descend a fixed distance to provide a constant impulse to the pendulum and it was this mechanism that Denison initially proposed for the Westminster clock.[162] In two papers to the Cambridge Philosophical Society in November 1848 and February 1849, Denison claimed that gravity escapements produced very little friction and only minor pendulum arc variations.[163]

It was not only their Cambridge Tripos performances which separated Airy and Denison: Denison had his own very specific views of science. Denison was sceptical of mathematicians who relied solely on calculations, devoid of any practical experience, when dealing with scientific problems.[164] Writing years after working with Airy, Denison boasted that while it was fashionable for 'scientific and moral philosophers of a certain school to emulate maxims or dogmas ... like the axioms of geometry or mechanics', he dealt only in 'purely scientific alternatives and probabilities'.[165] For Denison, to be 'inductive' was to be scientific. Pure mathematics was only of theoretical value and mere hypothesis.

[158] Ibid., p. 112.
[159] Edmund Beckett Denison, *A Rudimentary Treatise on Clocks and Watches, and the Bells; with a Full Account of the Westminster Clock and Bells*, (London, 1860), p. 261.
[160] Wilfrid Airy (ed.), *Autobiography of Sir George Biddell Airy*, p. 213.
[161] RGO 6/608, 'Letter from George Airy to Edmund Denison', (7 February, 1852), p. 146.
[162] McKay, *Big Ben*, pp. 151–53. [163] Ibid., p. 135.
[164] Edmund Beckett Denison, *Astronomy Without Mathematics*, (London, 1865), pp. 2–3.
[165] Edmund Beckett, *On the Origin of the Laws of Nature*, (London, 1879), pp. 1, 4.

Although not mentioning Airy specifically, Denison was critical of mathematical theory in natural philosophy, asserting that,

All that we can say of the well-known law of gravity is that it is shown to be immeasurably more probable than any other explanation of the motions of the universe. The undulating theory of light and heat is at present the most probable one because it explains all the known phenomena better than any other; but there is not the smallest direct proof of the luminiferous aether which it assumes.[166]

Denison believed that mathematics was a vital aspect of clock-building, but he emphasized the discipline's limits and, in a manner not dissimilar to Vulliamy, applauded the role of experience and observation in practical science, including horology.

Not long after commencing work together formally, Airy and Denison's differences surfaced. Denison was well known for his bad temper and quickness to disagree. During discussions surrounding the clock's remontoire action, their differing opinions erupted into argument. While Airy remained content that his mathematical escapement secured accuracy, Denison continued to develop his own designs. When in January 1852 Denison produced a set of drawings for Dent to follow, Airy raised several concerns which revealed very different conceptions of how the clock should be built. Airy noted that Denison's plans included a pendulum of five cwt which would drop through the centre of the clock tower. (Figure 6.5) This enormous weight was, Airy reckoned, excessive: such a pendulum might cause 'the whole tower to swing'.[167] Denison admitted it 'was perhaps chiefly a matter of fancy, to have it the heaviest in the world'. He wanted a 5-cwt pendulum because the Royal Exchange's pendulum weighed 4 cwt. Airy damned this pendulum envy as a moment of 'the ancient barbarism, whose people knew no other way . . . [than the] action of a simple train carrying very large hands'. Rather than vainly aspiring to size, precision and accuracy should be prioritized. He believed that 'Instead of setting up the *heaviest* pendulum in the world, it will be a far greater triumph to set up the *lightest* pendulum in the world'.[168] A light pendulum would make the clock easier to regulate. Airy likened this reduction in size to the accuracy of an intricate chronometer. However, Denison remained obstinate, reiterating that size mattered, and it mattered that Parliament have the world's biggest pendulum. Unable to resolve the weight by correspondence, Airy and Denison met in February and agreed on a heavier pendulum.

[166] Ibid., pp. 4–5; see Chapter 5 for Airy's view of the undulatory theory of light.
[167] RGO 6/608, 'Letter from George Airy to Edmund Denison', (30 January, 1852), p. 125.
[168] RGO 6/608, 'Letter from George Airy to Edmund Denison', (2 February, 1852), p. 129.

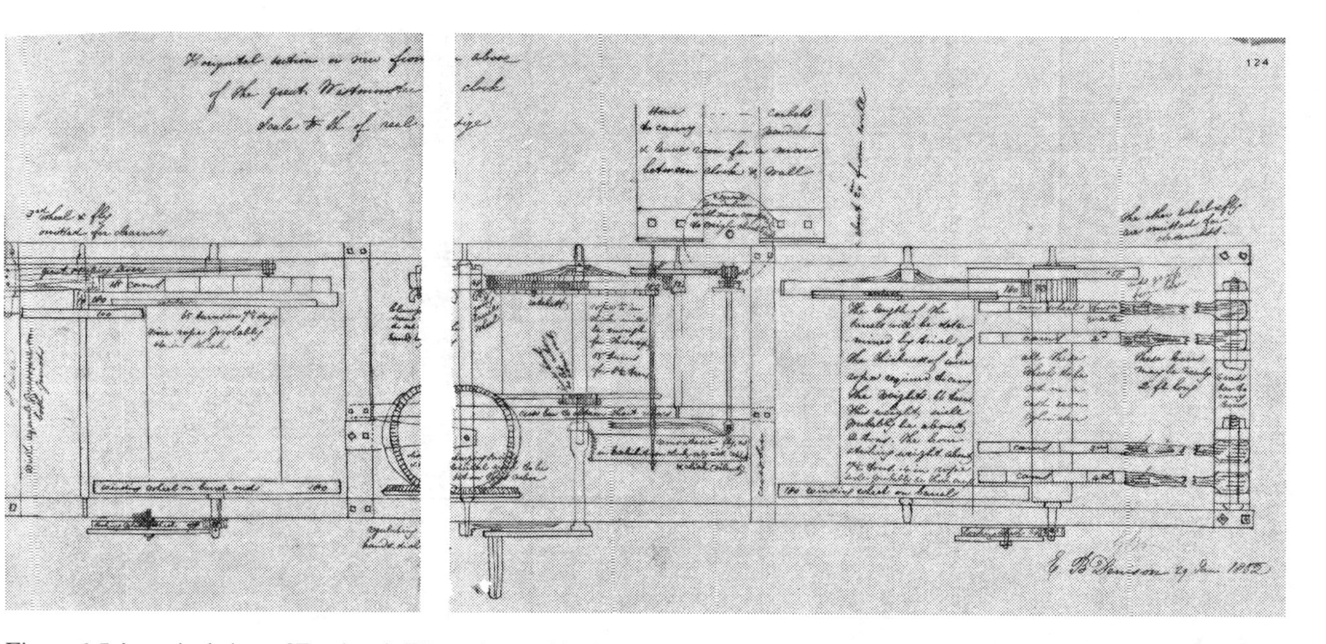

Figure 6.5 A vertical view of Denison's Westminster Clock, drawn in January 1852 for Airy's approval and Dent's instruction. Reproduced by kind permission of the Syndics of Cambridge University Library

Along with this disagreement the two designers also contested the means of regulating the pendulum. In order to alter the clock's rate daily for deviations of two or three seconds, Airy devised a rack-and-pinion mechanism which moved a sliding weight near the base of the pendulum.[169] When weight was applied, the pendulum's going increased, while reducing the weight retarded it. Denison though, disliked Airy's penchant for devising new mechanisms. What Airy proposed was a 'mere fanciful novelty'.[170] Although prepared to see a model of Airy's 'pendulum-retarding apparatus', Denison preferred a manual system of correction. He would only be convinced if he could witness the device in action. Otherwise he maintained that the pendulum should be corrected by adding or removing weights to the top of the pendulum bob.

Airy felt this disagreement was eroding both his own and Denison's authority. Dent, who was supposedly under their direction, was gaining too much 'power' in light of their estrangement.[171] Airy's solution was that when any difference of opinion occurred between Denison and himself, the matter would be settled 'by a superior authority' and that authority was not Dent, but the First Commissioner of Works, Lord Seymour.[172] This decision was crucial because it established that resolving horological disagreements was not an experienced clockmaker's business, but a politician's. Airy thus asserted that Seymour, despite lacking mechanical know-how, was a more suitable authority than Dent. Denison objected to this selection of Seymour as an adjudicator, believing it to be a most 'unpractical turn'. As he saw it, the only disagreement he had regarded the regulation of the pendulum. All Denison wanted was to see a trial of Airy's mechanism. Clocks were not built by resolutions but by 'argument or experiment'.[173] Denison recognized the problem of authority in managing the project, but preferred Dent as a judge instead of Seymour. Dent's opinion would be more valuable 'than the superior authority of any First Commissioner'. If Dent agreed with Airy on any point, Denison promised to concede. In this spirit, Denison urged Airy to see sense in 'throwing this auxiliary . . . train of correspondence out of gear'. They would do 'more good by discussing the flexibility of remontoire-springs than of resolutions'.[174] Yet although Denison advocated attention to mechanics rather than authority, he shared

[169] RGO 6/608, 'Letter from Edmund Denison to George Airy', (7 February, 1852), p. 143ii.
[170] RGO 6/608, 'Letter from George Airy to Edmund Denison', (9 February, 1852), p. 149ii.
[171] Ibid., p. 149. [172] Ibid., pp. 149ii–150.
[173] RGO 6/608, 'Letter from Edmund Denison to George Airy', (10 February, 1852), p. 151ii.
[174] Ibid., p. 152ii.

Airy's concerns over Dent assuming excessive autonomy. He was quick to assert that Dent could never be 'an umpire between us' and warned that Airy's obstinacy would leave the clockmaker 'at liberty to do exactly as he pleased'. In spite of this Airy and Denison agreed to cooperate without Seymour's judgement and in reference to Dent's advice.

This solution appeared at first to establish order in the project. Airy went abroad on Observatory business in April 1852, leaving Denison and Dent in charge of the clock. While he was away they agreed on a method of retarding and accelerating the pendulum. Denison informed Airy that in Dent's workshop measures had been taken for manually placing or removing weights to the pendulum bob (see Map A). However, Denison also reported that in Dent's opinion the most effective part of the pendulum on which to add or remove weights was at the middle of the rod. Dent spoke from experience on this point. Denison therefore proposed 'a broadish collar ... fixed on the pendulum rod at half its height'.[175] The clock's attendants would be provided with an appropriate set of weights. Removing weights from the collar would quickly slow the pendulum's swing. Denison had Dent try it with an existing pendulum in the workshop and found that all of Dent's workmen could perform the task with ease. What this was, was a simple practice which any workman could master, rather than an elaborate apparatus. Airy's response was not enthusiastic. He doubted that a humble workman could be trusted to regulate the clock and maintained that a well-developed mechanism which reduced the risk of human error was preferable.[176] He feared that indelicate workmen might shake the pendulum rod while placing or removing weights and cause inaccuracies in the clock.

Disagreements continued to dog the project. In November 1852, while observing the 'going on' of the Westminster clock's pendulum in Dent's factory near the Strand, Denison conceived of an improved '3 legged escapement converted into a gravity escapement with the additions of long stopping teeth'.[177] This reduced further the friction encountered in converting pendulum movement into clockwork motion. To Airy's protests against this new device, Denison argued that it had to be 'seen ... in real life'.[178] Denison's confidence in mechanical clockwork design from observation and practical experience lacked the mathematical credibility

[175] RGO 6/608, 'Letter from Edmund Denison to George Airy', (15 March, 1852), p. 168.
[176] RGO 6/608, 'Letter from George Airy to Edmund Denison', (16 March, 1852), pp. 170–170ii.
[177] RGO 6/608, 'Letter from Edmund Denison to George Airy', (27 November, 1852), pp. 218–19.
[178] RGO 6/608, 'Letter from Edmund Denison to George Airy', (18 December, 1852), p. 222; for Airy's protests, see RGO 6/608, 'Letter from George Airy to Edmund Denison', (18 December, 1852), p. 221.

Airy felt to be so important. The tensions between Airy and Denison continued into 1853 until eventually Airy withdrew from the project altogether. This time, however, problems emerged in relation to working with Barry. Since the competition which had seen Dent chosen ahead of Vulliamy, Barry had remained committed to his initial choice of clock builder. As a result, Denison was 'incurably suspicious of Barry as an inveterate jobber' and refused to cooperate with him.[179] Offended by Denison's manner of business, Airy withdrew from the project informing Denison that 'we cannot with advantage profess to act in concert'.[180] Although he was convinced of Denison's 'mechanical ingenuity and horological knowledge', the lawyer's character was in dispute. Working with him convinced Airy that 'our ideas of the mode of conducting public business are very different'.[181] Despite Denison's continual requests that Airy return to the project and offer advice, Airy refused, and only accepted a supervisory role after persuasion from the Board of Works.[182] Airy thus secured the right to examine the clock once complete and approve payment to the clockmaker. The Office of Works' reluctance to lose Airy's involvement with the clock is telling of its dependence on the Astronomer Royal.

Airy's effective role in the project after 1853 was limited. Denison took charge of the establishment of time at Parliament and soon asserted his own conceptions of accuracy to the endeavour, 'even while Mr Airy's nominal superintendence continued'.[183] Denison though, lacked Airy's conviction that galvanic-telegraphic apparatus provided the means of ensuring scientific improvement. While he wanted 'to be always exact within a second of Greenwich time', he felt Airy's daily reporting to Greenwich to be excessive.[184] Denison insisted a weekly check would be sufficient. He reduced the control of Greenwich over Westminster time and contended that he would make the clock so accurate that it would itself be an authority with minimal supervision. By 1860 Denison reported that 'Provision is made in the clock for reporting its own rate of going to the Greenwich Observatory at any convenient hour' by electric

[179] RGO 6/608, 'Letter from Edmund Denison to George Airy', (10 June, 1852), pp. 201–201ii.

[180] RGO 6/608, 'Letter from Edmund Denison to George Airy', (1 November, 1853), p. 254; RGO 6/608, 'Letter from George Airy to Edmund Denison', (7 November, 1853), p. 260.

[181] RGO 6/608, 'Letter from George Airy to William Molesworth', (7 November, 1853), pp. 76ii–77.

[182] RGO 6/608, 'Letter from Edmund Denison to George Airy', (21 November, 1853), p. 261; RGO 6/608, 'Letter from Edmund Denison to George Airy', (1 April, 1854), p. 264; RGO 6/608, 'Letter from George Airy to Edmund Denison', (4 April, 1854), p. 266; Denison to the Commissioners of Works, &c., in PP. 1854–55 (436), p. 19.

[183] Denison, *Clocks and Locks*, p. 113. [184] Ibid., pp. 125, 129.

telegraph, but that the system was still awaiting the government to finance a connection to a local railway station.[185] Denison had no sense of urgency about this. He argued that the manual practices of regulating the clock were sufficient. From the Clock Tower the recently erected time-ball in the Strand, regulated from Greenwich to drop every day at one o'clock, could be observed, while the striking of the Strand clock could be heard at Westminster. Denison calculated the sound travelled at four and half seconds a mile to reach the tower, meaning accurate regulation might be maintained by manual observation.[186] He suggested that this was 'sufficient without the other electrical connection'. Effectively Denison abandoned Airy's accuracy based on instantaneous galvanic time signalling.

Denison instead placed emphasis on the accuracy of the clock itself as a source of authority. The three main challenges for Denison were protecting the pendulum from external influences, avoiding losing time when winding the clock, and ensuring the clock could be accurately controlled from within the Clock Tower. Instead of Airy's remontoire escapement, Denison employed his own means of regulating the driving power of the clock. His 'double three-legged gravity escapement' provided a constant impulse to the pendulum to protect it from interference from the weather and this ensured the clock was driven with regularity.[187] (Figure 6.6) The face hands of the clock were liable to be disturbed by wind, rain, and snow, which threatened the integrity of the pendulum, but Denison was confident his escapement reduced this problem. This device consisted of two arms which were in turn raised before lowering, allowing a circular wheel to rotate which regulated the beating of the pendulum.[188]

To solve the difficulty of losing time while winding the clock, Denison considered using 'the tide of the river below', despite proposals to 'turn the spire of the tower into a windmill cupola'.[189] Instead he employed a winding wheel with a set of ratchet teeth which could be wound without stopping the clock.[190] Despite these measures, Denison accepted the clock might gain or lose time, and so equipped his clock with apparatus to compensate for error. This was the system which Dent had proposed as an alternative to Airy's rack-and-pinion apparatus for correcting deviations. Towards the top of the pendulum a 'collar' was fixed to carry small regulating weights.[191] This provided what Denison believed to be a mathematically grounded practice of securing accuracy. He explained that using differential calculus for the radius of oscillation, it could be

[185] Denison, *A Rudimentary Treatise*, pp. 260–61. [186] Ibid., p. 261.
[187] McKay, *Big Ben*, p. 130. [188] Ibid., pp. 151–52.
[189] Denison, *Clocks and Locks*, p. 128. [190] Denison, *A Rudimentary Treatise*, p. 249.
[191] Ibid., p. 246.

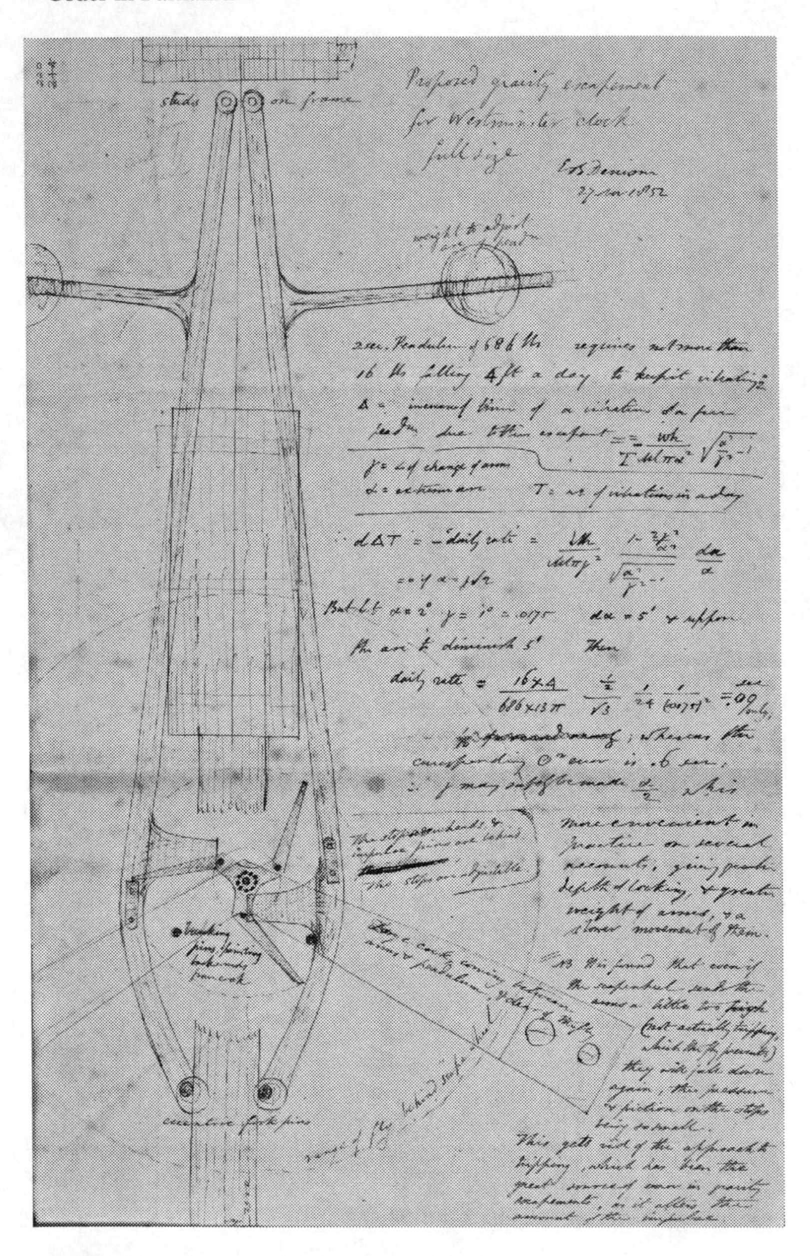

Figure 6.6 The full-scale drawing of Denison's three-legged gravity escapement for the Westminster Clock, dated 27 November 1852. Reproduced by kind permission of the Syndics of Cambridge University Library

calculated how much weight was to be added or removed from the pendulum collar to establish the correct time. Denison demonstrated the precision of this practice as he explained how,

> small weights can easily be put on or taken off . . . if you have a set of them marked ¼, ½, 1, 2, 4, according to the number of seconds a day by which they will accelerate, you can make any regulation of it for a range of errors amounting to 8 seconds a day.[192]

The practice of regulating the clock was therefore manually controlled at Parliament, rather than galvanically from Greenwich.

As for disseminating time around Parliament's rooms, Denison was not interested in galvanic apparatus, instead focusing his efforts on the construction of the clock's bells. Without galvanic clock regulation, the five Westminster bells became the primary technique of projecting the time of the Westminster clock. Denison believed that the production of five well-toned bells, including the largest ever to be cast in England (14 tonnes) called for experiment.[193] He felt that as a bell of such scale was itself a radical trial, its construction required extensive research and investigation. The result of Denison's 'observations and experiments' was a bell he believed to be of good sound and strength.[194] To compensate for the large striker used to generate enough sound to echo through Parliament and allow government's clocks and politicians' watches to be set to Westminster time, it was important that the bell had a great 'power both of bearing blows and of giving out sound'.[195] Denison strove to calculate the ideal size, shape, and chemical composition for a bell to regulate government. The cost of £100 for experiments on these characteristics was considered slight, 'compared with the importance of success or failure in a national work of this kind'.[196]

The Chief Commissioner of the Board of Works sent Barry and Wheatstone to examine the bells on display at the 1855 Paris Industrial Exhibition, but little useful information was obtained. The Board then commissioned Wheatstone to gather details on bell founding for Denison 'respecting the most esteemed chimes in France and Belgium'.[197] Wheatstone, however, concluded that Continental bell founders failed to employ the 'modern discoveries of science' in their work.[198] This compelled Denison to do his own research to guide his designs. After experiments on several of Europe's most esteemed bells, he settled on a

[192] Ibid., p. 66.
[193] Edmund Beckett Denison, 'On the great bell of Westminster', *Royal Institution of Great Britain*, (Friday 6 March, 1857), p. 1.
[194] Ibid., p. 2. [195] Ibid., p. 3. [196] Ibid., p. 3.
[197] Wheatstone to John Thornborrow, in PP. 1854–55 (436), p. 26. [198] Ibid., p. 26.

shape 'something between ... the great bell of Notre Dame ... [and] the great bell of Bow ... the same as that of St Paul's, York'.[199] Noticeably absent from Denison's plan was the use of mathematical theorems. Instead of 'geometric rules', he favoured 'purely empirical' considerations in the design of the great bell.[200] He was pleased to note the figure of the diameter of his bell was based on 'no kind of *a priori* reason'. His bell was a construct of empirical 'trial'.

As for the composition of the bells, Denison boasted that by working alongside Wheatstone, they had been made referencing chemical knowledge. Generally, bells were cast with a five to three copper to tin ratio.[201] Denison agreed with this compound for his bells, but questioned the correct ratio. Through experiment he claimed to have shown a three to one ratio produced an alloy 'like glass', yet a low tin content reduced the sweetness of sound produced. 'By trial' he asserted that the ideal copper-tin proportion was twenty-two to seven exactly, and produced chemical tables comparing the Westminster Bell to those at York and Lincoln.[202] Denison's bell was strong, euphonically pleasing, and seemingly 'elastic'.[203] Above all it was grounded in observations of existing bells and experiments on the properties of various chemical compositions. If Denison could propound no 'scientific theory of bell-founding', he could at least argue that the manner of projecting time at Parliament was grounded in experiment. Despite these efforts, Denison's great bell ('Big Ben'), cast in 1856, cracked in 1857 after being transported from Messers Warner and Sons of Cripplegate to London (Figure 6.7). Denison arranged for Mears of Whitechapel to recast the bell, having blamed Warner's casting skill for the original's failure.[204] Using the same alloy, the bell was recast in April 1858.

In 1860 Airy reported to the Board of Works on Denison's completed clock, generally applauding it as a machine of appropriate accuracy. Although testifying to Denison's 'horological skill', Airy was critical of his choice to replace his remontoire action. Airy did not feel Denison's escapement would prevent 'sudden gusts of wind acting on the hands' from causing the clock to vibrate and lose regularity.[205] Publicly, however, the clock was applauded as a spectacular work of horology. As the clock neared completion, it attracted increasing public interest. In 1857, *The*

[199] Denison, 'On the great bell of Westminster', p. 4; for a comparison with bells in nineteenth-century rural France, see Alain Corbin, *Village Bells: Sound and Meaning in the Nineteenth-Century French Countryside*, (London, 1999).
[200] Denison, 'On the great bell of Westminster', p. 7. [201] Ibid., p. 8. [202] Ibid., p. 11.
[203] Ibid., p. 3. [204] McKay, *Big Ben*, p. 104.
[205] Airy in, *The Bell (new Palace, Westminster). Copy of reports on the condition of the great bell of the new Palace of Westminster*, PP. 1860 (553), p. 6.

Figure 6.7 The spectacular recasting of the Great Bell for the clock tower attracted a considerable crowd in Whitechapel

Times reported triumphantly that the clock's modernity would ensure that 'the present Houses of Parliament will always remain an enduring record of our civilization, wealth, and greatness' (Figure 6.8). A telegraphic link to Greenwich promised to ensure accurate time, which the publication asserted to be a huge concern for the nation, but was confident the link was far from essential when the predicted deviation of the timepiece was not even two seconds in two years. The whole project was portrayed to readers as being at the very forefront of science. *The Times* informed audiences of the clock's huge pendulum and intricate workings, which to the untrained eye might appear more as 'a double-action mangle than the largest, strongest, most carefully finished, and accurate clock that has probably ever yet been made since time began'.[206] The publication continued to note that one Continental observer had observed that while the new Parliament, including its clock, excited 'the admiration of all Europe', the British public knew little of the building. Yet *The Times* was sure that if there was one work at Westminster which should be understood well by all British audiences, it was the clock encased in the

[206] (Anon.), 'The tower and clock at Westminster', *The Times*, (London, England), 18 November, 1857; p. 6; Issue 22841.

Figure 6.8 The completed Westminster Clock mechanism under trial in Dent's workshop

architecturally splendid Clock Tower. It asserted that 'since the machinery for measuring time was first invented no clock-case so magnificent and elaborate has yet been conceived'.

Sound and Vision

Although not implemented to the extent of Airy's proposals, government time was eventually regulated from Greenwich. From 1862 the Westminster clock was in contact with Greenwich twice a day and by 1864 there was an hourly signal in place, but the clock itself remained independent of direct Greenwich control.[207] This practice was far from the level of regulation Airy envisaged. Airy made bids to introduce a system of synchronized clocks into Parliament in 1859 and 1860, but the Board of Works objected to any more of the Astronomer Royal's intrusions into daily politics.[208] The building's individual clocks would

[207] Howse, *Greenwich Time and the Longitude*, pp. 100, 104.

[208] Yuto Ishibashi, 'In pursuit of accurate timekeeping: Liverpool and Victorian electrical horology', *Annals of Science*, Vol. 71, No. 4 (2014), pp. 474–96, 493.

be set traditionally by a human carrying time from room to room kept by a chronometer (Figure 6.9). Airy's demand for Greenwich-controlled order in Parliament was not discarded, but it was quietly adulterated. Indeed, this response was typical from Parliament at this time. While there was a great awareness that the efficient management of time in the Palace was an imperative, reactions to this problem were restrained. There was little desire within Parliament to change the way business was regulated during

Figure 6.9 The House of Lords under the control of electrically distributed Greenwich time

the 1840s and 1850s, especially after the revolutions which swept through Europe in 1848. Increasing the speed of change or efficiency of business carried with it radical, even revolutionary, connotations.[209]

Despite this, Parliament distributed time with impressive authority. The most obvious monument to Parliament's command over 'new' time was its clock tower (Figure 6.10). Though shorter than the Victoria Tower, this was a dramatic symbol of the authority of Parliament's time. It was a structure to emphasize the science behind Britain's new legislative palace. As late as the 1880s few public clocks were accurate to Greenwich time, so what was dispensed at Parliament was a rare commodity.[210] Careful attention was paid to the face of the clock. While Barry produced several designs, eventually settling on one of twelve sections and cast-iron dials, Denison was largely unconcerned by the face's appearance. He assured Barry that observers cared not for the dials, but rather for the position of the hands.[211] More significant was the elaborate system of lighting installed to illuminate the dials. Lit by gas, pipes and burners were installed in lighting chambers behind each side of the clock face. The face itself was glass, glazed to look like porcelain, allowing a measured amount of light to pass. Gas piping ran up through the tower's air shaft to the dials, with the lamps mounted on the walls behind the face. Initially the clock was lit until midnight.[212]

In a sense, this illumination united values of reason and enlightenment with discipline and precision. As London's premier time distributer, the combination of these values presented an impressive image of science's role in modern industrial society, but this message was not only to be seen; it was there to be heard. The Westminster chimes became the most public symbol of Parliamentary authority over time. This is not to say that the clock and its chimes met universal approbation. The clock was put in motion in May 1859, with the chime of the bell following in July, and the quarter bells from September. Yet the initial reaction was far from harmonious. Complaints of the excessive noise of 'Big Ben' were soon lodged, first from ear-sore MPs in the Commons, and then by public residents around Westminster. Between 1859 and 1863 there were hopes that the bell would be permanently silenced.[213] Yet the sound of the bells

[209] Vieira, *Time and Politics*, pp. 69–71.

[210] Ryan A. Vieira, *Time and Politics: Parliament and the Culture of Modernity in Britain and the British World*, (Oxford, 2015), pp. 64–65; also see John Darwin, *The Triumphs of Big Ben*, (London, 1986), p. 107.

[211] Chris McKay, *Big Ben: The Great Clock and the Bells at the Palace of Westminster*, (Oxford, 2010), pp. 121–23.

[212] Ibid., pp. 124–26.

[213] Caroline Shenton, *Mr Barry's War: Rebuilding the Houses of Parliament after the Great Fire of 1834*, (Oxford, 2016), pp. 243, 248.

Figure 6.10 The iconic Westminster Tower, housing the Westminster Clock and the great bell, 'Big Ben'

was a crucial element in the clock's display of accuracy and the government's authority to dispense time. Airy preferred visual time-signals as these were, he argued, more accurate. A sound signal took time to travel and the precision of an observation was contingent on an observer's distance from the clock. However, Airy acknowledged that a sound, such as a chime or gunshot, was a popular form of disseminating time. In 1863, after being consulted over the establishment of a time-gun in Newcastle, Airy explained that to 'secure public sympathy' a sound signal was preferable.[214] The advantage sound held was that it could permeate the homes and workplaces of those who made no effort to observe a visual signal. It was a much harder signal to ignore, and potentially of greater social impact.

From 1861, a time-gun sounded 1pm each day in Edinburgh; Newcastle followed this practice two years later.[215] Yet in London, 'Big Ben' was the premier sound signal for Greenwich time. In 1863 *The Reader* noted that a national division had arisen between the popular sound signals of the north and the visual signals of the south. It complained that the church bells of London lacked authority and the only accurate, trusted sound was the peals of the Westminster Clock's bells. It was an exception to the 'repeated lying tongues' of the city's churches.[216] When Airy himself read this article, probably in December 1863, he must have been both pleased that the authority of the Westminster Clock was acknowledged, but disappointed that the public had such a preference for sound signals. The chimes of the Westminster Clock were a powerful symbol of Parliament's authority and reliability, and accessible to a broad public. Sound was the medium by which the clock would shape the activities of those members of society with little motivation to observe Greenwich time. Although he considered them relatively inaccurate, Airy recognized that the trustworthy chiming of Parliament's bells was important to fashioning attitudes to time in London. Parliament's scientific basis was to be presented by both sight and sound.

A Time for Order in Parliament

In mid-Victorian Britain keeping time was an urgent concern, seemingly implicit to an increasingly industrial society. As the legislature of this society, it is unsurprising that Parliament was a battleground of contrasting ideas of time management. It is significant that Airy found what he

[214] RGO 6/615, 'Letter from Airy to James Mather', (7 September, 1863), pp. 115–16.
[215] RGO 6/615, 'Letter from C. F. Varley to Airy', (10 June, 1864), p. 137.
[216] RGO 6.615 'Cutting captioned 'Science: on time and time-guns', from *The Reader*', (12 December, 1863), p. 239.

deemed to be a suitable time regime in the factory environment of a drapery in St Paul's. Parliament was at the heart of a culture increasingly controlled by 'factory' time and it was perhaps appropriate that science, controlled at Greenwich, brought ordered regulation to the place of government. Scientific knowledge, in the form of galvanic apparatus and mechanical instruments, provided ways of ordering and disciplining government business. At Parliament, science could be constructed, replicated, and contested, but it could also be a powerful form of control.

It has been shown how fundamentally the Westminster clock was conceived of as a symbol of Parliament's authority, and that to establish this authority the clock should employ science to secure accuracy. The clock was to embody the science of precise readings of astronomical time recorded at Greenwich and therefore project a standard time throughout London which was uncontested and authoritative. Parliament time was to be emblematic of modern society. However, this analysis has also shown how problematic establishing accurate time at Westminster was. Crucially, the conception of accuracy was contested. The question of a time system to regulate Parliament engendered dilemmas over human error and new apparatus. It involved choices between trusting instantaneous galvanic telegraphic time-signalling, seemingly non-human, yet experimental, or relying on overtly human regimes which, though tried, risked error. The time system thus raised profound concerns over the relationship between instruments and man. It embodied differing perceptions of what it was to be accurate. This extended to the horological discourse itself, over the roles of mathematical theory and practical experience. Airy's mathematical, galvanic, telegraphic time regime was one definition of an accurate system, shaped by his own scientific values and repetitive of his work at Greenwich to implement controlled precision.

Parliament was a vital site of contesting time and authority. Airy's efforts to implement time from the controlled space of the Royal Observatory to the Houses of Parliament present an insightful analysis of the difficulties of establishing national time. In Conrad's *The Secret Agent* the First Secretary of an unspecified foreign embassy, Mr Vladimir, calculates that the 'blowing up of the first meridian' would provoke the British state into a spate of reactionary measures, clamping down on anarchistic dissent. He perceives that the most effective way to invoke fear of disorder and anarchy in Britain is to attack the middle-class fetish of 'science'. Vladimir speculates that 'if one could throw a bomb into pure mathematics', British political stability would appear under threat.[217] As

[217] Joseph Conrad, *The Secret Agent: A Simple Tale*, (London: 2008), p. 27–29.

such an act was impossible he demands the destruction of the Observatory. The Observatory was the centre of imperial astronomy and world time; it was at the heart of the ordered political, commercial, and social life which contrasted so sharply to Vladimir's demonstration of anarchistic chaos. Over the preceding forty years, Greenwich time had spread rapidly through London, Britain, the empire, and the world. One of the earliest places where Greenwich time was introduced was not a town or city, but an individual building. While Conrad's anarchistic plot targeted Greenwich to provoke disorder, half a century earlier, the Astronomer Royal plotted to establish Greenwich time in Parliament and bring order to the daily business of governance. Science could govern Parliament.

Conclusion: The House of Experiment

On opening, Britain's new Palace of Westminster was a network of scientific works. As visitors were informed, in the place where the most important affairs of Britain, and indeed 'all the world ... [were] deliberated upon and settled', science shaped the light in politicians' eyes, the air in their lungs, and the time in their pockets.[1] Except it did not quite work out like this. When Barry died on 12 May 1860, he left behind a building instructive not only of science's value, but of its limitations. As late as 1885, *The Builder* complained that Barry's architectural plans for the Palace had never been completed.[2] Furthermore, despite the extensive geological and chemical efforts to select a durable stone, within seven years of its erection Parliament was showing signs of rapid decay. By 1860, this was so extensive that Charles Dickens publicly declared Parliament's stone to be the 'the worst ever used in the metropolis'.[3] While the Anston stone of Parliament's structure crumbled, science's performance on the inside was equally disappointing. In 1852 the government sacked Reid after seventeen years of experiment. He failed to establish a ventilation system which could satisfy MPs. Gurney's Bude Light remained controversial too and was not chosen for the permanent Palace. In the decade following Gurney's experiments, his lighting was replaced with a trial of David Boswell Reid's design, before Barry devised a system which was employed until April 1852 when MPs demanded its removal.[4] To cap it all, Denison's bell cracked shortly after its recasting in 1858, having been positioned in the Westminster Tower. Taken all together it would seem that the application of science at Parliament did not present a particularly triumphant narrative.

[1] See (Anon.), *The New Palace of Westminster*, (London, 1852), pp. 12–14, 40–41; (Anon.), *A Description of the New Palace of Westminster*, (London, 1848).

[2] (Anon.), *Architectural Problems for Parliament*, (London, 1885), p. 15.

[3] Dickens quoted in Graham K. Lott and Christine Richardson, 'Yorkshire stone for building the Houses of Parliament (1839–c.1852)', *Proceedings of the Yorkshire Geological Society*, Vol. 51 (1997), pp. 265–72, 271.

[4] Denis Smith, 'The building services', in M. H. Port (ed.), *The Houses of Parliament*, (New Haven, 1976), pp. 218–31, 229.

What is remarkable, however, is the government's response to this concatenation of calamities. With the stone disintegrating, the architecture unfinished, the air and light unsatisfactory, and the bell cracked, the Board of Works remained committed to its cohort of scientific advisors. Efforts were made throughout the 1850s to save Parliament's stone, including several attempts to apply chemical solutions to forestall decay. In spring 1857 Nicholas Charles Szerelmey, a former Hungarian army engineer, applied Egyptian asphaltum from the Dead Sea dissolved in water to Parliament's exterior. He cited ancient Egypt as evidence of its efficaciousness, recalling how 'You can see it on the Pyramids, for instance ... you will find that the stone has a preparation of some such kind of composition, not all, but a great part'.[5] This approach to architectural salvation was not uncommon in this period. As surveyor to Westminster Abbey, George Gilbert Scott had, in 1849, applied a solution of shellac dissolved in spirits of wine to some of the abbey's internal stonework. He persisted with this solution throughout the 1850s and 1860s with mixed results.[6] Szerelmey's remedy likewise did little to save Parliament's stone.[7] The Commons subsequently demanded an enquiry into the decay in 1860. William Francis Cowper (1811–1888), the Liberal First Commissioner of Works, appointed a commission into Anston's poor performance in March 1861, consisting of a selection of geologists, chemists, and architects. With Roderick Murchison, now De la Beche's successor as Director General of the British Geological Survey, in the chair, architects Sydney Smirke, William Tite (President of the RIBA), and George Gilbert Scott, geologists David Ansted and James Tennant, and chemists Augustus Hofmann and Edward Frankland formed the committee.[8] Charles Smith and Charles Barry's son, Edward Middleton Barry (1830–1880), also participated. The enquiry asserted that the cause of the rapid decay was the poor selection of individual cuts of stone rather than the initial choice of magnesian

[5] William Cowper, *Report of the committee on the decay of the stone of the new palace at Westminster*, PP. 1861 (504), p. 6.

[6] Gavin Stamp, 'Scott, Sir George Gilbert (1811–1878)', *Oxford Dictionary of National Biography*, Oxford University Press, 2004 [http://ezproxy.ouls.ox.ac.uk:2204/view/article/24869, accessed 23 August 2014]; Christine Reynolds (ed.), *Surveyors of the Fabric of Westminster Abbey, 1827–1906: Reports and Letters*, (Woodbridge, 2011), pp. 37–38, 46–47; for Scott see, P. S. Barnwell, Geoffrey Tyack, and William Whyte (eds.), *George Gilbert Scott, 1811–1878: An Architect and His Influence*, (Donington, 2014); Gavin Stamp, *Gothic for the Steam Age: An Illustrated Biography of George Gilbert Scott*, (London, 2015).

[7] 'Letter 3778: Faraday to Alfred Austin', (12 May, 1860), in Frank A. J. L. James (ed.), *The Correspondence of Michael Faraday: Volume 5: Nov., 1855–Oct., 1860, Letters 3033–3873*, (London, 2008), pp. 682–83, 682.

[8] PP. 1861 (504), p. iii.

limestone.[9] The commission attributed Parliament's decay to the stone contractors who oversaw the cutting of stone in the Anston quarry. Nevertheless, it recommended Portland stone be used in future for public buildings in London. Portland was not as 'hard as the Anston', but had 'a power of resisting' the influence of the London atmosphere.[10]

The commission did not resolve the problems of Parliament's stone, but it did represent the extent to which the government continued to place its faith in individuals possessing scientific knowledge. This commitment was not just evident in matters of geology. In 1854 a committee reappointed Gurney to superintend Parliament's lighting, which he did until retiring in 1863. Gurney also took control of ventilating Parliament. He reversed Reid's upcast system, in which air entered through the floor and exited via the roof, but this caused MPs so much discomfort that he reverted to Reid's original practice.[11] It is no coincidence that when the Whigs lost power in 1843, the government ceased financing Gurney's Westminster experiments, and that when Edwin Chadwick reformed the Office of Woods and Forests after returning to power in 1846, Gurney returned too as a trusted government authority.[12] Gurney's eventual reappointment to administer the lighting and ventilation also coincided with the fall of the Earl of Derby's 1852 Conservative administration.[13]

The trust that both Whig and Conservative administrations had in science was also apparent in their efforts to rescue the cracked bell. In December 1859 Charles May informed Airy of the bell's failings. May had 'no doubt that the cracking is caused by too brittle alloy & too heavy a hammer'.[14] He implored Airy to apply his mathematical theory to assess the problem. Airy explained to May how mathematics could resolve the problem: 'there is a certain numerical quantity, which ... we will call "impact-momentum" which is proportional to the product of the falling weight by the *square* of its velocity'.[15] Airy used formula to compare 'the

[9] Ibid., p. vi. [10] Ibid., p. viii.

[11] See *First report from the select committee on ventilation of the House; together with the proceedings of the committee, and minutes of evidence*, PP. 1854 (149); *Second report from the select committee on ventilation of the House; with the minutes of evidence taken before them*, PP. 1854 (270); *Report from the select committee of the House of Lords, appointed to inquire into the possibility of improving the ventilation and the lighting of the House, and the contiguous chambers, galleries and passages; and to report thereon to the House, together with the minutes of evidence, and an appendix*, PP. 1854 (384); T. R. Harris, *Sir Goldsworthy Gurney, 1793–1875*, (Penzance, 1975), p. 88.

[12] Dale H. Porter, *The Life and Times of Sir Goldsworthy Gurney: Gentleman Scientist and Inventor, 1793–1875*, (Bethlehem: 1998), pp. 174–75.

[13] Ibid., p. 176.

[14] RGO 6/599, 'Correspondence on chronometers', Letter from Charles May to Airy, (26 December, 1859), p. 435.

[15] RGO 6/599, Letter from Airy to Charles May, (29 December, 1859), p. 437.

destructive effect' of 'Big Ben' to the 'Great Tom' bell of Oxford. He calculated the resistance of each bell to the impact of the hammer to produce 'the true measure of the destructive effect of the hammer'.[16] He claimed that the Westminster Bell had been subjected to five and a half times the destructive effect to that which the Oxford Bell endured on a regular basis. Along with Airy's calculations, a series of examinations were carried out to ascertain the cause of the bell's failings. In late 1859, John Tyndall examined 'by direct experiment, the condition of the bell while in a state of vibration'.[17] He divided the circumference of the sound-bow of the bell into twenty-eight equal parts to compare the amplitudes of vibrations at different sections. Depending on where the bell was struck, Tyndall found certain segments vibrated more violently than others, and it was at these points that the natural weaknesses in the bell's metal were exposed.[18] To remedy this problem, the government asked Airy to calculate a revised position for the bell. Airy's 'theory of a bell's sound vibrations' identified the four points of the bell's circumference which remained stationary during ringing and explained that if the bell were turned two feet and eight inches in relation to the hammer, the cracks would be placed at the point of least strain.[19] The bell has remained in this position ever since.

What then can we conclude from all this scientific activity? First, that the building of the new Palace of Westminster revealed the interrelated nature of science and politics. Within a context of political reform and with the old Palace in ruins, building Parliament made obvious the role of science in political culture at a time when it really mattered. This relationship was materially demonstrated through architecture to national audiences. Various Whig and Conservative administrations, officials from the Office of Woods and Forests, and more radical elements of Parliament all looked to science as something to emulate and utilize. Science could be venerated, either as an apparently objective manner of knowledge production, or drawn on from individuals with specialist knowledge to solve technical problems. Seeking specialist knowledge and using science as a model of how to behave as a statesman were ways of governing. This book has built on this understanding of the interaction between science and politics in the mid-nineteenth century. However, it has shown that this relationship involved more than ideals of governance and rhetoric, but took shape materially in the form of the Houses of Parliament. The adoption of scientific manners and learning was not just promised in words, but built in stone. The Palace was the architectural expression of

[16] Ibid., p. 439.
[17] Tyndall in *The Bell (new Palace, Westminster). Copy of reports on the condition of the great bell of the new Palace of Westminster*, PP. 1860 (553), p. 1.
[18] Ibid., pp. 1–2. [19] Ibid., p. 9.

Figure C.1 The completed Palace of Westminster as it stands today

this interaction between science and politics. It fits as part of wider efforts, from varying administrations, to govern in a manner considered scientific.

Interpreting the Palace of Westminster as a work of science has important historical implications. Despite its shortcomings, the opening of the new Parliament represented a moment where the relationship between science and politics was made visible (Figure C.1). While this involved a particular network of individuals and a specific context, thinking of the Houses of Parliament as more than a question of aesthetics and style raises questions over how we should interpret other buildings. Looking beyond the use of Gothic at Westminster provides valuable insights into early Victorian science and politics; insights which should encourage us to do the same with other architectural projects in the years following Parliament's construction. Moving beyond questions of aesthetics could yield new interpretations on the architecture of state buildings, like George Gilbert Scott's Foreign Office, as well as town halls and structures for local government. Attention to these buildings' technical arrangements, in addition to their stylistic characteristics, could transform the way we perceive such works. This move away from questions of style has international ramifications. Melbourne's Parliament House in Australia, constructed in the late 1850s, received even greater attention to time keeping than the Parliament at Westminster, while projects on the scale of Edwin Lutyens' (1869–1944) work around the Rajpath in New Delhi, including the Viceroy's House and Secretariat Building, entailed similar problems over how to build appropriate architecture for government and administration.[20] Looking beyond matters of style for such buildings could prove every bit as fruitful as it has done for the Palace of Westminster. To understand the architecture of governments, and indeed all architecture, we have to cast off architectural history's past fixation with style.

The architecture of science was not, in nineteenth-century Britain, limited to laboratories and museums, but was present in buildings as diverse as churches, colleges, and even the nation's Parliament. There is, however, a particular importance to the relationship between science and politics as embodied in the Palace's architecture. The links between politics and science continued long after Parliament's opening, but the Palace was the product of a particular group of ideas and individuals; what was built was the result of a specific moment. In the 1830s and 1840s there was a consensus, among Whigs, Tories, radicals, bureaucrats, and men of science that building in reference to scientific knowledge was a way of ensuring the new Parliament could be implicated as a harbinger of progress. Science offered control over the building's materials, air, light, and time, and this was believed to be crucial to making a modern architectural organ of governance. At the location from where an empire was governed, science promised radical new ways of controlling nature.

Recognizing how science and politics came together has consequences for our conception of the Palace today. The fittingness of having a Gothic home for a modern Parliament is a subject of much debate. Does the character of the building sustain a model of politics detached from the cares of modern society? Happily this book does not wade into this minefield, but as questions surround the building's future, increasingly tense in light of the growing costs of maintaining it, it surely matters that we realize exactly what the Houses of Parliament meant when they were built. Science's role was about more than the technical, but a much bigger idea; a fantastic idea in which it played a grandiose role in politics. It is clear that the building was supposed to reflect an enlightened society. The problem was that there were many types of science in the 1830s. Science was a double-edged product. It had the potential to enhance the credibility of a government, but also played the part of a radical. In the wrong hands it could blind, asphyxiate, disable, or even kill honourable members. Science was easily characterized as a dangerous resource, capable of wrecking the political

[20] Gavin Stamp, 'Lutyens, Sir Edwin Landseer (1869–1944)', *Oxford Dictionary of National Biography*, Oxford University Press, 2004; online edn, May 2012 [http://www.oxforddnb.com/view/article/34638, accessed 30 July 2015].

system. What the building embodies is the kind of knowledge early Victorian politicians were prepared to trust; the sort of science which they believed appropriate to guide government. The question of reliable knowledge is constantly faced by societies and it was a problem overcome during the building of the Palace. A building like Parliament can reflect a society's relationship with knowledge, as much as its religious values, artistic tastes, or political system. It is the way in which a science, considered reliable, is produced that tells us the most about the relationship between a political system and its favoured bodies of knowledge. The difficulties of manufacturing and managing knowledge which societies face can be resolved through the architecture they build. Yet today the science of the building has become almost invisible; we no longer see its truly revolutionary character. In a way, Charles Barry's work has become an analogy of the very science used to build it. As scientific knowledge becomes generally accepted, the marks of its production become less noticeable. As a consensus builds over its validity, the way it was made becomes of less concern. So too with Parliament, that as it has become the recognized home of British politics, the knowledge used to build it has been forgotten. We cannot evaluate what the Palace of Westminster means today if we ignore the science which made it.

This turning to science to guide the building of Britain's national assembly was symptomatic of a political system increasingly unsure of its position and identity. The 1832 Reform Act and destruction of the old Parliament in 1834 represented a challenge, both politically and architecturally, which created continued doubt during the 1830s and 1840s over how a modern legislature should operate and what it was for. At this moment, rebuilding Parliament was a vast project to address these uncertainties. The scheme was about maintaining Parliament's position in British society. When we recognize this, it becomes apparent that questions of style represent just one part of a much larger plan to build an appropriate Palace. This involved an aesthetics of science, as much as it did a science of aesthetics. We have to look beyond Barry and Pugin, and investigate the efforts of Airy, Reid, Faraday, Gurney, and De la Beche. The content of their work was as important as Pugin's attention to Gothic detail. The knowledge employed, be it chemistry, geology, or mathematics, had to be reliable and ordered for it to be of value to Britain's political establishment. This was why the longevity of the building's stone mattered. This was why Reid's ventilation experiments aroused so much concern. It explains how Gurney's lighting took on so much political significance. And it is why Airy deemed the establishment of a disciplined system of timekeeping to be such an imperative. Science had to be represented in a credible fashion for it to contribute to the political identity of the country's political representation. Knowledge could only enhance Parliament's authority if it was itself credible. Achieving this was a way of bringing order to government amidst the social turmoil of the 1830s. Science and politics were powerfully united though Parliament's architecture.

Select Bibliography

Publication is in London unless otherwise stated

Primary Material

I. *Manuscripts*

Cambridge, Cambridgeshire

Royal Greenwich Observatory Archives, Cambridge University Library
RGO 6/599, Papers of George Biddell Airy, 'Correspondence on Chronometers, 1849–1859'.
RGO 6/607, Papers of George Biddell Airy, 'Correspondence on New Palace Clock, 1845–1848'.
RGO 6/608, Papers of George Biddell Airy, 'Clock for New Palace Westminster, 1851–1856'.
RGO 6/609, Papers of George Biddell Airy, 'Clock for New Palace Westminster, 1857–1861'.
RGO/6/615, Papers of George Biddell Airy, 'Galvanic Time Balls, time signals and sympathetic clocks regulated by the Royal Observatory, 1861 Jan. to 1866 Feb.'.

Edinburgh

Centre for Research Collections, Edinburgh University
P.89.16, David Boswell Reid, 'Remarks on Dr Hope's "Summary," presented to the patrons of the University'.
P.89.19, David Boswell Reid, 'Testimonials regarding Dr D. B. Reid's qualifications as a lecturer on chemistry, and as a teacher of practical chemistry' (1833).
P.137.25, Thomas Charles Hope, 'Summary of a Memorial to be presented to the Right Honourable the Lord Provost, Magistrates, and Council, respecting the institution of a professorship of practical chemistry in the University of Edinburgh'.
PS86.8, David Boswell Reid, 'Extracts from official documents, reports, and papers, referring to the progress of Dr. Reid's plans for ventilation'.

S.B.5404/2*13, David Boswell Reid, 'The Study of Chemistry: its nature, and influence on the progress of society: importance of introducing it as an early branch of education in all schools and academies'.
New College Library, Edinburgh University
CHA4, 'Letters to Thomas Chalmers', 1831–1834.

Keyworth, Nottinghamshire

British Geological Survey Archives
GSM/DC/A/C/1, Papers of Henry Thomas De la Beche, 'Entry Book of In-and-out Letters, 1835–1842'.
GSM/DC/A/C/11, Papers of Henry Thomas De la Beche, 'Correspondence to the Director General, 1836–1847'.
GSM/MG/C/20, Papers of Henry Thomas De la Beche, C. H. Smith, 'Building Stones' (5 January 1848).

London

British Library
British Museum, Additional. Ms. 40,413, Peel Papers, Vol. CCXXXIII, 'General Correspondence'.
British Museum Additional Ms. 46,126, Murchison Papers, Vol. II.
Royal Institute of British Architects Archives
SmC/1/1, Papers of C. H. Smith, 'Abstracts of notes made by C. H. Smith during his travels with the commissioners in search of stone for the Houses of Parliament in the autumn of the year 1838 and spring of 1839'.
SmC/1/2, Papers of C. H. Smith, 'Various papers on stone for building, and on some other materials'.
King's College London Archives
Papers of John Frederic Daniell, 5/1, 'John Frederic Daniell – Ms Life – Lectures' (c.1845–50).
National Archives
WORK (Records of the Board of Works), 11/17/5, 'Stone selection: correspondence and papers' (1836–1843).
Parliamentary Archives
ARC/PRO/WORK11/26/6, 'Report on St Stephen's Chapel, 1835–57'.
BLY/58, 'Sketch book of Henry Bailey, 1840–52'.
MOU/Box7/327, 'The clock tower' (1836).
OOW/23, 'Ground plan of Dr D. B. Reid's premises'.
8/L/10/2(10), 'Barry/Hardman Correspondence, 1845–1860'.

Oxford, Oxfordshire

Special Collections, Bodleian Library
Ms Dep. Papers of the British Association for the Advancement of Science 1, 'Correspondence of John Phillips'.

Ms Dep. Papers of the British Association for the Advancement of Science 5, 'Miscellaneous Papers, 1831–1869'.

Oxford University Museum of Natural History

Papers of William Smith, Box 44 (Ms Journal), 'Journal of William Smith LL. D., geologist, civil engineer, and mineral surveyor, on a tour of observation on the principal freestone quarries in England and Scotland, in company with Mr Barry the architect, Mr De la Beche, and Mr C H Smith, for the purpose of finding stone best calculated for the construction of the new Houses of Parliament' (18 August to 23 October, 1838).

Papers of William Smith, Box 44, Folder 1, 'Miscellaneous notes relating to survey for building stone for Houses of Parliament in 1838'.

Papers of William Smith, Box 44, Folder 5, 'Papers on economic geology' (1837–1839).

II. State Papers

Report from the Select Committee on House of Commons Buildings; together with the minutes of evidence taken before them, Parliamentary Paper (PP.) 1833 (17).

Report from the Select Committee on the House of Commons' Buildings; with the minutes of evidence taken before them, PP. 1833 (269).

Report from the Select Committee on Rebuilding Houses of Parliament; with the minutes of evidence, and an appendix, PP. 1835 (262).

Report from the Select Committee on the Admission of Ladies to the Strangers' Gallery; with the minutes of evidence, PP. 1835 (437).

Report from Select Committee on the Ventilation of the Houses of Parliament; with the minutes of evidence, PP. 1835 (583).

Houses of Parliament Plans. Report of commissioners appointed to consider the plans for building the new Houses of Parliament, PP. 1836 (66).

Report from Select Committee on Houses of Parliament; with the minutes of evidence, PP. 1836 (245).

Ventilation of the House. Copy of a letter from Dr. Reid to Lord Duncannon, dated February 4th, 1837, relative to the acoustic and ventilating arrangements lately made in the House of Commons, PP. 1837 (21).

Report from the Select Committee on the Thames Tunnel; with the minutes of evidence, PP. 1837 (499).

Ventilation of the House. Copy of a letter from Sir Frederick Trench to Lord Viscount Duncannon, on the subject of ventilating the House of Commons, with Lord Duncannon's answer, PP. 1837–38 (204).

Ventilation of the House. Copy of a letter from Dr. Reid to the Viscount Duncannon, in reply to observations addressed to His Lordship by Sir Frederick Trench, PP. 1837–38 (277).

Ventilation and lighting of the House. Letters from Sir Frederick Trench to Lord Duncannon, on the subject of ventilation and lighting the House of Commons, PP. 1837–38 (358).

Ventilation of the House. Return of the detailed expenses incurred in experiments for improving the ventilation, &c. of the House of Commons, in the experiment of lighting with gas, also in lighting with candles, ending with the present lustres and shades, PP. 1837–38 (725).

Hume, Joseph, *Report from the select committee on lighting the House; together with the minutes of evidence, appendix and index*, PP. 1839 (501).

Barry, Charles, *Report of the Commissioners appointed to visit the Quarries, and to inquire into the Qualities of the Stone to be used in Building the New Houses of Parliament*, PP. 1839 (574).

Report from Select Committee on Ventilation of the New Houses of Parliament; with the minutes of evidence, and appendix, PP. 1841 (51).

Lemon, Charles, *Report from the select committee on lighting the House; together with the minutes of evidence, and appendix*, PP. 1842 (251).

Report from the select committee on ventilation of the new houses of Parliament; with the minutes of evidence, PP. 1842 (536).

Report from the select committee on Houses of Parliament; together with the minutes of evidence taken before them, PP. 1844 (448).

First Reports of the Commissioners for inquiring into the state of large towns and populous districts, PP. 1844 (572).

Brought from the Lords, 9 August 1844. Second Report from the select committee of the House of Lords appointed to inquire into the progress of the building of the Houses of Parliament, and to report thereon to the house; with the minutes of evidence taken before the committee, PP. 1844 (629).

Report from the Select Committee on Westminster Bridge and new Palace, PP. 1846 (177).

Second report from the Select Committee on Westminster Bridge and new Palace, PP. 1846 (349).

Third Report from the Select Committee on Westminster Bridge and new Palace; together with the minutes of evidence, appendix, and index, PP. 1846 (574).

Reports from the Select Committee of the House of Lords appointed to inquire into the progress of the building of the Houses of Parliament, and to report thereon to the house: together with the minutes of evidence taken before the said committee, PP. 1846 (719).

First report from the select committee on new House of Commons, PP. 1850 (650, 650-II).

Second Report from the Select Committee on Ventilation and Lighting of the House; together with the proceedings of the committee, minutes of evidence, appendix and index, PP. 1852 (402).

Atkins, S. Elliott, *Copy of the Memorial presented to Her Majesty's Commissioners of Works and Public Buildings by the Clockmakers' Company of London, respecting the great clock to be erected at the new Palace at Westminster; together with the answer thereto*, PP. 1852 (415).

Clocks (New Houses of Parliament), PP. 1852 (500).

Clocks (New Houses of Parliament), PP. 1852 (500-I).

Ventilation and lighting of the House committee. Report of the standing committee on the ventilating and lighting the House of Commons, PP. 1852–53 (570).

First report from the select committee on ventilation of the House; together with the proceedings of the committee, and minutes of evidence, PP. 1854 (149).

Second report from the select committee on ventilation of the House; with the minutes of evidence taken before them, PP. 1854 (270).

Report from the select committee of the House of Lords, appointed to inquire into the possibility of improving the ventilation and the lighting of the House, and the contiguous chambers, galleries and passages; and to report thereon to the House, together with the minutes of evidence, and an appendix, PP. 1854 (384).

Westminster New Palace. Copies of all papers and correspondence relating to the great clock and bells for the new Palace at Westminster (in continuation of Parliamentary Paper, no. 500, of session 1852), PP. 1854–5 (436).

The Bell (new Palace, Westminster). Copy of reports on the condition of the great bell of the new Palace of Westminster, PP. 1860 (553).

Cowper, William, Report of the committee on the decay of the stone of the new palace at Westminster, PP. 1861 (504).

III. *Periodicals*

Architectural Magazine (1834–1838)

Blackwood's Edinburgh Magazine (1817–1905)

The Bude Light (1841)

Builder (1842–1966)

Building News (1855–1926)

Caledonian Mercury (1720–1867)

The Civil Engineer and Architect's Journal (1837–1868)

Cornwall Royal Gazette, Falmouth Packet and Plymouth Journal (1801–1854)

Edinburgh Review (1802–1929)

The Friend of Africa; by the society for the extinction of the slave trade, and for the civilization of Africa (1841–1843)

Hansard (1803–2005)

Illustrated London News (1842–1900)

John Bull (1820–1892)

The Lady's Newspaper (1847–1863)

Literary Gazette and Journal of Belles Lettres, Arts, Sciences, &c. (1817–1860)

Mechanics' Magazine (1823–1872)

The Mirror of Literature, Amusement, and Instruction (1822–1847)

Morning Chronicle (1770–1865)

Morning Post (1772–1937)

The Musical World: a journal and record of science, criticism, literature, & intelligence, connected with the art (1836–1890)
Pictorial Times (1843–1848)
Preston Chronicle (1812–1854)
Punch (1841–1992)
Quarterly Review (1809–1967)
St Paul's Magazine (1867–1874)
The Scotsman (1817–)
The Times (1785–)
Westminster Review (1824–1914)

IV. Printed

Adkins, Kathleen (ed.), *Travel Diaries (1817–1820) of Sir Charles Barry (1795–1860) (Personal Extracts)*, (Privately Published, 1986).

Ainger, Alfred, *On Ventilation, in Reference to the Houses of Parliament*, (London, 1835).

Airy, George Biddell, *Mathematical Tracts on Physical Astronomy, the Figure of the Earth, Precession and Nutation, and the Calculus of Variations*, (Cambridge, 1826).

Airy, George Biddell, *On the Undulatory Theory of Optics*, (London, 1866).

Airy, Wilfrid (ed.), *Autobiography of Sir George Biddell Airy*, (Cambridge, 1896).

Alison, Archibald, 'The British School of Architecture', *Blackwood's Edinburgh Magazine*, Vol. XL, No. CCL (August, 1836), pp. 227–38.

Allen, William, *A Narrative of the Expedition Sent by Her Majesty's Government to the River Niger, in 1841: under the Command of Captain H. D. Trotter, R.N.*, (London, 1848).

Anderson, John Wilson, *Letter to the Right Honourable the Lord Provost, Magistrates, and Town Council of Edinburgh, as patrons of the University, in reference to the contemplated establishment of a lectureship of practical chemistry*, (Edinburgh, 1834).

Anderson, John Wilson, *Postscript of a letter to the Right Honourable the Lord Provost, Magistrates, and Town Council of Edinburgh, as patrons of the University, in reply to a pamphlet by Dr. Reid, lecturer on chemistry, in reference to the contemplated establishment of a lectureship of practical chemistry*, (Edinburgh, 1834).

(Anon.), *Architectural Problems for Parliament*, (London, 1885).

Arnott, Neil, *On Warming and Ventilating: with Directions for Making and Using the Thermometer-Stove, or Self-Regulating Fire, and Other New Apparatus*, (London, 1838).

(Anon.), 'Arts and Sciences. British Association', *The Literary Gazette and Journal of Belles Lettres, Arts, Sciences, &c.*, 28 August, 1841 (London), pp. 561–63.

Bagehot, Walter, *The English Constitution*, (ed.) Miles Taylor, (Oxford, 2001).

Barry, Alfred, *The Life and Works of Sir Charles Barry*, (London, 1867).

Barry, Alfred, *The Architect of the New Palace at Westminster. A Reply to a Pamphlet by E. W. Pugin, Esq., Entitled "Who Was the Art-Architect of the Houses of Parliament?"*, (London, 1868).
Becker, Bernard H., *Scientific London*, (London, 1874).
Beckett, Edmund, *On the Origin of the Laws of Nature*, (London, 1879).
Belcher, Margaret (ed.), *The Collected Letters of A. W. N. Pugin*, Vol. 2, 1843–1845, (Oxford, 2003).
Brayley, Edward Wedlake and Britton, John, *The History of the Ancient Palace and Late Houses of Parliament at Westminster*, (London, 1836).
Brewster, David, *The Martyrs of Science, or the Lives of Galileo, Tycho Brahe, and Kepler*, (London, 1841).
Brougham, Henry, 'The New Houses of Parliament', *The Edinburgh Review*, Vol. LXV, No. CXXXI, (April, 1837), pp. 174–79.
Brougham, Henry, *The Life and Times of Henry Lord Brougham written by himself*, Vol. 3, (Edinburgh, 1871).
(Anon.), *The Bude Light: A Social, Satirical, Farcical, Fashionable, Personal, Political, Musical, Poetical, Attical, Dramatical, Tart, Smart, Courting, Sporting, Literary, Skiterary, Monthly Illuminator*, (London, 1841).
Carlyle, Thomas, 'Signs of the Times', in Maurice Cross (ed.), *Selections from the Edinburgh Review*, (Paris, 1835), pp. 91–106.
Carlyle, Thomas (ed.), *Latter-Day Pamphlets*, (New York, 1850).
Chalmers, Thomas, *The Bridgewater Treatises: on the Power Wisdom and Goodness of God as Manifested in the Adaption of External Nature to the Moral and Intellectual Constitution of Man*, Vol. I, (London, 1833).
Chalmers, Thomas, *The Bridgewater Treatises: on the Power Wisdom and Goodness of God as Manifested in the Adaption of External Nature to the Moral and Intellectual Constitution of Man*, Vol. II, (London, 1833).
Christison, Robert, *The Life of Sir Robert Christison, Bart*, (Edinburgh, 1885).
Cochrane, John, *A Treatise on the Game of Chess*, (London, 1822).
Cole, Henry, 'Parliaments of our Ancestors', *Westminster Review*, Vol. XXI, (October, 1834), pp. 319–34.
'The Conductor', 'A new site for the Houses of Parliament suggested, and the fundamental principles on which they ought to be designed pointed out', *Architectural Magazine*, Vol. III, No. 25, (March, 1836) pp. 100–03.
Conolly, M. F., *Biographical Dictionary of Eminent Men of Fife, of Past and Present Times*, (Edinburgh, 1866).
Conrad, Joseph, *The Secret Agent: A Simple Tale*, (London: 2008).
Cross, Maurice (ed.), *Selections from the Edinburgh Review*, (Paris, 1835).
Cust, Edward, *A letter to the Right Honourable Sir Robert Peel, Bart. M. P. on the expedience of a better system of control over buildings erected at the public expense; and on the subject of rebuilding the Houses of Parliament*, (London, 1835).
Daniell, John Frederic, *An Introduction to the Study of Chemical Philosophy: Being a Preparatory View of the Forces Which Concur to the Production of Chemical Phenomena*, 2nd ed., (London, 1843).
De la Beche, Henry T., *A Geological Manual*, 2nd ed., (London, 1832).
De la Beche, Henry T., *Report on the Geology of Cornwall, Devon, and West Somerset*, (London, 1839).

De la Beche, Henry Thomas, 'Some Account of Myself', in Richard Morris (ed.), *A Journal of Sir Henry De la Beche: Pioneer Geologist (1796–1855)*, (Royal Institution of South Wales: 2013), pp. 22–36.

Denison, Edmund Beckett, *Clocks and Locks. From the 'Encyclopaedia Britannica'. Second Edition; with a Full Account of the Great Clock at Westminster*, (Edinburgh, 1857).

Denison, Edmund Beckett, 'On the Great Bell of Westminster', *Royal Institution of Great Britain*, Weekly Evening Meeting, (Friday 6 March, 1857), p. 1.

Denison, Edmund Beckett, *On some of the grounds of dissatisfaction with modern gothic architecture. A lecture delivered at the Royal Institution of Great Britain*, (London, 1859).

Denison, Edmund Beckett, *A Rudimentary Treatise on Clocks and Watches, and the Bells; with a Full Account of the Westminster Clock and Bells*, (London, 1860).

Denison, Edmund Beckett, *Astronomy without Mathematics*, (London, 1865).

(Anon.), *A Description of the New Palace of Westminster*, (London, 1848).

Faraday, Michael, *The Chemical History of a Candle: a Course of Lectures Delivered by Michael Faraday*, (London, 1904).

Fowler, Charles, *Remarks on the resolutions adopted by the committees of the Houses of Lords and Commons for rebuilding the Houses of Parliament, particularly with reference to their dictating the style to be adopted, (reprinted from the Architectural Magazine*, September, 1835).

Fowler, Charles, *On the Proposed Site of the New Houses of Parliament*, (London, 1836).

Gracie, John Black (ed.), *The Grey Festival; Being a Narrative of the Proceedings Connected with the Dinner Given to Earl Grey, at Edinburgh, on Monday, the 15th September, 1834, and a Corrected Report of the Speeches*, (Edinburgh, 1834).

Gregory, William, *Observations on the proposed appointment of a teacher of practical chemistry in the University; with remarks on some passages in Dr Reid's letter to the council on the subject*, (Edinburgh, c.1834).

Gwilt, Joseph, *An Encyclopaedia of Architecture, Historical, Theoretical, and Practical*, (London, 1842).

Hamilton, W. R., *Letter from W. R. Hamilton, to the Earl of Elgin, on the New Houses of Parliament*, (London, 1836).

Hamilton, W. R., *Second letter from W. R. Hamilton, esq. to the Earl of Elgin, on the propriety of adopting the Greek style of architecture in the construction of the New Houses of Parliament*, (London, 1836).

Hamilton, W. R., *Third letter from W. R. Hamilton, esq. to the Earl of Elgin, on the propriety of adopting the Greek style of architecture in preference to the Gothic, in the construction of the New Houses of Parliament*, (London, 1837).

Harris, Elisha, 'An introductory outline of the progress of improvement in ventilation', in David Boswell Reid, *Ventilation in American Dwellings; with a Series of Diagrams, Presenting Examples in Different Classes of Habitations*, (New York, 1858), pp. iii–xxxv.

Harwood, J., *Memoirs of the Right Honourable Henry, Lord Brougham*, (London, 1840).

Hopper, Thomas, *Designs for the Houses of Parliament*, (London, c.1840).

Hopper, Thomas, *Hopper versus Cust, on the Subject of Rebuilding the Houses of Parliament*, (London, 1837).

Jackson, J. R., *Observations on a letter from W. R. Hamilton, esq. to the Earl of Elgin, on the New Houses of Parliament*, (London, 1837).

James, Frank A. J. L. (ed.), *The Correspondence of Michael Faraday: Vol. 1, 1811–Dec., 1831, Letters 1–524*, (London, 1991).

James, Frank A. J. L. (ed.), *The Correspondence of Michael Faraday: Vol. 2, 1832–Dec., 1840, Letters 525–1333*, (London, 1993).

James, Frank A. J. L. (ed.), *The Correspondence of Michael Faraday: Vol. 3, 1841–Dec., 1848: Letters 1334–2145*, (London, 1996).

James, Frank A. J. L. (ed.), *The Correspondence of Michael Faraday: Vol. 4, Jan.,1849–Oct., 1855, Letters 2146–3032*, (London, 1996).

James, Frank A. J. L. (ed.), *The Correspondence of Michael Faraday: Vol. 5: Nov., 1855–Oct., 1860, Letters 3033–3873*, (London, 2008).

James, Frank A. J. L. (ed.), *The Correspondence of Michael Faraday: Vol. 6, Nov., 1860–Aug., 1867, Undated Letters, and Additional Letters for Volumes 1–5: Letters 3874–5053*, (London, 2012).

"J.H.F.", 'New books', *The Mirror of Literature, Amusement, and Instruction*, Vol. 37, No. 1062, (5 June, 1841), p. 367.

Juvara, T., *Strictures on architectural monstrosities, and suggestions for an improvement in the direction of public works*, (London, 1835).

Kelsall, C., *A letter to the Society of the Dilettanti, on the works in progress at Windsor*, (London, 1827).

Lyell, Charles, *Principles of Geology: Being an Inquiry How Far the Former Changes of the Earth's Surface Are Referable to Causes Now in Operation*, (London, 1834).

Lyell, Charles, *Elements of Geology*, (London, 1838).

Mercer, Vaudrey, *The Life and Letters of Edward John Dent: Chronometer Maker and Some Accounts of His Successors*, (London, 1977).

Mill, John Stuart, *System of Logic Ratiocinative and Inductive: Being a Connected View of the Principles of Evidence and the Methods of Scientific Investigation*, (London, 1886).

Morris, Richard (ed.), *A Journal of Sir Henry De la Beche: Pioneer Geologist (1796–1855)*, (Royal Institution of South Wales: 2013).

Morritt, John S., 'Review of Hamilton, etc. on architecture', *The Quarterly Review*, Vol. 58, (February, 1837), pp. 61–82.

M'William, James Ormiston, *Medical History of the Expedition to the Niger during the Years 1841–2 Comprising an Account of the Fever Which Led to Its Abrupt Termination*, (London, 1843).

(Anon.), *The New Palace of Westminster*, (London, 1852).

(Anon.), *The New Palace of Westminster*, (London, 1867).

(Anon.), 'Notices and abstracts of communications to the British Association for the advancement of science', in *Report of the Eighth Meeting of the British Association for the Advancement of Science; Held at Newcastle in August 1838*, (London, 1839), pp. 131–32.

(Anon.), 'Notices of the press', *The Musical World: A Journal and Record of Science, Criticism, Literature, & Intelligence, Connected with the Art*, Vol. XVI – New Series, Vol. IX, (1 July, 1841), p. 16.

Phillips, John, *Memoirs of William Smith, LL.D., Author of the 'Map of the Strata of England and Wales'*, (London, 1844).

Phillips, Richard, *A letter to Dr. David Boswell Reid, experimental assistant to Professor Hope, in answer to his pamphlet, entitled 'An Exposure Of The Misrepresentations In The Philosophical Magazine And Annals', &c.*, (London, 1831).

(Anon.), *A portion of the papers relating to the Great Clock for the New Palace at Westminster, printed by order of the House of Lords, (*London, 1848*).*

Pugin, A. Welby, *A letter to A. W. Hakewill, architect, in answer to his reflections on the style for rebuilding the Houses of Parliament*, (Salisbury, 1835).

Pugin, A. Welby, *Ornaments of the 15th and 16th Centuries*, (London, 1836).

Pugin, A. Welby, *The True Principles of Pointed or Christian Architecture: Set Forth in Two Lectures Delivered at St. Marie's, Oscott*, (London, 1841).

Pugin, A. Welby, *An Apology for the Revival of Christian Architecture in England*, (London, 1843).

Pugin, A. Welby, *Glossary of Ecclesiastical Ornament and Costume, Compiled and Illustrated from Antient Authorities and Examples*, (London, 1844).

Pugin, E. Welby, *Who Was the Art Architect of the Houses of Parliament. A Statement of Facts, Founded on the Letters of Sir Charles Barry and the Diaries of Augustus Welby Pugin*, (London, 1867).

Pugin, E. Welby, *Notes on the reply of the Rev. Alfred Barry D. D. principal of Cheltenham College, to the "Infatuated Statements" made by E. W. Pugin, on the Houses of Parliament*, (London, 1868).

Reid, David Boswell, *Elements of Practical Chemistry*, (Edinburgh, 1830).

Reid, David Boswell, *An exposure of the continued misrepresentations by Richard Phillips, esq ... in his attempt to vindicate himself from Dr. Reid's first exposure of his misrepresentation in that journal*, (Edinburgh, 1831).

Reid, David Boswell, *An exposure of the misrepresentations in the Philosophical Magazine and Annals, for December, 1830, in its attack upon the author's Elements of Practical Chemistry*, (Edinburgh, 1831).

Reid, David Boswell, *A Memorial to the patrons of the University on the present state of practical chemistry*, (Edinburgh, 1833).

Reid, David Boswell, *A letter to the Right Honourable the Lord Provost, magistrates, and council, patrons of the University Of Edinburgh, on the present state of practical chemistry: with remarks on some statements in a pamphlet by Dr Anderson, assistant to Dr Hope*, (Edinburgh, 1834).

Reid, David Boswell, *A letter to the Right Honourable the Burgh Commissioners, on the evidence of Dr Christiston, Professor of Materia Medica in the University of Edinburgh*, (Edinburgh, 1835).

Reid, David Boswell, *Brief outlines illustrative of the alterations in the House of Commons, in reference to the acoustic and ventilating arrangements*, (Edinburgh, 1837).

Reid, David Boswell, *Testimonials regarding Dr D.B. Reid's qualifications as a lecturer on chemistry and teacher of practical chemistry, April 1837*, (London, 1837).

Reid, David Boswell, *Professorship of Chemistry in the University of Edinburgh: Testimonials in Favour of Dr D. B. Reid, F. R. S. E.*, (Edinburgh, 1843).

Reid, David Boswell, *Illustrations of the Theory and Practice of Ventilation, with Remarks on Warming, Exclusive Lighting, and the Communication of Sound*, (London, 1844).
Reid, David Boswell, *Ventilation. A reply to misstatements made by 'The Times' and by 'The Athenaeum' in reference to ships and buildings ventilated by the author*, (London, 1845).
Reid, David Boswell, *Narrative of Facts as to the New Houses of Parliament*, (London, 1849).
Reid, David Boswell, *Rudiments of Chemistry; with Illustrations of the Chemical Phenomena of Daily Life*, 4th ed., (London, 1851).
Reid, David Boswell, *Ventilation in American Dwellings; with a Series of Diagrams, Presenting Examples in Different Classes of Habitations*, (New York, 1858).
Reid, Hugo, *Memoir of the Late David Boswell Reid*, (Edinburgh, 1863).
Report of the Eighth Meeting of the British Association for the Advancement of Science; Held at Newcastle in August 1838, (London, 1839).
Reynolds, Christine (ed.), *Surveyors of the Fabric of Westminster Abbey, 1827–1906: Reports and Letters*, (Woodbridge, 2011).
Richardson, Benjamin Ward, *The Health of Nations: A Review of the Works of Edwin Chadwick, with a Biographical Dissertation*, Vol. II, (London, 1887).
Ruskin, John, *The Stones of Venice*, (London, 2000).
Ryde, Henry T., *Illustrations of the New Palace of Westminster*, (London, 1849).
Sanders, Lloyd C. (ed.), *Lord Melbourne's Papers*, (London, 1889).
Sharpe, Tom, 'New insights into the early life of Henry Thomas De la Beche (1796–1855)', in Richard Morris (ed.), *A Journal of Sir Henry De la Beche: Pioneer Geologist (1796–1855)*, (Royal Institution of South Wales: 2013), pp. 5–21.
Sharpe, T., and McCartney, P. J. (eds.), *The Papers of H. T. De la Beche (1796–1855) in the National Museum of Wales*, (Cardiff, 1998).
Smith, William, *A Delineation of the Strata of England and Wales: with Part of Scotland; Exhibiting the Collieries and Mines, the Marshes and Fen Land Originally Overflowed by the Sea, and the Varieties of Soil According to the Variations in the Substrata, Illustrated by the Most Descriptive Names*, (London, 1815).
Smith, William, *Stratigraphical System of Organised Fossils, with Reference to the Specimens of the Original Geological Collection in the British Museum: explaining Their State of Preservation and Their Use in Identifying the British Strata*, (London, 1817).
Douglas, Stair, *The Life and Selections from the Correspondence of William Whewell, D.D. Late Master of Trinity College Cambridge*, (London, 1881).
Sylvester, Charles, *The Philosophy of Domestic Economy*, (Nottingham, 1819).
Symonds, Arthur, 'New House of Commons', *Westminster Review*, Vol. XXII, (January, 1835), pp. 163–72.
Tredgold, Thomas, *Principles of Warming and Ventilating Public Buildings, Dwelling-Houses, Manufactories, Hospitals, Hot-Houses, Conservatories, & c.*, (London, 1824).
Vitruvius Pollio, Marcos, *De Architectura*, (trans.) Richard Schofield, (London, 2009).

'W. E. H.', 'Mr. Barry's design for the new Houses of Parliament', *Westminster Review*, Vol. 3, (25 July, 1836), pp. 409–24.

Wheatstone, Charles, 'Description of the electro-magnetic clock', *Abstracts of the Papers Printed in the Philosophical Transactions of the Royal Society of London*, Vol. 4, (1837–1843), pp. 249–50.

Whewell, William, *History of the Inductive Sciences, from the Earliest to the Present Times*, Vol. I, (1837).

Willis, Robert, *An Attempt to Analyse the Automaton Chess Player, of Mr. De Kempelen*, (London, 1821).

Willis, Robert, *Remarks on the Architecture of the Middle Ages, Especially of Italy*, (Cambridge, 1835).

Willis, Robert, *Principles of Mechanism, Designed for the Use of Students in the Universities, and for Engineering Students Generally*, (London, 1841).

Xylopolist, *A few remarks on the style and execution of the New Houses of Parliament, the insertion of which was refused by a scientific journal for unknown reasons. With some additional observations, occasioned by the debate on the subject in the House of Commons, the 14th February, 1848*, (London, 1848).

Secondary Material

I. Published

Alexander, Jennifer Karns, 'Thinking again about science in technology', *Isis*, Vol. 103, No. 3 (September, 2012), pp. 518–26.

Alberti, Samuel J. M. M., 'The status of museums: authority, identity, and material culture', in David N. Livingstone and Charles W. J. Withers (eds.), *Geographies of Nineteenth-Century Science*, (Chicago, 2011), pp. 51–72.

Allen, J. R. L., *Late Churches and Chapels in Berkshire: A Geological Perspective from the Late Eighteenth Century to the First World War*, (Oxford, 2007).

Allen, J. R. L., *Building Late Churches in North Hampshire: A Geological Guide to Their Fabrics and Decoration from the Mid-Eighteenth Century to the First World War*, (Oxford, 2009).

Anderson, R. E., 'Hamilton, William Richard (1777–1859)', rev. R. A. Jones, *Oxford Dictionary of National Biography*, Oxford University Press, 2004; online edn, May 2006 [www.oxforddnb.com/view/article/12147, accessed 7 March 2014].

Arkell, W. J., *Oxford Stone*, (London, 1947).

Arnold, Dana, 'Burton, Decimus (1800–1881)', *Oxford Dictionary of National Biography*, Oxford University Press, 2004; online edn, May 2012 [www.oxforddnb.com/view/article/4125, accessed 22 February 2015].

Ashworth, William J., 'John Herschel, George Airy, and the roaming eye of the state', *History of Science*, 36:2, (1 June, 1998), pp. 151–78.

Atterbury, Paul, and Wainwright, Clive (eds.), *Pugin: A Gothic Passion*, (New Haven, 1994).

Baigent, Elizabeth, 'Jackson, Julian (1790–1853)', *Oxford Dictionary of National Biography*, Oxford University Press, 2004 [www.oxforddnb.com/view/article/14540, accessed 12 September 2014].

Banham, Reyner, *The Architecture of the Well-Tempered Environment*, (London, 1969).

Barker, G. F. R., 'Hall, Benjamin, Baron Llanover (1802–1867)', rev. H. C. G. Matthew, *Oxford Dictionary of National Biography*, Oxford University Press, 2004; online edn, January 2012 [www.oxforddnb.com/view/article/11945, accessed 22 February 2015].

Barnes, Barry, and Edge, David (eds.), *Science in Context: Readings in the Sociology of Science*, (Milton Keynes, 1982).

Barnwell, P. S., Geoffrey Tyack, and William Whyte (eds.), *George Gilbert Scott, 1811–1878: An Architect and His Influence*, (Donington, 2014).

Bate, David G., 'Sir Henry Thomas De la Beche and the founding of the British Geological Survey', *Mercian Geologist*, 2010, 17 (3), pp. 149–65.

Bebbington, David W., *Evangelicalism in Modern Britain: A History from the 1730s to the 1980s*, (London, 1989).

Bebbington, David W., 'Science and evangelical theology in Britain from Wesley to Orr', in David N. Livingstone, D. G. Hart, and Mark A. Noll (eds.), *Evangelicals and Science in Historical Perspective*, (Oxford, 1999), pp. 120–41.

Becher, Harvey W., 'Radicals, Whigs and Conservatives: the middle and lower classes in the analytical revolution at Cambridge in the age of aristocracy', *British Journal for the History of Science*, Vol. 28, No. 4 (December, 1995), pp. 405–26.

Becher, Harvey W., 'William Whewell's odyssey: from mathematics to moral philosophy', in Menachem Fisch and Simon Schaffer (eds.), *William Whewell: A Composite Portrait*, (Oxford, 1991), pp. 1–29.

Bennett, J. A., 'George Biddell Airy and horology', *Annals of Science*, Vol. 37, Issue 3 (1980), pp. 269–85.

Bennett, J. A., *The Mathematical Science of Christopher Wren*, (Cambridge, 1982).

Bennett, J. A., *The Divided Circle: A History of Instruments for Astronomy, Navigation and Surveying*, (Oxford, 1987).

Bennett, Tony, and Joyce, Patrick (eds.), *Material Powers: Cultural Studies, History and the Material Turn*, (London, 2010).

Bergdoll, Barry, 'Of crystals, cells, and strata: natural history and debates on the form of a new architecture in the nineteenth century', *Architectural History*, Vol. 50 (2007), pp. 1–29.

Bijker, Weibe E., Thomas P. Hughes, and Trevor J. Pinch (eds.), *The Social Construction of Technological Systems: New Directions in the Sociology and History of Technology*, (Cambridge, Massachusetts, 1987).

Blau, Eve, *Ruskinian Gothic: The Architecture of Deane and Woodward, 1845–1861*, (Princeton, 1982).

Boase, G. C., 'Dent, Edward John (1790–1853)', rev. Anita McConnell, *Oxford Dictionary of National Biography*, Oxford University Press, 2004; online edn, May 2007 [http://ezproxy.ouls.ox.ac.uk:2117/view/article/7512, accessed 24 August 2013].

Bonython, Elizabeth, *King Cole: A Picture Portrait of Sir Henry Cole, KCB 1808–1882*, (London, 1982).

Bord, Joe, 'Whiggery, science and administration: Grenville and Lord Henry Petty in the Ministry of All the Talents, 1806–7', *Historical Research*, Vol. 76, No. 191 (February, 2003), pp. 108–27.

Bord, Joe, *Science and Whig Manners: Science and Political Style in Britain, c.1790–1850*, (Basingstoke, 2009).

Bourdieu, Pierre, *Language and Symbolic Power*, (trans.) Gino Raymond and Matthew Adamson, (Cambridge, 1991).

Bowdler, Roger, 'Gwilt, Joseph (1784–1863)', *Oxford Dictionary of National Biography*, Oxford University Press, 2004; online edn, October 2007 [www.oxforddnb.com/view/article/11811, accessed 4 December 2013].

Brand, Vanessa (ed.), *The Study of the Past in the Victorian Age*, (Oxford, 1998).

Brown, Stewart J., 'Chalmers, Thomas (1780–1847)', *Oxford Dictionary of National Biography*, Oxford University Press, 2004; online edn, October 2007 [www.oxforddnb.com/view/article/5033, accessed 26 December 2012].

Brown, Stewart J., *Providence and Empire: Religion, Politics and Society in the United Kingdom, 1815–1914*, (Harlow, 2008).

Bruegmann, Robert, 'Central heating and forced ventilation: origins and effects on architectural design', *Journal of the Society of Architectural Historians*, Vol. 37, No. 3 (October, 1978), pp. 143–60.

Buchanan, Alexandrina, *Robert Willis (1800–1875) and the Foundation of Architectural History*, (Woodbridge, 2013).

Buchanan, R. A., 'Engineers and government in nineteenth-century Britain', in Roy MacLeod (ed.), *Government and Expertise: Specialists, Administrators and Professionals, 1860–1919*, (Cambridge, 1988), pp. 41–58.

Buchdahl, Gerd, 'Deductive versus inductivist – Mill vs Whewell', in Menachem Fisch and Simon Schaffer (eds.), *William Whewell: A Composite Portrait*, (Oxford, 1991), pp. 311–44.

Bud, Robert, '"Applied Science": a phrase in search of a meaning', *Isis*, Vol. 103, No. 3 (September 2012), pp. 537–45.

Bud, Robert, 'Introduction', *Isis*, Vol. 103, No. 3 (September, 2012), pp. 515–17.

Bud, Robert and Cozzens, Susan E. (eds.), *Invisible Connections: Instruments, Institutions, and Science*, (Washington, 1992).

Bud, Robert and Robert, Gerrylynn K., *Science Versus Practice: Chemistry in Victorian Britain*, (Manchester, 1984).

Burchell, Graham, Colin Gordon, and Peter Miller (eds.), *The Foucault Effect: Studies in Governmentality with Two Lectures by and an Interview with Michel Foucault*, (Chicago, 1991).

Burke, John G. (ed.), *The Uses of Science in the Age of Newton*, (Berkeley, 1983).

Burrow, J. W., *Whigs and Liberals: Continuity and Change in English Political Thought*, (Oxford, 1988).

Bynum, W. F., 'Wakley, Thomas (1795–1862)', *Oxford Dictionary of National Biography*, Oxford University Press, 2004 [http://ezproxy.ouls.ox.ac.uk:2117/view/article/28425, accessed 5 April 2013].

Cantor, Geoffrey, 'The changing role of Young's ether', *British Journal for the History of Science*, Vol. 5, No. 1 (June, 1970), pp. 44–62.

Cantor, Geoffrey, 'Henry Brougham and the Scottish methodological tradition', *Studies in History and Philosophy of Science*, 2, no. 1 (1971), pp. 69–89.

Cantor, Geoffrey, 'The reception of the wave theory of light in Britain: a case study illustrating the role of methodology in scientific debate', *Historical Studies in the Physical Sciences*, Vol. 6 (1975), pp. 109–32.

Cantor, Geoffrey, *Optics after Newton: Theories of Light in Britain and Ireland, 1704–1840*, (Manchester, 1983).

Cantor, Geoffrey N., 'Reading the book of nature: the relation between Faraday's religion and his science', in David Gooding and Frank A. J. L. James (eds.), *Faraday Rediscovered: Essays on the Life and Work of Michael Faraday, 1791–1867*, (Basingstoke, 1985), pp. 69–81.

Cantor, Geoffrey, *Michael Faraday: Sandemanian and Scientist. A Study of Science and Religion in the Nineteenth-Century*, (Basingstoke, 1991).

Cantor, Geoffrey, David Gooding, and Frank A. J. L. James, *Michael Faraday*, (New York, 1991).

Carlyle, E. I., 'Smith, Charles Harriot (1792–1864)', rev. M. A. Goodall, *Oxford Dictionary of National Biography*, Oxford University Press, 2004 [http://ezproxy.ouls.ox.ac.uk:2204/view/article/25787, accessed 31 August 2014].

Casey, Edward S., 'How to get from space to place in a fairly short stretch of time: phenomenological prolegomena', in Steven Feld and Keith H. Basso (eds.), *Senses of Place*, (Santa Fe, 1996), pp. 13–52.

Chaldecott, John A., 'Platinum and the Greenwich system of time-signals in Britain: the work of George Biddell Airy and Charles Vincent Walker from 1849 to 1870', *Platinum Metals Review*, 30 (1), pp. 29–37.

Chancellor, Valerie, *The Political Life of Joseph Hume, 1777–1855*, (London, 1986).

Chancellor, V. E., 'Hume, Joseph (1777–1855)', *Oxford Dictionary of National Biography*, Oxford University Press, 2004; online edn, January 2008 [www.oxforddnb.com/view/article/14148, accessed 7 March 2014].

Chapman, Allan, 'Sir George Airy (1801–1892) and the concept of international standards in science, timekeeping and navigation', *Vistas in Astronomy, 28* (London, 1985), pp. 321–28.

Chapman, Allan, 'Private research and public duty: George Biddell Airy and the search for Neptune', *Journal for the History of Astronomy*, 19 (Cambridge, 1988), pp. 121–39.

Chapman, Allan, 'George Biddell Airy, F. R. S. (1801–1892): a centenary commemoration', *Notes and Records of the Royal Society of London*, Vol. 46, No. 1 (January, 1992), pp. 103–10.

Chapman, Allan, 'The pit and the pendulum: G. B. Airy and the determination of gravity', *Antiquarian Horology* (Autumn, 1993), pp. 70–78.

Chapman, Allan, *Dividing the Circle: The Development of Critical Angular Measurement in Astronomy, 1500–1850*, 2nd ed., (Chichester, 1995).

Chapman, Allan, *The Victorian Amateur Astronomer: Independent astronomical research in Britain, 1820–1920*, (Chichester, 1998).

Chapman, Allan, 'Airy, Sir George Biddell (1801–1892)', *Oxford Dictionary of National Biography*, Oxford University Press, 2004; online edn, January 2011 [http://ezproxy.ouls.ox.ac.uk:2117/view/article/251, accessed 24 Aug 2013].

Christie, William, *The Edinburgh Review in the Literary Culture of Romantic Britain: Mammoth and Megalonyx*, (London, 2009).

Clark, Kenneth, *The Gothic Revival: An Essay in the History of Taste*, (Harmondsworth, 1962).

Clark, William, Jan Golinski, and Simon Schaffer (eds.), *The Sciences in Enlightened Europe*, (Chicago, 1999).

Clarke, Jonathan, 'Pioneering yet peculiar: John Nash's contribution to late Georgian building technology', in Geoffrey Tyack (ed.), *John Nash: Architect of the Picturesque*, (Swindon, 2013), pp. 153–68.

Clifton-Taylor, Alec, *The Pattern of English Building*, (London, 1987).

Collini, Stefan, Donald Winch and John Burrow, *That Noble Science of Politics: A Study in Nineteenth-Century Intellectual History*, (Cambridge, 1983).

Colvin, Howard, *A Biographical Dictionary of British Architects, 1600–1840*, 3rd ed., (New Haven, 1995).

Cooper, Ann, 'Cole, Sir Henry (1808–1882)', *Oxford Dictionary of National Biography*, Oxford University Press, 2004; online edn, January 2008 [www.oxforddnb.com/view/article/5852, accessed 7 March 2014].

Cooter, Roger and Pumfrey, Stephen, 'Separate spheres and public places: reflections on the history of science popularization and science in popular culture', *History of Science*, Vol. 32, No. 3 (1994), pp. 237–67.

Corbin, Alain, *Village Bells: Sound and Meaning in the Nineteenth-Century French Countryside*, (London, 1999).

Cotter, Charles H., 'George Biddell Airy and his mechanical correction of the magnetic compass', *Annals of Science*, Vol. 33, Issue 3 (1976), pp. 263–74.

Cotter, Charles H., 'The early history of ship magnetism: the Airy-Scoresby controversy', *Annals of Science*, Vol. 34, Issue 6 (1977), pp. 589–99.

Craik, A. D. D., 'Victorian "applied mathematics"', in Raymond Flood, Adrian Rice, and Robin Wilson (eds.), *Mathematics in Victorian Britain*, (Oxford, 2011), pp. 177–98.

Crimmins, James E., *Utilitarian Philosophy and Politics: Bentham's Later Years*, (London, 2011).

Crook, J. Mordaunt, 'The pre-Victorian architect: professionalism & patronage', *Architectural History*, Vol. 12 (1969), pp. 62–78.

Crook, J. Mordaunt, *The Dilemma of Style: Architectural Ideas from the Picturesque to the Post-Modern*, (London, 1987).

Crook, J. Mordaunt and Port, M. H. (eds.), *The History of the King's Works*, Vol. VI: 1782–1851, (London, 1973).

Crook, Tom, 'Secrecy and liberal modernity in Victorian and Edwardian England', in Simon Gunn and James Vernon (eds.), *The Peculiarities of Liberal Modernity in Imperial Britain*, (Berkeley, 2011), pp. 72–90.

Cruickshank, Dan (ed.), *Timeless Architecture: 1*, (London, 1985).

(Anon.), 'Cust, Sir Edward, baronet (1794–1878)', rev. James Lunt, *Oxford Dictionary of National Biography*, Oxford University Press, 2004 [www.oxforddnb.com/view/article/6973, accessed 7 March 2014].

Cust, L. H., 'Meeson, Alfred (1808–1885)', rev. Susie Barson, *Oxford Dictionary of National Biography*, Oxford University Press, 2004 [www.oxforddnb.com/view/article/18511, accessed 4 Dec 2013].

Daniels, Rebecca and Brandwood, Geoff (eds.), *Ruskin and Architecture*, (Reading, 2003).
Darwin, John, *The Triumphs of Big Ben*, (London, 1986).
Daunton, Martin, *Progress and Poverty: An Economic and Social History of Britain, 1700–1850*, (Oxford, 1995).
Daunton, Martin, *Trusting Leviathan: The Politics of Taxation in Britain, 1799–1914*, (Cambridge, 2001).
Daunton, Martin (ed.), *The Organization of Knowledge in Victorian Britain*, (Oxford, 2005).
Davies, David I., 'John Frederic Daniell, 1791–1845', *Chemistry in Britain* (October, 1990), pp. 946–60.
Davison, Graeme, 'The city as a natural system: theories of urban society in early nineteenth-century Britain', in Derek Fraser and Anthony Sutcliffe (eds.), *The Pursuit of Urban History*, (London, 1983), pp. 349–70.
Desmond, Adrian, *The Politics of Evolution: Morphology, Medicine, and Reform in Radical London*, (Chicago, 1989).
Droth, Martina, Jason Edwards, and Michael Hatt (eds.), *Sculpture Victorious: Art in an Age of Invention, 1837–1901*, (Yale, 2014).
Elliott, Brent, 'Loudon, John Claudius (1783–1843)', *Oxford Dictionary of National Biography*, Oxford University Press, 2004; online edn, May 2010 [www.oxforddnb.com/view/article/17031, accessed 7 March 2014].
Fanning, A. E., *Steady as She Goes: A History of the Compass Department of the Admiralty*, (London, 1986).
Feld, Steven and Basso, Keith H. (eds.), *Senses of Place*, (Santa Fe, 1996).
Ferriday, Peter, *Lord Grimthorpe, 1816–1905*, (London, 1957).
Ferriday, Peter (ed.), *Victorian Architecture*, (London, 1963).
Fisch, Menachem, *William Whewell: Philosopher of Science*, (Oxford, 1991).
Fisch, Menachem and Schaffer, Simon (eds.), *William Whewell: A Composite Portrait*, (Oxford, 1991).
Fleetwood-Hesketh, Peter, 'Sir Charles Barry', in Peter Ferriday (ed.), *Victorian Architecture*, (London, 1963), pp. 125–35.
Flett, John Smith, *The First Hundred Years of the Geological Survey of Great Britain*, (London, 1937).
Flood, Raymond, Adrian Rice, and Robin Wilson (eds.), *Mathematics in Victorian Britain*, (Oxford, 2011).
Forgan, Sophie, 'From servant to savant: the institutional context', in David Gooding and Frank A. J. L. James (eds.), *Faraday Rediscovered: Essays on the Life and Work of Michael Faraday, 1791–1867*, (Basingstoke, 1985), pp. 51–67.
Forgan, Sophie, 'Context, image and function: a preliminary enquiry into the architecture of scientific societies', *British Journal for the History of Science*, Vol. 19, Issue 1, (March, 1986), pp. 89–113.
Forgan, Sophie, 'The architecture of display: museums, universities and objects in nineteenth-century Britain', *History of Science*, Vol. 32, Issue 2 (1994), pp. 139–62.
Forgan, Sophie, '"But indifferently lodged . . . ": perception and place in building for science in Victorian London', in Crosbie Smith and Jon Agar (eds.), *Making*

Space for Science: Territorial Themes in the Shaping of Knowledge, (Basingstoke, 1998), pp. 195–215.

Forgan, Sophie, 'Building the museum: knowledge, conflict and the power of place', *Isis*, Vol. 96, No. 4 (2005), pp. 572–85.

Foucault, Michel, *Archaeology of Knowledge*, (trans.) A. M. Sheridan Smith, (London, 1969).

Foucault, Michel, *Discipline and Punish: The Birth of the Prison*, (trans.) Alan Sheridan, (Harmondsworth, 1977).

Foucault, Michel, 'Governmentality', in Graham Burchell, Colin Gordon, and Peter Miller (eds.), *The Foucault Effect: Studies in Governmentality with Two Lectures by and an Interview with Michel Foucault*, (Chicago, 1991), pp. 87–104.

Fox, Celina (ed.), *London – World City, 1800–1840*, (New Haven, 1992).

Fraser, Derek and Sutcliffe, Anthony (eds.), *The Pursuit of Urban History*, (London, 1983).

Fredericksen, Andrea, 'Parliament's genius loci: the politics of place after the 1834 fire', in Christine Riding and Jacqueline Riding (eds.), *The Houses of Parliament: History, Art, Architecture*, (London, 2000), pp. 99–111.

Fry, Michael, 'Alison, Sir Archibald, first baronet (1792–1867)', *Oxford Dictionary of National Biography*, Oxford University Press, 2004 [www.oxforddnb.com/view/article/349, accessed 7 March 2014].

Galison, Peter and Thompson, Emily (eds.), *The Architecture of Science*, (Cambridge, Massachusetts: 1999).

Garnham, Trevor, *Oxford Museum: Deane and Woodward*, (London, 1992).

Gerbino, Anthony and Johnston, Stephen, *Compass and Rule: Architecture as Mathematical Practice in England, 1500–1700*, (New Haven, 2009).

Gieryn, Thomas F., 'Boundary-work and the demarcation of science from non-science: strains and interests in professional ideologies of scientists', *American Sociological Review*, Vol. 48, No. 6 (December, 1983), pp. 781–95.

Gillin, Edward John, 'Gothic fantastic: Parliament, Pugin, and the architecture of science', *True Principles*, Vol. 4, No. 5, (Winter, 2015), pp. 382–89.

Gillin, Edward John, 'Prophets of progress: authority in the scientific projections and religious realizations of the *Great Eastern* steamship', *Technology and Culture*, Vol. 56, No. 4 (October, 2015), pp. 928–56.

Gillin, Edward John, 'Stones of science: Charles Harriot Smith and the importance of geology in architecture, 1834–64', *Architectural History*, Vol. 59 (2016), pp. 281–310.

Gillin, Edward John, 'Reid, David Boswell (1805–1863)', *Oxford Dictionary of National Biography*, Oxford University Press, April 2016 [http://ezproxy-prd.bodleian.ox.ac.uk:2167/view/article/23327, accessed 17 September 2016].

Girouard, Mark, 'Charles Barry: a centenary assessment', *Country Life*, Vol. CXXVIII, No. 3319 (13 October, 1960), pp. 796–97.

Girouard, Mark, *The Victorian Country House*, (Oxford, 1971).

Gleadle, Kathryn, *Borderline Citizens: Women, Gender, and Political Culture in Britain, 1815–1867*, (Oxford, 2009).

Gleich, Moritz, 'Architect and service architect: the quarrel between Charles Barry and David Boswell Reid.', *Interdisciplinary Science Reviews*, Vol. 37, No. 4 (December, 2012), pp. 332–44.

Glennie, Paul and Thrift, Nigel, *Shaping the Day: A History of Timekeeping in England and Wales, 1300–1800*, (Oxford, 2009).
Goldman, Lawrence, *Science, Reform, and Politics in Victorian Britain: The Social Science Association, 1857–1886*, (Cambridge, 2002).
Golinski, Jan, *Science as Public Culture: Chemistry and Enlightenment in Britain, 1760–1820*, (Cambridgc, 1992).
Gooday, Graeme J. N., *The Morals of Measurement: Accuracy, Irony, and Trust in Late Victorian Electrical Practice*, (Cambridge, 2004).
Gooday, Graeme J. N., 'Liars, experts and authorities', *History of Science*, Vol. 46, Issue 4 (2008), pp. 431–56.
Gooday, Graeme, 'Placing or replacing the laboratory in the history of science?', *Isis*, Vol. 99, No. 4 (December, 2008), pp. 783–95.
Gooday, Graeme, '"Vague and artificial": the historically elusive distinction between pure and applied science', *Isis*, Vol. 103, No. 3 (September, 2012), pp. 546–54.
Gooding, David, '"In nature's school": Faraday as an experimentalist', in David Gooding and Frank A. J. L. James (eds.), *Faraday Rediscovered: Essays on the Life and Work of Michael Faraday, 1791–1867*, (Basingstoke, 1985), pp. 105–35.
Gooding, David, and James, Frank A. J. L. (eds.), *Faraday Rediscovered: Essays on the Life and Work of Michael Faraday, 1791–1867*, (Basingstoke, 1985).
Gooding, David, Trevor Pinch, and Simon Schaffer (eds.), *The Uses of Experiment: Studies in the Natural Sciences*, (Cambridge, 1989).
Gooding, David, Trevor Pinch, and Simon Schaffer, 'Introduction: some uses of experiment', in David Gooding, Trevor Pinch, and Simon Schaffer (eds.), *The Uses of Experiment: Studies in the Natural Sciences*, (Cambridge, 1989), pp. 1–27.
Goodsell, Charles T., 'The architecture of parliaments: legislative houses and political culture', *British Journal of Political Science*, Vol. 18, No. 3, (July, 1988) pp. 287–302.
Grattan-Guinness, I., 'Mathematics and mathematical physics from Cambridge, 1815–40: a survey of the achievements and of the French influences', in P. M. Harman (ed.), *Wranglers and Physicists: Studies on Cambridge Mathematical Physics in the Nineteenth Century*, (Manchester, 1985), pp. 84–111.
Guldi, Jo, *Roads to Power: Britain Invents the Infrastructure State*, (Cambridge, Massachusetts, 2012).
Gunn, Simon and Vernon, James (eds.), *The Peculiarities of Liberal Modernity in Imperial Britain*, (Berkeley, 2011).
Gunn, Simon and Vernon, James, 'Introduction: what was liberal modernity and why was it peculiar in imperial Britain', in Simon Gunn and James Vernon (eds.), *The Peculiarities of Liberal Modernity in Imperial Britain*, (Berkeley, 2011), pp. 1–18.
Gunnis, Rupert, *Dictionary of British Sculptors, 1660–1851*, (London, 1953).
Habibi, Don A., *John Stuart Mill and the Ethic of Human Growth*, (London, 2001).
Haile, Neville, 'Buckland, William (1784–1856)', *Oxford Dictionary of National Biography*, Oxford University Press, 2004; online edn, October 2007 [http://ezproxy.ouls.ox.ac.uk:2204/view/article/3859, accessed 23 August 2014].

Hall, Michael, 'What do Victorian churches mean? Symbolism and sacramentalism in Anglican church architecture, 1850–1870', *Journal of the Society of Architectural Historians*, Vol. 59, No. 1 (March, 2000), pp. 78–95.

Hall, Michael, 'G. F. Bodley and the response to Ruskin in ecclesiastical architecture in the 1850s', in Rebecca Daniels and Geoff Brandwood (eds.), *Ruskin and Architecture*, (Reading, 2003), pp. 249–76.

Hall, Michael, *George Frederick Bodley: And the Later Gothic Revival in Britain and America*, (New Haven, 2014).

Hardy, Anne, 'Lyon Playfair and the idea of Progress: science and medicine in Victorian parliamentary politics', in Dorothy Porter and Roy Porter (eds.), *Doctors, Politics and Society: Historical Essays*, (Amsterdam, 1993), pp. 81–106.

Harman, P. M. (ed.), *Wranglers and Physicists: Studies on Cambridge Mathematical Physics in the Nineteenth Century*, (Manchester, 1985).

Harris, Jose, 'Mill, John Stuart (1806–1873)', *Oxford Dictionary of National Biography*, Oxford University Press, 2004; online edn, January 2012 [www.oxforddnb.com/view/article/18711, accessed 7 March 2014].

Harris, T. R., *Sir Goldsworthy Gurney, 1793–1875*, (Penzance, 1975).

Hatt, Michael, Martina Droth, and Jason Edwards, 'Sculpture victorious', in Martina Droth, Jason Edwards, and Michael Hatt (eds.), *Sculpture Victorious: Art in an Age of Invention, 1837–1901*, (Yale, 2014), pp. 15–55.

Hawkins, Angus, *The Forgotten Prime Minister: The 14th Earl of Derby. Volume II: Achievement, 1851–1869*, (Oxford, 2008).

Headrick, Daniel R., *The Tools of Empire: Technology and European Imperialism in the Nineteenth Century*, (Oxford, 1981).

Headrick, Daniel R., *Power over Peoples: Technology, Environments, and Western Imperialism, 1400 to the Present*, (Princeton, 2010).

Heilbroner, Robert L., 'Do machines make history?', in Merritt Roe Smith and Leo Marx (eds.), *Does Technology Drive History? The Dilemma of Technological Determinism*, (Cambridge, Massachusetts, 1994), pp. 53–65.

Hicks, Dan, 'The material-cultural turn: event and effect', in Dan Hicks and Mary C. Beaudry (eds.), *The Oxford Handbook of Material Culture Studies*, (Oxford, 2010), pp. 25–98.

Hicks, Dan and Beaudry, Mary C. (eds.), *The Oxford Handbook of Material Culture Studies*, (Oxford, 2010).

Hill, Rosemary, 'Butterfield, William (1814–1900)', *Oxford Dictionary of National Biography*, Oxford University Press, 2004 [www.oxforddnb.com/view/article/4228, accessed 29 November 2014].

Hill, Rosemary, *God's Architect: Pugin and the Building of Romantic Britain*, (London, 2008).

Hilton, Boyd, *The Age of Atonement: The Influence of Evangelicalism on Social and Economic Thought, 1785–1865*, (Oxford, 1986).

Hix, John, *The Glass House*, (London, 1974).

Hoppen, K. Theodore, *The Mid-Victorian Generation, 1846–1886*, (Oxford, 1998).

Hoppen, K. Theodore, 'Ponsonby, John William, fourth earl of Bessborough (1781–1847)', *Oxford Dictionary of National Biography*, Oxford University

Press, 2004; online edn, January 2008 [www.oxforddnb.com/view/article/22500, accessed 4 December 2013].

Howell-Thomas, Dorothy, *Duncannon: Reformer and Reconciler, 1781–1847*, (Norwich, 1992).

Howse, Derek, *Greenwich Time and the Discovery of the Longitude*, (Oxford, 1980).

Howse, Derek, *Greenwich Time and the Longitude*, (London, 1997).

Huch, Ronald K. and Ziegler, Paul R., *Joseph Hume: The People's M.P.*, (Philadelphia, 1985).

Hughes, Thomas P., 'Technological momentum', in Merritt Roe Smith and Leo Marx (eds.), *Does Technology Drive History? The Dilemma of Technological Determinism*, (Cambridge, Massachusetts, 1994),101–13.

Hughes, Thomas P., *Human-Built World: How to Think about Technology and Culture*, (Chicago, 2004).

Hunt, Bruce J., 'The Ohm is where the art is: British telegraph engineers and the development of electrical standards', *Osiris*, 2nd Series, Vol. 9, Instruments (1994), pp. 48–63.

Hunt, Bruce, 'Scientists, engineers and wildman whitehouse: measurement and credibility in early cable telegraphy', *British Journal for the History of Science*, Vol. 29, Issue 2, (June, 1996), pp. 155–69.

Hunt, Bruce J., 'Doing science in a global empire: cable telegraphy and electrical physics in Victorian Britain', in Bernard Lightman (ed.), *Victorian Science in Context*, (Chicago, 1997), pp. 312–33.

Ihalainen, Pasi, 'The sermon, court, and Parliament, 1689–1789', in Keith A. Francis and William Gibson (eds.), *Oxford Handbook of The British Sermon, 1689–1901*, (Oxford, 2012), pp. 229–44.

Inkster, Ian and Morrell, Jack (eds.), *Metropolis and Province: Science in British Culture, 1780–1850*, (London, 1983).

Ishibashi, Yuto, 'In pursuit of accurate timekeeping: Liverpool and Victorian electrical horology', *Annals of Science*, Vol. 71, Issue 4 (2014), pp. 474–96.

Jackson, Neil, 'Clarity or camouflage? The development of constructional polychromy in the 1850s and early 1860s', *Architectural History*, Vol. 47 (2004), pp. 201–26.

Jacyna, L. S., *Philosophic Whigs: Medicine, Science and Citizenship in Edinburgh, 1789–1848*, (London, 1994).

James, Frank A. J. L., '"The optical mode of investigation": light and matter in Faraday's natural philosophy', in David Gooding and Frank A. J. L. James (eds.), *Faraday Rediscovered: Essays on the Life and Work of Michael Faraday, 1791–1867*, (Basingstoke, 1985), pp. 137–61.

James, Frank A. J. L., '"The civil-engineer's talent": Michael Faraday, science, engineering and the English lighthouse service, 1836–1865', *Transactions of the Newcomen Society*, 70 (1998–99), pp. 153–60.

James, Frank A. J. L., 'Brande, William Thomas (1788–1866)', *Oxford Dictionary of National Biography*, Oxford University Press, 2004 [http://ezproxy.ouls.ox.ac.uk:2117/view/article/3258, accessed 7 April 2013].

James, Frank A. J. L., 'Daniell, John Frederic (1790–1845)', *Oxford Dictionary of National Biography*, Oxford University Press, 2004 [http://ezproxy.ouls.ox.ac.uk:2117/view/article/7124, accessed 21 July 2013].

James, Frank A. J. L., 'Faraday, Michael (1791–1867)', *Oxford Dictionary of National Biography*, Oxford University Press, 2004; online edn, January 2011 [http://ezproxy.ouls.ox.ac.uk:2117/view/article/9153, accessed 21 September 2013].

Jenkins, Frank, 'The Victorian architectural profession', in Peter Ferriday (ed.), *Victorian Architecture*, (London, 1963), pp. 39–49.

Jenkins, T. A., *Parliament, Party and Politics in Victorian Britain*, (Manchester, 1996).

Joyce, Patrick, *The Rule of Freedom: Liberalism and the Modern City*, (London, 2003).

Joyce, Patrick, 'Filing the Raj: political technologies of the imperial British state', in Tony Bennett and Patrick Joyce (eds.), *Material Powers: Cultural Studies, History and the Material Turn*, (London, 2010),102–23.

Joyce, Patrick, *The State of Freedom: A Social History of the British State since 1800*, (Cambridge, 2013).

Joyce, Patrick and Bennett, Tony, 'Material powers: introduction', in Tony Bennett and Patrick Joyce (eds.), *Material Powers: Cultural Studies, History and the Material Turn*, (London, 2010), pp. 1–21.

Kennedy, David, 'Dr. D. B. Reid and the Teaching of Chemistry', *Studies: An Irish Quarterly Review*, Vol. 31, No. 123, (September, 1942), pp. 343–50.

Kern, Stephen, *The Culture of Time and Space, 1880–1918*, (London, 1983).

Knell, Simon J., *The Culture of English Geology, 1815–1851: A Science Revealed through Its Collecting*, (Aldershot, 2000).

Kostof, Spiro (ed.), *The Architect: Chapters in the History of the Profession*, (New York, 1977).

Kriegel, Abraham D., 'Liberty and whiggery in early nineteenth-century England', *Journal of Modern History*, Vol. 52, No. 2 (June, 1980), pp. 253–78.

Latour, Bruno, *We Have Never Been Modern*, (trans.) Catherine Porter, (Cambridge, Massachusetts, 1993).

Latour, Bruno and Woolgar, Steve, *Laboratory Life: The Construction of Scientific Facts*, (Princeton, 1986).

Latour, Bruno, and Woolgar, Steve, 'The cycle of credibility', in Barry Barnes and David Edge (eds.), *Science in Context: Readings in the Sociology of Science*, (Milton Keynes, 1982), pp. 35–43.

Leach, Peter, 'Fowler, Charles (1792–1867)', rev. *Oxford Dictionary of National Biography*, Oxford University Press, 2004 [www.oxforddnb.com/view/article/37426, accessed 7 March 2014].

Leathlean, Howard, 'Loudon's architectural magazine and the Houses of Parliament competition', *Victorian Periodicals Review*, Vol. 26, No. 3 (Fall, 1993), pp. 145–53.

Lee, Matthew Lee, 'Birkbeck, George (1776–1841)', *Oxford Dictionary of National Biography*, Oxford University Press, 2004 [http://ezproxy.ouls.ox.ac.uk:2117/view/article/2454, accessed 7 April 2013].

Lefebvre, Henri, *The Production of Space*, (trans.) Donald Nicholson-Smith, (Oxford, 1991).

Leggett, Don, 'Spectacle and witnessing: constructing readings of Charles Parsons's marine turbine', *Technology and Culture*, Vol. 52, No. 2 (April, 2011), pp. 287–309.

Leggett, Don, 'Naval architecture, expertise and navigating authority in the British Admiralty, c.1885–1906', *Journal for Maritime Research*, Vol. 16, No. 1 (May, 2014), pp. 73–88.

Levine, Neil, *Modern Architecture: Representation and Reality*, (New Haven, 2009).

Lightman, Bernard (ed.), *Victorian Science in Context*, (Chicago, 1997).

Lightman, Bernard, 'The visual theology of Victorian popularisers of science: from reverent eye to chemical retina', *Isis*, Vol. 91, No. 4 (December, 2000), pp. 651–80.

Lightman, Bernard, *Victorian Popularizers of Science: Designing Nature for New Audiences*, (Chicago, 2007).

Lightman, Bernard, 'Refashioning the Spaces of London Science: Elite Epistemes in the Nineteenth Century', in David N. Livingstone and Charles W. J. Withers (eds.), *Geographies of Nineteenth-Century Science*, (Chicago, 2011), pp. 25–50.

Lightman, Bernard and Zon, Bennett (eds.), *Evolution and Victorian Culture*, (Cambridge, 2014).

Livingstone, David N., D. G. Hart, and Mark A. Noll (eds.), *Evangelicals and Science in Historical Perspective*, (Oxford, 1999).

Livingstone, David N. and Withers, Charles W. J. (eds.), *Geographies of Nineteenth-Century Science*, (Chicago, 2011).

Lobban, Michael, 'Brougham, Henry Peter, first Baron Brougham and Vaux (1778–1868)', *Oxford Dictionary of National Biography*, Oxford University Press, 2004; online edn, January 2008 [http://ezproxy.ouls.ox.ac.uk:2117/view/article/3581, accessed 28 March 2013].

Lott, Graham K., 'The development of the Victorian stone industry', The English Stone Forum Conference, (York, 15–17 April, 2005), pp. 44–56.

Lott, Graham K. and Richardson, Christine, 'Yorkshire stone for building the Houses of Parliament (1839–c.1852)', *Proceedings of the Yorkshire Geological Society*, Vol. 51 (1997), pp. 265–72.

Lounsbury, Carl R., 'Architecture and cultural history', in Dan Hicks and Mary C. Beaudry (eds.), *The Oxford Handbook of Material Culture Studies*, (Oxford, 2010), pp. 484–501.

Lucas, Caroline, *Honourable Friends?* (London, 2015).

Luckin, Bill, 'Arnott, Neil (1788–1874)', *Oxford Dictionary of National Biography*, Oxford University Press, 2004 [http://ezproxy.ouls.ox.ac.uk:2117/view/article/694, accessed 7 April 2013].

Maclean, Allan, *Telford's Highland Churches: The Highland Churches and Manses of Thomas Telford*, (Inverness, 1989).

MacLeod, Roy M., 'Science and Government in Victorian England: Lighthouse Illumination and the Board of Trade, 1866–1886', *Isis*, Vol. 60, No. 1 (Spring, 1969), pp. 4–38.

Macleod, Roy, 'Introduction: On the Advancement of Science', in Roy Macleod and Peter Collins (eds.), *The Parliament of Science: The British Association for the Advancement of Science, 1831–1981*, (Northwood, 1981), pp. 17–42.

MacLeod, Roy M., 'Whigs and savants: reflections on the reform movement in the Royal Society, 1830–48', in Ian Inkster and Jack Morrell (eds.), *Metropolis and Province: Science in British culture, 1780–1850*, (London, 1983), pp. 55–90.

MacLeod, Roy (ed.), *Government and Expertise: Specialists, Administrators and Professionals, 1860–1919*, (Cambridge, 1988).

Macleod, Roy, and Collins, Peter (eds.), *The Parliament of Science: the British Association for the Advancement of Science, 1831–1981*, (Northwood, 1981).

Mainardi, Patricia, *Art and Politics of the Second Empire: The Universal Expositions of 1855 and 1867*, (New Haven, 1987).

Mandler, Peter, *Aristocratic Government in the Age of Reform: Whigs and Liberals, 1830–1852*, (Oxford, 1990).

Mandler, Peter, 'Lamb, William, second Viscount Melbourne (1779–1848)', *Oxford Dictionary of National Biography*, Oxford University Press, 2004; online edn, January 2008 [www.oxforddnb.com/view/article/15920, accessed 7 March 2014].

Marsden, Ben, 'Engineering science in Glasgow: economy, efficiency and measurement as prime movers in the differentiation of an academic discipline', *British Journal for the History of Science*, Vol. 25, No. 3 (September, 1992), pp. 319–46.

Marsden, Ben, 'Blowing hot and cold: reports and retorts on the status of the air-engine as success or failure, 1830–1855', *History of Science*, Vol. 36, Issue 4 (1998), pp. 373–420.

Marsden, Ben, '"The progeny of these two "fellows"": Robert Willis, William Whewell and the sciences of mechanism, mechanics and machinery in early Victorian Britain', *British Journal for the History of Science*, Vol. 37, No. 4 (December, 2004), pp. 401–34.

Marsden, Ben, 'Willis, Robert (1800–1875)', *Oxford Dictionary of National Biography*, Oxford University Press, 2004; online edn, October 2009 [www.oxforddnb.com/view/article/29584, accessed 7 March 2014].

Marsden, Ben and Smith, Crosbie, *Engineering Empires: A Cultural History of Technology in Nineteenth-Century Britain*, (Basingstoke, 2005).

Martin, Ged, 'Hawes, Sir Benjamin (1797–1862)', *Oxford Dictionary of National Biography*, Oxford University Press, 2004; online edn, May 2009 [http://ezproxy.ouls.ox.ac.uk:2117/view/article/12643, accessed 5 April 2013].

Matthew, H. C. G., 'Warburton, Henry (1784–1858)', *Oxford Dictionary of National Biography*, Oxford University Press, 2004; online edn, May 2009 [http://ezproxy.ouls.ox.ac.uk:2117/view/article/28672, accessed 8 April 2013].

McCartney, Paul J., *Henry De la Beche: Observations on an Observer*, (Cardiff, 1977).

McKay, Charles, *Big Ben: The Great Clock and the Bells at the Palace of Westminster*, (Oxford, 2010).

Meadows, A. J., *Greenwich Observatory, Volume 2: Recent History, (1836–1975)*, (London, 1975).

Metcalf, Thomas R., 'Canning, Charles John, Earl Canning (1812–1862)', *Oxford Dictionary of National Biography*, Oxford University Press, 2004; online edn, January 2008 [http://ezproxy.ouls.ox.ac.uk:2117/view/article/4554, accessed 24 August 2013].

Miele, Chris, 'Real antiquity and the ancient object: the science of Gothic architecture and the restoration of medieval buildings', in Vanessa Brand (ed.), *The Study of the Past in the Victorian Age*, (Oxford, 1998), pp. 103–24.

Mitchell, Timothy, *Rule of Experts: Egypt, Techno-Politics, Modernity*, (Berkeley, 2002).
Morrell, Jack B., 'Practical chemistry in the University of Edinburgh, 1799–1843', *Ambix*, 16 (1969), pp. 66–80.
Morrell, Jack, 'Economic and ornamental geology: the Geological and Polytechnic Society of the West Riding of Yorkshire, 1837–53', in Ian Inkster and Jack Morrell (eds.), *Metropolis and Province: Science in British Culture, 1780–1850*, (London, 1983), pp. 231–56.
Morrell, Jack, 'Hope, Thomas Charles (1766–1844)', *Oxford Dictionary of National Biography*, Oxford University Press, 2004 [http://ezproxy.ouls.ox.ac.uk:2117/view/article/13738, accessed 5 April 2013].
Morrell, Jack, *John Phillips and the Business of Victorian Science*, (Aldershot, 2005).
Morrell, Jack and Thackray, Arnold, *Gentlemen of Science: Early Years of the British Association for the Advancement of Science*, (Oxford, 1981).
Morrison-Low, A. D., 'Brewster, Sir David (1781–1868)', *Oxford Dictionary of National Biography*, Oxford University Press, 2004; online edn, October 2005 [http://ezproxy.ouls.ox.ac.uk:2117/view/article/3371, accessed 21 September 2013].
Morton, John L., *Strata: The Remarkable Life Story of William Smith, 'The Father of English Geology'*, (Horsham, 2004).
Morus, Iwan Rhys, 'Different experimental lives: Michael Faraday and William Sturgeon', *History of Science*, Vol. 30, Issue 1 (March, 1992), pp. 1–28.
Morus, Iwan Rhys, *Frankenstein's Children: Electricity, Exhibition, and Experiment in Early-Nineteenth-Century London*, (Princeton, 1998).
Morus, Iwan Rhys, '"The nervous system of Britain": space, time and the electric telegraph in the Victorian age', *British Journal for the History of Science*, Vol. 33, No. 4, On Time: history, science and commemoration, (December, 2000), pp. 455–75.
Morus, Iwan Rhys, *When Physics Became King*, (Chicago, 2005).
Morus, Iwan Rhys, 'Worlds of wonder: sensation and the Victorian scientific performance', *Isis*, Vol. 101, No. 4 (December, 2010), pp. 806–16.
Morus, Iwan, Simon Schaffer, and Jim Secord, 'Scientific London', in Celina Fox (ed.), *London – World City, 1800–1840*, (New Haven, 1992), pp. 129–42.
Newbould, Ian, *Whiggery and Reform, 1830–41: The Politics of Government*, (Basingstoke, 1990).
O'Dwyer, Frederick, *The Architecture of Deane and Woodward*, (Cork, 1997).
Ogden, James, 'D'Israeli, Isaac (1766–1848)', *Oxford Dictionary of National Biography*, Oxford University Press, 2004; online edn, May 2008 [http://ezproxy.ouls.ox.ac.uk:2117/view/article/7690, accessed 21 September 2013].
Oliver, Stuart, 'The Thames embankment and the discipline of nature in modernity', *The Geographical Journal*, Vol. 166, No. 3 (September, 2000), pp. 227–38.
Olley, John, 'The Reform Club: Charles Barry', in Dan Cruickshank (ed.), *Timeless Architecture: 1*, (London, 1985), pp. 23–46.
Ophir, Adi and Shapin, Steven, 'The place of knowledge: a methodological survey', *Science in Context*, 4, 1 (1991), pp. 3–21.

Otter, Chris, *The Victorian Eye: A Political History of Light and Vision in Britain, 1800–1910*, (Chicago, 2008).
Parry, Jonathan, *The Rise and Fall of Liberal Government in Victorian Britain*, (New Haven, 1993).
Perkin, Harold, *The Rise of Professional Society: England since 1880*, (London, 1989).
Pevsner, Nikolaus, *Robert Willis*, (Northampton, Massachusetts, 1970).
Phillips, John A. and Wetherell, Charles, 'The Great Reform Act of 1832 and the political modernization of England', *American Historical Review*, Vol. 100, No. 2 (April, 1995), pp. 411–36.
Pickering, Andrew, 'Material culture and the dance of agency', in Dan Hicks and Mary C. Beaudry (eds.), *The Oxford Handbook of Material Culture Studies*, (Oxford, 2010), pp. 191–208.
Pickstone, John, 'Science in nineteenth-century England: plural configurations and singular politics', in Martin Daunton (ed.), *The Organization of Knowledge in Victorian Britain*, (Oxford, 2005), pp. 29–60.
Pike, David L., '"The greatest wonder of the world": Brunel's tunnel and the meanings of underground London', *Victorian Literature and Culture*, Vol. 33, No. 2 (2005), pp. 341–67.
Pinch, Trever J. and Bijker, Wiebe E., 'The social construction of facts and artefacts: or how the society of science and the sociology of technology might benefit each other', in Weibe E. Bijker, Thomas P. Hughes, and Trevor J. Pinch (eds.), *The Social Construction of Technological Systems: New Directions in the Sociology and History of Technology*, (Cambridge, Massachusetts, 1987), pp. 17–50.
Port, M. H., 'Parliamentary scrutiny and treasury stringency', in J. Mordaunt Crook and M. H. Port (eds.), *The History of the King's Works*, Vol. VI: 1782–1851, (London, 1973), pp. 157–78.
Port, M. H., 'The failure of experiment', in J. Mordaunt Crook and M. H. Port (eds.), *The History of the King's Works*, Vol. VI: 1782–1851, (London, 1973), pp. 209–49.
Port, M. H., 'Buckingham Palace', in J. Mordaunt Crook and M. H. Port (eds.), *The History of the King's Works*, Vol. VI: 1782–1851, (London, 1973), pp. 263–93.
Port, M. H., 'The New Houses of Parliament', in J. Mordaunt Crook and M. H. Port (eds.), *The History of the King's Works*, Vol. VI: 1782–1851, (London, 1973), pp. 573–626.
Port, M. H. (ed.), *The Houses of Parliament*, (New Haven, 1976).
Port, M. H., 'The old Houses of Parliament', in M. H. Port (ed.), *The Houses of Parliament*, (New Haven, 1976), pp. 5–19.
Port, M. H., 'The Houses of Parliament competition', in M. H. Port (ed.), *The Houses of Parliament*, (New Haven, 1976), pp. 20–52.
Port, M. H., 'Problems of building in the 1840s', in M. H. Port (ed.), *The Houses of Parliament*, (New Haven, 1976), pp. 97–121.
Port, M. H., 'Barry's last years: the new Palace in the 1850s', in M. H. Port (ed.), *The Houses of Parliament*, (New Haven, 1976), pp. 142–72.
Port, M. H., 'Trench, Sir Frederick William (*c.* 1777–1859)', *Oxford Dictionary of National Biography*, Oxford University Press, 2004 [http://ezproxy.ouls.ox.ac.uk:2117/view/article/27699, accessed 5 April 2013].

Port, M. H., *600 New Churches: The Church Building Commission, 1818–1856*, (Reading, 2006).

Port, M. H., 'Barry, Sir Charles (1795–1860)', *Oxford Dictionary of National Biography*, Oxford University Press, 2004; online edn, October 2008 [www.oxforddnb.com/view/article/1550, accessed 4 December 2013].

Port, M. H., 'Founders of the Royal Institute of British Architects (*act.* 1834–1835)', *Oxford Dictionary of National Biography*, Oxford University Press, May 2013 [www.oxforddnb.com/view/theme/97265, accessed 7 December 2013].

Porter, Dale H., *The Life and Times of Sir Goldsworthy Gurney: Gentleman Scientist and Inventor, 1793–1875*, (Bethlehem: 1998).

Porter, Dale H., *The Thames Embankment: Environment, Technology, and Society in Victorian London*, (Akron, 1998).

Porter, Roy, *The Making of Geology: Earth Science in Britain, 1660–1815*, (Cambridge, 1977).

Porter, Roy, 'Gentlemen and Geology: The Emergence of a Scientific Career, 1660–1920', *The Historical Journal*, Vol. 21, No. 4 (December, 1978), pp. 809–36.

Porter, Dorothy and Porter, Roy (eds.), *Doctors, Politics and Society: Historical Essays*, (Amsterdam, 1993),

Prest, John, 'Peel, Sir Robert, second baronet (1788–1850)', *Oxford Dictionary of National Biography*, Oxford University Press, 2004; online edn, May 2009 [www.oxforddnb.com/view/article/21764, accessed 7 March 2014].

Quinault, Roland, 'Westminster and the Victorian constitution', *Transactions of the Royal Historical Society*, Vol. 2 (1992), pp. 79–104.

Richards, Joan L., 'Mathematics in Victorian Britain by Raymond Flood; Adrian Rice; Robin Wilson. Review', *Isis*, Vol. 104, No. 4 (December, 2013), pp. 853–55.

Richardson, Christine, *Yorkshire Stone to London: To Create the Houses of Parliament*, (Sheffield, 2007).

Riddell, Richard, 'Smirke, Sir Robert (1780–1867)', *Oxford Dictionary of National Biography*, Oxford University Press, 2004; online edn, May 2010 [www.oxforddnb.com/view/article/25763, accessed 12 September 2014].

Riding, Christine and Riding, Jacqueline (eds.), *The Houses of Parliament: History, Art, Architecture*, (London, 2000).

Rooney, David, *Ruth Belville: The Greenwich Time lady*, (London, 2008).

Rope, H. E. G., *Pugin*, (Hassocks, 1935).

Rorabaugh, W. J., 'Politics and the architectural competition for the Houses of Parliament, 1834–1837', *Victorian Studies*, Vol. 17, No. 2 (1973), pp. 155–75.

Rudwick, Martin J. S., *The Great Devonian Controversy: The Shaping of Scientific Knowledge among Gentlemanly Specialists*, (Chicago, 1985).

Rudwick, Martin J. S., *The New Science of Geology: Studies in the Earth Sciences in the Age of Reform*, (Aldershot, 2004).

Rudwick, Martin J. S., 'Travel, travel, travel: geological fieldwork in the 1830s', in Martin J. S. Rudwick (ed.), *The New Science of Geology: Studies in the Earth Sciences in the Age of Reform*, (Aldershot, 2004), pp. 1–10.

Rudwick, Martin, 'Lyell, Sir Charles, first baronet (1797–1875)', *Oxford Dictionary of National Biography*, Oxford University Press, 2004; online edn, May 2012 [http://ezproxy.ouls.ox.ac.uk:2204/view/article/17243, accessed 23 August 2014].

Rudwick, Martin J. S., *Worlds Before Adam: The Reconstruction of Geohistory in the Age of Reform*, (Chicago, 2008).

Saint, Andrew, *The Image of the Architect*, (New Haven, 1983).

Saint, Andrew, *Architect and Engineer: A Study in Sibling Rivalry*, (New Haven, 2007).

Salmon, Frank, *Building on Ruins: The Rediscovery of Rome and English Architecture*, (Aldershot, 2000).

Sanders, L. C., 'Beckett, Edmund , first Baron Grimthorpe (1816–1905)', rev. Catherine Pease-Watkin, *Oxford Dictionary of National Biography*, Oxford University Press, 2004; online edn, May 2007 [http://ezproxy.ouls.ox.ac.uk:2117/view/article/30665, accessed 24 August 2013].

Sawyer, Sean, 'Delusions of national grandeur: reflections on the intersection of architecture and history at the Palace of Westminster, 1789–1834', *Transactions of the Royal Historical Society*, Vol. 13, (2003), pp. 237–50.

Schaffer, Simon, 'Natural philosophy and public spectacle in the eighteenth century', *History of Science*, Vol. 21, Issue 1 (1983), pp. 1–43.

Schaffer, Simon, 'Astronomers mark time: discipline and the personal equation', *Science in Context*, Vol. 2, Issue 1 (March, 1988), pp. 115–45.

Schaffer, Simon, 'Glass works: Newton's prisms and the uses of experiment', in David Gooding, Trevor Pinch, and Simon Schaffer (eds.), *The Uses of Experiment: Studies in the Natural Sciences*, (Cambridge, 1989), pp. 67–104.

Schaffer, Simon 'The history and geography of the intellectual world: Whewell's politics of language', in Menachem Fisch and Simon Schaffer (eds.), *William Whewell: A Composite Portrait*, (Oxford, 1991), pp. 201–31.

Schaffer, Simon, 'Late Victorian metrology and its instrumentation: a manufactory of Ohms', in Robert Bud and Susan E. Cozzens (eds.), *Invisible Connections: Instruments, Institutions, and Science*, (Washington, 1992), pp. 23–56.

Schaffer, Simon, 'Babbage's Intelligence: Calculating Engines and the Factory System', *Critical Inquiry*, Vol. 21, No. 1 (Autumn, 1994), pp. 203–27.

Schaffer, Simon, 'Babbage's dancer and the impresarios of mechanism', in Francis Spufford and Jenny Uglow (eds.), *Cultural Babbage: Technology, Time and Invention*, (London, 1996), pp. 53–80.

Schaffer, Simon, 'Metrology, metrication, and Victorian values', in Bernard Lightman (ed.), *Victorian Science in Context*, (Chicago, 1997), pp. 438–74.

Schaffer, Simon, 'Physics laboratories and the Victorian country house', in Crosbie Smith and Jon Agar (eds.), *Making Space for Science: Territorial Themes in the Shaping of Knowledge*, (Basingstoke, 1998), pp. 149–80.

Schaffer, Simon, 'Enlightened automata', in William Clark, Jan Golinski, and Simon Schaffer (eds.), *The Sciences in Enlightened Europe*, (London, 1999), pp. 126–65.

Schoenefeldt, Henrik, 'The Crystal Palace, environmentally considered', *Architectural Research Quarterly*, Vol. 12, Issue 3–4 (December, 2008), pp. 283–94.

Schoenefeldt, Henrik, 'The temporary Houses of Parliament and David Boswell Reid's architecture of experimentation', *Architectural History*, 57 (2014), pp. 173–213.

Secord, James A., 'King of Siluria: Roderick Murchison and the imperial theme in nineteenth-century British geology', *Victorian Studies*, Vol. 25, No. 4 (Summer, 1982), pp. 413–42.

Secord, James A., 'The geological survey of Great Britain as a research school, 1839–1855', *History of Science*, Vol. 24 Issue 3 (September, 1986), pp. 223–75.

Secord, James A., *Controversy in Victorian Geology: The Cambrian-Silurian Dispute*, (Princeton, 1986).

Secord, James A., *Victorian Sensation: The Extraordinary Publication, Reception, and Secret Authorship of Vestiges of the Natural History of Creation*, (Chicago, 2000).

Secord, J. A., 'Beche, Sir Henry Thomas De la (1796–1855)', *Oxford Dictionary of National Biography*, Oxford University Press, 2004 [http://ezproxy.ouls.ox.ac .uk:2117/view/article/1891, accessed 21 July 2013].

Secord, James A., *Visions of Science: Books and Readers at the Dawn of the Victorian Age*, (Oxford, 2014).

Sennett, Richard, *Flesh and Stone: The Body and the City in Western Civilization*, (London, 1994).

Shapin, Steven, '"Nibbling at the teats of science": Edinburgh and the diffusion of science in the 1830s', in Ian Inkster and Jack Morrell (eds.), *Metropolis and Province: Science in British Culture, 1780–1850*, (London, 1983), pp. 151–78.

Shapin, Steven, 'The house of experiment in seventeenth-century England', *Isis*, Vol. 79, No. 3, (September, 1988), pp. 373–404.

Shapin, Steven, 'The invisible technician', *American Scientist*, Vol. 77, No. 6 (November –December, 1989), pp. 554–63.

Shapin, Steven, *A Social History of Truth: Civility and Science in Seventeenth-Century England*, (Chicago, 1994).

Shapin, Steven, 'Placing the view from nowhere: historical and sociological problems in the location of science', *Transactions of the Institute of British Geographers, New Series*, Vol. 23, No. 1 (1998), pp. 5–12.

Shapin, Steven, *The Scientific Life: A Moral History of a Late Modern Vocation*, (Chicago, 2008).

Shapin, Steven, *Never Pure: Historical Studies of Science as if It Was Produced by People with Bodies, Situated in Time, Space, Culture, and Society, and Struggling for Credibility and Authority*, (Baltimore, 2010).

Shapin, Steven and Schaffer, Simon, *Leviathan and the Air-Pump: Hobbes, Boyle, and the Experimental Life*, (Princeton, 1985).

Sharpe, Tom, 'Slavery, sugar, and the survey', *Open University Geological Society Journal*, Vol. 29, No. 2 (symposium edition, 2008), pp. 88–94.

Shattock, Joanne, *Politics and Reviewers: The Edinburgh and the Quarterly in the Early Victorian Age*, (London, 1989).

Shenton, Caroline, *The Day Parliament Burned Down*, (Oxford, 2012).

Shenton, Caroline, *Mr Barry's War: Rebuilding the Houses of Parliament after the Great Fire of 1834*, (Oxford, 2016).

Smith, Crosbie, 'From design to dissolution: Thomas Chalmers' debt to John Robison', *British Journal for the History of Science*, Vol. 12, No. 1 (March, 1979), pp. 59–70.

Smith, Crosbie, '"Nowhere but in a great town": William Thomson's spiral of classroom credibility', in Crosbie Smith and Jon Agar (eds.), *Making Space for Science: Territorial Themes in the Shaping of Knowledge*, (Basingstoke, 1998), pp. 118–46.

Smith, Crosbie, *The Science of Energy: A Cultural History of Energy Physics in Victorian Britain*, (London, 1998).

Smith, Crosbie, '"The 'crinoline' of our steam engineers": reinventing the marine compound engine, 1850–1885', in David N. Livingstone and Charles W. J. Withers (eds.), *Geographies of Nineteenth-Century Science*, (Chicago, 2011), pp. 229–54.

Smith, Crosbie and Agar, Jon (eds.), *Making Space for Science: Territorial Themes in the Shaping of Knowledge*, (Basingstoke, 1998).

Smith, Crosbie and Agar, Jon, 'Introduction: making space for science', in Crosbie Smith and Jon Agar (eds.), *Making Space for Science: Territorial Themes in the Shaping of Knowledge*, (Basingstoke, 1998), pp. 1–23.

Smith, Crosbie, Ian Higginson, and Phillip Wolstenholme, '"Avoiding equally extravagance and parsimony": the moral economy of the ocean steamship', *Technology and Culture*, Vol. 44, No. 3 (July, 2003), pp. 443–69.

Smith, Crosbie, Ian Higginson, and Phillip Wolstenholme, '"Imitations of God's own works": making trustworthy the ocean steamship', *History of Science*, Vol. 41, (2003), pp. 379–426.

Smith, Crosbie and Scott, Anne, '"Trust in Providence": building confidence into the Cunard line of steamers', *Technology and Culture*, Vol. 48, No. 3 (July, 2007), pp. 471–96.

Smith, Denis, 'The techniques of the building', in M. H. Port (ed.), *The Houses of Parliament*, (New Haven, 1976), pp. 195–217.

Smith, Denis, 'The building services', in M. H. Port (ed.), *The Houses of Parliament*, (New Haven, 1976), pp. 218–231.

Smith, G. B., 'Gurney, Sir Goldsworthy (1793–1875)', rev. Anita McConnell, *Oxford Dictionary of National Biography*, Oxford University Press, 2004 [http://ezproxy.ouls.ox.ac.uk:2117/view/article/11764, accessed 21 Sept 2013].

Smith, Merritt Roe and Marx, Leo (eds.), *Does Technology Drive History? The Dilemma of Technological Determinism*, (Cambridge, Massachusetts, 1994).

Snyder, Laura J., *Reforming Philosophy: A Victorian Debate on Science and Society*, (Chicago, 2006).

Spufford, Francis and Uglow, Jenny (eds.), *Cultural Babbage: Technology, Time and Invention*, (London, 1996).

Stamp, Gavin, 'Hamilton, Thomas (1784–1858)', *Oxford Dictionary of National Biography*, Oxford University Press, 2004 [http://ezproxy.ouls.ox.ac.uk:2117/view/article/12131, accessed 7 April 2013].

Stamp, Gavin, 'Lutyens, Sir Edwin Landseer (1869–1944)', *Oxford Dictionary of National Biography*, Oxford University Press, 2004; online edn, May 2012 [www.oxforddnb.com/view/article/34638, accessed 30 July 2015].

Stamp, Gavin, 'Scott, Sir George Gilbert (1811–1878)', *Oxford Dictionary of National Biography*, Oxford University Press, 2004 [http://ezproxy.ouls.ox.ac.uk:2204/view/article/24869, accessed 23 August 2014].

Stamp, Gavin, *Gothic for the Steam Age: An Illustrated Biography of George Gilbert Scott*, (London, 2015).

Standage, Tom, *The Victorian Internet: The Remarkable Story of the Telegraph and the Nineteenth Century's Online Pioneers*, (London: 1998).

Stanton, Phoebe, *Pugin*, (London, 1971).

Stanton, Phoebe B., 'Barry and Pugin: a collaboration', in M. H. Port (ed.), *The Houses of Parliament*, (New Haven, 1976), pp. 53–72.

St Clair, William, 'Bruce, Thomas, seventh earl of Elgin and eleventh earl of Kincardine (1766–1841)', *Oxford Dictionary of National Biography*, Oxford University Press, 2004; online edn, May 2013 [www.oxforddnb.com/view/article/3759, accessed 7 March 2014].

Stewart, Robert, *Henry Brougham, 1778–1868: His Public Career*, (London, 1985).

Stewart, Robert, *Party and Politics, 1830–1852*, (Basingstoke, 1989).

Sturrock, Neil and Lawson-Smith, Peter, 'The grandfather of air-conditioning: The work and influence of David Boswell Reid, physician, chemist, engineer (1805–63)', *Proceedings of the Second International Congress on Construction History*, 3, (2006), pp. 2981–98.

Temperley, Howard, *White Dreams, Black Africa: The Antislavery Expedition to the River Niger, 1841–1842*, (New Haven, 1991).

Thomas, William, *The Philosophic Radicals: Nine Studies in Theory and Practice, 1817–1841*, (Oxford, 1979).

Thompson, E. P., 'Time, work-discipline, and industrial capitalism', *Past and Present*, No. 38 (December, 1967), pp. 56–97.

Thompson, Paul, *William Butterfield*, (London, 1971).

Thompson, S. P., 'Wheatstone, Sir Charles (1802–1875)', rev. Brian Bowers, *Oxford Dictionary of National Biography*, Oxford University Press, 2004; online edn, January 2011 [http://ezproxy.ouls.ox.ac.uk:2204/view/article/29184, accessed 1 September 2014].

Topham, Jonathan R., 'Science, natural theology, and evangelicalism in early nineteenth-century Scotland: Thomas Chalmers and the evidence controversy', in David N. Livingstone, D. G. Hart, and Mark A. Noll (eds.), *Evangelicals and Science in Historical Perspective*, (Oxford, 1999), pp. 142–74.

Tresch, John, *The Romantic Machine: Utopian Science and Technology after Napoleon*, (Chicago, 2012).

Tyack, Geoffrey, *Sir James Pennethorne and the Making of Victorian London*, (Cambridge, 1992).

Tyack, Geoffrey, 'William Butterfield and Oxford: adapted from a lecture by Geoffrey Tyack', in Geoffrey Tyack and Marjory Szurko (eds.), *William Butterfield and Keble College*, (Oxford, 2002).

Tyack, Geoffrey, 'Gilbert Scott and the Chapel of Exeter College, Oxford', *Architectural History*, Vol. 50 (2007), pp. 125–48.

Tyack, Geoffrey, 'Nash, John (1752 1835)', *Oxford Dictionary of National Biography*, Oxford University Press, 2004; online edn, May 2009 [www.oxforddnb.com/view/article/19786, accessed 22 Feb 2015].

Tyack, Geoffrey (ed.), *John Nash: Architect of the Picturesque*, (Swindon, 2013).

Tyack, Geoffrey and Szurko, Marjory (eds.), *William Butterfield and Keble College*, (Oxford, 2002).

Vaughan, Denys, 'Whitehurst, John (1713–1788)', *Oxford Dictionary of National Biography*, Oxford University Press, 2004; online edn, September 2013 [http://ezproxy-prd.bodleian.ox.ac.uk:2167/view/article/29295, accessed 2 September 2014].

Vernon, James, *Politics and the People: A Study in English Political Culture, c.1815–1867*, (Cambridge, 1993).

Vieira, Ryan A., *Time and Politics: Parliament and the Culture of Modernity in Britain and the British World*, (Oxford, 2015).

Vincent, David, 'Government and the management of information, 1844–2009', in Simon Gunn and James Vernon (eds.), *The Peculiarities of Liberal Modernity in Imperial Britain*, (Berkeley, 2011), pp. 165–81.

Warwick, Andrew, *Masters of Theory: Cambridge and the Rise of Mathematical Physics*, (Chicago, 2003).

Waters, David W., 'Nautical astronomy and the problem of longitude', in John G. Burke (ed.), *The Uses of Science in the Age of Newton*, (Berkeley, 1983), pp. 143–69.

Watkin, David, 'Kelsall, Charles (1782–1857)', rev. *Oxford Dictionary of National Biography*, Oxford University Press, 2004 [www.oxforddnb.com/view/article/37627, accessed 22 February 2015].

Watkin, David, 'Soane, Sir John (1753–1837)', *Oxford Dictionary of National Biography*, Oxford University Press, 2004; online edn, January 2008 [www.oxforddnb.com/view/article/25983, accessed 22 February 2015].

Watkin, David, *Thomas Hope, 1769–1831: And the Neo-Classical Idea*, (London, 1968).

Webster, Christopher, *R. D. Chantrell (1793–1872) and the Architecture of a Lost Generation*, (Reading, 2010).

Wedgwood, Alexandra, 'Pugin, Auguste Charles (1768/9–1832)', *Oxford Dictionary of National Biography*, Oxford University Press, 2004; online edn, January 2008 [www.oxforddnb.com/view/article/22868, accessed 4 December 2013].

Wedgwood, Alexandra, 'Pugin, Augustus Welby Northmore (1812–1852)', *Oxford Dictionary of National Biography*, Oxford University Press, 2004; online edn, January 2008 [www.oxforddnb.com/view/article/22869, accessed 4 December 2013].

Wedgwood, Alexandra, 'The new Palace of Westminster', in Christine Riding and Jacqueline Riding, *The Houses of Parliament: History, Art, Architecture*, (London, 2000), pp. 113–35.

Weitzman, George H., 'The Utilitarians and the Houses of Parliament', *Journal of the Society of Architectural Historians*, Vol. 20, No. 3, (October, 1961) pp. 99–107.

White, Brenda M., 'Christison, Sir Robert, first baronet (1797–1882)', *Oxford Dictionary of National Biography*, Oxford University Press, 2004; online edn, October 2009 [http://ezproxy.ouls.ox.ac.uk:2117/view/article/5370, accessed 8 April 2013].

Wheeler, Katherine, *Victorian Perceptions of Renaissance Architecture*, (Farnham, 2014).

Whyte, William, 'How do buildings mean? Some issues of interpretation in the history of architecture', *History and Theory*, Vol. 45, No. 2 (May, 2006), pp. 153–77.

Whyte, William, *Oxford Jackson: Architecture, Education, Status, and Style, 1835–1924*, (Oxford, 2006).

Whyte, William, 'Building the nation in the town: architecture and national identity in Britain', in William Whyte and Oliver Zimmer (eds.), *Nationalism and the Reshaping of Urban Communities in Europe, 1848–1914*, (Basingstoke, 2011), pp. 204–33.

Whyte, William, and Zimmer, Oliver (eds.), *Nationalism and the Reshaping of Urban Communities in Europe, 1848–1914*, (Basingstoke, 2011).

Williams, L. Pearce, *Michael Faraday: A Biography*, (London, 1965).

Williams, W. R., 'Vernon, Robert, first Baron Lyveden (1800–1873)', rev. H. C. G. Matthew, *Oxford Dictionary of National Biography*, Oxford University Press, 2004; online edn, January 2008 [www.oxforddnb.com/view/article/25898, accessed 25 February 2015].

Wilson, David B., 'The educational matrix: physics education at early-Victorian Cambridge, Edinburgh and Glasgow Universities', in P. M. Harman (ed.), *Wranglers and Physicists: Studies on Cambridge Mathematical Physics in the Nineteenth Century*, (Manchester, 1985), pp. 12–48.

Wilton-Ely, John, 'The rise of the professional architect in England', in Spiro Kostof (ed.), *The Architect: Chapters in the History of the Profession*, (New York, 1977), pp. 180–208.

Winter, Alison, '"Compasses all awry": the iron ship and the ambiguities of cultural authority in Victorian Britain', *Victorian Studies*, Vol. 38, No. 1 (Autumn, 1994), pp. 69–98.

Winter, Alison, *Mesmerized: Powers of Mind in Victorian Britain*, (Chicago, 1998).

Winter, Emma L., 'German fresco painting and the new Houses of Parliament at Westminster, 1834–1851', *Historical Journal*, Vol. 47, No. 2 (June, 2004), pp. 291–329.

Wise, M. Norton (ed.), *The Values of Precision*, (Princeton, 1995).

Wise, M. Norton, 'Introduction', in M. Norton Wise (ed.), *The Values of Precision*, (Princeton, 1995), pp. 3–13.

Wise, M. Norton and Smith, Crosbie, 'Measurement, work and industry in Lord Kelvin's Britain', *Historical Studies in the Physical and Biological Sciences*, Vol. 17, No. 1 (1986), pp. 147–73.

Withers, Charles W. J., 'Place and the "spatial turn" in geography and in history', *Journal of the History of Ideas*, Vol. 70, No. 4 (October, 2009), pp. 637–58.

Withers, Charles W. J., *Geography and Science in Britain, 1831–1939: A Study of the British Association for the Advancement of Science*, (Manchester, 2010).

Withers, Charles W. J. and Livingstone, David N., 'Thinking geographically about nineteenth-century science', in David N. Livingstone and Charles W. J. Withers (eds.), *Geographies of Nineteenth-Century Science*, (Chicago, 2011), pp. 1–19.

Yanni, Carla, 'Nature and nomenclature: William Whewell and the production of architectural knowledge in early Victorian Britain', *Architectural History*, Vol. 40 (1997), pp. 204–21.

Yanni, Carla, *Nature's Museums: Victorian Science and the Architecture of Display*, (Baltimore, 1999).

Yanni, Carla, 'Development and display: progressive evolution in British Victorian architecture and architectural theory', in Bernard Lightman and Bennett Zon (eds.), *Evolution and Victorian Culture*, (Cambridge, 2014), pp. 227–60.

Yeo, Richard, 'Science and intellectual authority in mid-nineteenth-century Britain: Robert Chambers and vestiges of the natural history of creation', *Victorian Studies*, Vol. 28, No. 1 (Autumn, 1984), pp. 5–31.

Yeo, Richard, 'An idol of the market-place: Baconianism in nineteenth-century Britain', *History of Science*, Vol. 23, No. 3, (September, 1985), pp. 251–98.

Yeo, Richard, *Defining Science: William Whewell, Natural Knowledge, and Public Debate in Early Victorian Britain*, (Cambridge, 1993).

Yeo, Richard, 'Whewell, William (1794–1866)', *Oxford Dictionary of National Biography*, Oxford University Press, 2004; online edn, May 2009 [www.oxforddnb.com/view/article/29200, accessed 7 March 2014].

II. Unpublished

Aspin, Philip, 'Architecture and Identity in the English Gothic Revival, 1800–1850', (DPhil Thesis: University of Oxford, 2013).

Thorne, Robert, 'House of Lords roof', privately commissioned research document, (27 November, 1991).

Thorne, Robert, 'The galvanising of the iron roof plates at the Palace of Westminster', privately commissioned research document, (1991).

Index

Printed in the United States
By Bookmasters